The Art of Science

Richard Hamblyn is the author of *The Invention of Clouds*, which won the 2002 *Los Angeles Times* Book Prize and was shortlisted for the Samuel Johnson Prize; *Terra: Tales of the Earth*, a study of natural disasters; *Data Soliloquies*, co-written with the digital artist Martin John Callanan; and *The Cloud Book*, published in association with the Met Office. He teaches creative writing at Birkbeck College, University of London.

RICHARD HAMBLYN

The Art of Science

A Natural History of Ideas

PICADOR

First published 2011 by Picador

First published in paperback 2012 by Picador
an imprint of Pan Macmillan, a division of Macmillan Publishers Limited
Pan Macmillan, 20 New Wharf Road, London N1 9RR
Basingstoke and Oxford
Associated companies throughout the world
www.panmacmillan.com

ISBN 978-0-330-49076-4

1 3 5 7 9 8 6 4 2

A CIP catalogue record for this book is available from the British Library.

Printed and bound by CPI Group (UK) Ltd, Croydon, CR0 4YY

Visit **www.picador.com** to read more about all our books
and to buy them. You will also find features, author interviews and
news of any author events, and you can sign up for e-newsletters
so that you're always first to hear about our new releases.

Art is I; science is we.

Claude Bernard

Contents

List of Illustrations xi

Introduction xiii

Babylonian Awakening 1

Ancient Egyptian Mathematics 4

Zeno's Paradox and Other Presocratic Fragments 8

The Saltness of the Sea 15

Atoms and Infinity 19

Whirlwinds 24

The Penalty of Rust 30

The Ferris Wheel Universe 34

The Colour of Shame 38

Imposters 41

The Gates of Knowledge 48

Dream Pool Essays 52

Empirical Rainbows 56

The Art of Numbring 60

Copernican Revolutions 63

Supernova 67

The Moons of Jupiter 71

The Circulation of Blood 76

A Quire of Echoes 80

The Air-Pump 84

The Discovery of Plant Cells 88

Mayflies 94

Sunspots 98

Temperature Scales 101

Newtonian Apples 104

Newton for the Ladies 110

Inoculation 115

Walking Plants 121

Scientific Style 127

Insiders and Outsiders 133

Glaciers 138

Electricity 144

Blunt *versus* Pointed 152

Seismic Waves 157

Classification 163

The Cure for Scurvy 175

Animal Magnetism 178

On His Colour-Blindness 182

Comet Sweeper 185

Botanical Poems 193

Phosphorus 199

The Miner's Safety Lamp 204

Brownian Motion 210

The Velocity of Meteorites 213

First Use of the Word 'Scientist' 216

The First Computer 219

The Silurian System 228

The Humboldt Current 232

The Temple of Serapis 236

Darwin's Dangerous Idea 242

Butterfly Hunting in Malaya 250

Drops of Water 253

The Physics of a Wet Towel 257

Pasteurization 260

Stammering 264

Dinosaur Teeth 268

Carnivorous Plants 273

Lunar Cosmorama 278

Krakatoa Dust 282

Edison's Carbon Telephone 287

Fidgeting 290

Surface Tension 293

The Household Laboratory 298

In Search of Krypton 304

Radioactivity 309

N-Rays 315

Paranoid Schizophrenia 319

Ripples in the Sand 325

Snowflakes 330

Science *versus* Religion 335

Planck Time 340

The Quantum Theory of Light 342

The Force of Gravity Does Not Exist 346

Bombarding the Atom 350

Nuclear Fission 354

The Impact of the War on Scientists 360

Anti-Science 367

The Two Cultures 373

The Double Helix 377

Continental Drift 380

Fossil Collecting 384

Compass-Building Bacteria 391

Exploding Ants 394

The Human Reserve 397

The Evolution of the Eye 400

Lichenometrics 405

The Turtle's Shell 408

Hebridean Seal Survival 411

The Mathematics of the ISBN 414

Brain Power 416

The Bends 419

Rounding Errors and the Butterfly Effect 421

Pulsars 425

The First Three Minutes 431

Black Holes 434

Climate Change 437

Fuzzy Sets and Beauty Maps 443

The Human Genome 448

Swimming with *E. coli* 452

Coffee Stains 456

The God Particle 459

Afterword 464

Further Reading 469

Acknowledgements 473

Index 477

List of Illustrations

1. A problem from the Rhind Mathematical Papyrus, c. 1650 BC. Photo Researchers/Science Photo Library. p. 5
2. Engraving of a tornado from John P. Finley, *Tornadoes: What they are and how to observe them* (1887). Image courtesy History of Science Collections, University of Oklahoma Libraries. p. 27
3. Epicycles from a page of Ptolemy's *Almagest* (2nd century AD). Science Museum/SSPL. p. 35
4. Page from William Harvey's *De Motu Cordis* (1628). Science Museum/SSPL. p. 78
5. Cork cells, from Robert Hooke's *Micrographia* (1665). Science Museum/SSPL. p. 89
6. Abraham Trembley's microscope. Science Photo Library. p. 123
7. Chinese Seismograph, c. AD 132. Science Museum/SSPL. p. 161
8. The three main cloud types, from Luke Howard, *On the Modifications of Clouds*, 1803. Science Museum/SSPL. p. 170
9. William and Caroline Herschel at their telescope, early 1780s. Mary Evans Picture Library. p. 188
10. Illustration from Humphry Davy, *On the Safety Lamp for Coal Miners* (1818). Science Photo Library. p. 208
11. Pascal's calculating machine, 1642. Science Museum/SSPL. p. 220
12. The Temple of Serapis at Puzzuoli, from Charles Lyell's *The Principles of Geology* (1830). © Natural History Museum, London. p. 238
13. Cartoon depicting Charles Darwin as an ape. From *The Hornet* (1871). Mary Evans Picture Library/INTERFOTO AGENTUR. p. 247
14. Drop I from Agnes Catlow's *Drops of Water* (1851). p. 256
15. Teeth of the iguanodon, from Gideon Mantell, *The Wonders of Geology* (1838). The Royal Society. p. 269
16. Model of lunar craters photographed by James Nasmyth, c. 1871. Science Museum/SSPL. p. 280
17. The periodic table. p. 308
18. A Ripple-forming Vortex with Generating Ridge, from Hertha Ayrton, 'The Origin and Growth of Ripple-Mark' (1910). The Royal Society. p. 329
19. A page of snowflake photographs by Wilson Bentley, c. 1920. NOAA. p. 332
20. Continental drift maps by Alfred Wegener, 1924. © Natural History Museum, London. p. 381

21. An amateur geologist's field kit, from Herbert S. Zim, *Rocks and Minerals* (1957). © St Martin's Press, Inc. p. 385

22. Turtle skeleton, from Sarah Cooper, *Animal Life in the Sea and On Land* (1887). Author's collection. p. 409

23. Computer-generated Mandelbrot fractal. Victor Habbick Visions / Science Photo Library. p. 446

Introduction

In the course of putting this anthology together I kept being reminded of a T-shirt slogan that was popular in the mid-1970s. THERE ARE ONLY 10 TYPES OF PEOPLE IN THE WORLD, it read: THOSE WHO UNDERSTAND BINARY NUMBERS AND THOSE WHO DON'T. I still recall the troubling sensation of knowing, aged about twelve or thirteen, that the joke was probably funny, but not understanding why. I hadn't the faintest idea what it meant, but I remember liking the way that it looked and sounded, like a coded message from some unfamiliar world in which words and numbers had different applications to those in everyday life. And not only was the computer-card script exciting in itself, but the strange word 'binary' sounded a lot like 'bionic', the coolest word in the English language, as far as I was concerned. So it didn't really matter that I didn't get the joke – the not knowing was part of the appeal.

A year or so later, when the binary system came up on the school mathematics curriculum, the penny finally dropped: '10' in binary was '2' in decimal! 'There are only 2 types of people in the world . . .', the joke being predicated on whether or not you could translate the '10', which of course you could do only if you were the type who understood binary: QED! Though it was hardly the greatest one-liner in history, it had a powerful impact on my adolescent mind, for I can clearly remember the complex pleasure of finally understanding something that I had assumed would always be beyond me. Solving the message had not diminished the appeal of its curious geeky poetry, in fact quite the reverse; for though there was a certain kind of pleasure to be had in not understanding the code – and this was true of the actual poetry we were also required to study – there was more and better pleasure to be had in grasping its hidden meaning.

It soon became apparent, however, that mathematics wasn't for me – binary notation was about as far as my not-very-numerate brain could go – but in common with the majority of the unmathematically inclined, I'm a little in awe of its mysterious authority. $E = mc^2$, for example, has a poetic resonance quite distinct from whatever explanatory content it might have, and for every person who understands it, who is at ease with its finely tuned symbolic language, there must be hundreds who can recite it, who could even write it down correctly, without having a clue what it means. And that's just the easy stuff: how many visitors to Scientists' Corner in the nave of Westminster Abbey can make head or tail of the inscription carved into Paul Dirac's memorial stone – just a few feet away from Newton's great monument – which reads: $i\gamma.\partial\psi = m\psi$? By accounting for the spin of the electron, this celebrated equation (one of the most beautiful mathematical statements ever made, according to its admirers) reconciled Einstein's special theory of relativity with Erwin Schrödinger's wave equation, thereby laying the foundations of quantum mechanics, one of the few genuine revolutions in the history of human thought. Yet the Dirac equation – the only one in the Abbey: perhaps the only equation in any abbey – remains a closed book to all but a handful of initiates able to translate its compact hieroglyphics into a statement about the nature of the universe.* Everyone else, myself included, irrespective of how well educated we might be, goes through life more or less completely disconnected from the world of such recondite ideas. As C. P. Snow observed more than half a century ago, in the course of his famous 'Two Cultures' essay, 'the great edifice of modern physics goes up, and the majority of the cleverest people in the western world have about as much insight into it as their neolithic ancestors would have had.'

True enough, but does it matter? So what if Dirac's equation can only be appreciated by a few hundred physicists and mathematicians: the rest of us seem to get along perfectly well without it, unencumbered by the brain-aching mathematical training that would be the price of possessing such knowledge. In the same way that very few people can read ancient Aramaic, or have ever made it all the way through to the end of *Finnegans Wake*, the constituency for difficult mathematical ideas is bound to be selectively small. Yet the comparison is not quite right, for the inability to read an ancient language or an experimental novel rarely leads to a blanket rejection of all other languages or all other literature – apart from the really challenging stuff – whereas there is a tendency among non-scientists to characterize the whole of Western science as uniformly reductive and difficult, as though all of it was as offputting and apparently inhuman as Dirac's unpronounceable $i\gamma.\partial\psi = m\psi$.

This outlook has a long history, as a number of pieces in this anthology attest. D. H. Lawrence, for example, writing in the late 1920s, just as Dirac was completing his equation, argued that the cold, rational, scientific view of the world was incompatible with the religious and poetical embrace of Life with a capital 'l'. 'The Universe is dead for us, and how is it to come alive again?' he asked, now that ' "knowledge" has killed the sun, making it a ball of gas, with spots' (see p. 371). A century earlier, William Blake had raised similar

---

* The only other ecclesiastical equation of which I am aware is a stained-glass window in the Grace Cathedral, San Francisco, that incorporates Einstein's celebrated formulation $E = mc^2$.

objections to the rise of Newtonian mechanics – 'Art is the Tree of Life. Science is the Tree of Death', he wrote, in a Romantic refashioning of an ancient anti-science theology that is still in the air today, though these days it tends towards more secular frames of reference: 'I don't get on with scientists because I'm not on the autistic spectrum', as one of my colleagues once said to me, not entirely facetiously, to which I wish I'd had the wit to reply that if it's a *spectrum*, sunshine, then everybody's on it.

But this is the point at which the cultural disengagement from science does begin to matter: the point at which it shades into hostility. In fact, over the past few decades, as John Carey pointed out in his introduction to *The Faber Book of Science* (a ground-breaking anthology that appeared in 1995), 'ignorance of science has acquired a degree of political correctness', in the light of which scientists find themselves accused of being, at best, unimaginative, unemotional and laughably precise, and at worst, power-hungry, inhumane and oblivious to danger – the familiar caricature popularized by Mary Shelley's *Frankenstein* (1818), and which has haunted the cultural landscape ever since.

One of the aims of this collection is to prove such accusations to be groundless. Science, however one might want to define it, is first and foremost a product of the human imagination; it is a cultural endeavour like any other, albeit one with certain unique procedures and conventions that have built up around it over time. Take those conventions out of the frame, and 'the whole of science is nothing more than a refinement of everyday thinking', as Albert Einstein described it in 1936, and his abiding belief that even the most esoteric idea can be explained in ordinary language to anyone has been a guiding principle of this book. Einstein's own technical writings were, of course, impenetrable to most non-physicists, but in popular works such as *The Evolution of Physics* (an extract from which appears on pp. 342–5), he ensured that his and others' theoretical achievements became known around the world through an insightful mix of imagery and expression that had a shaping influence on the public understanding of science. I can, for example, think of no better introduction to the light wave/particle conundrum that so vexed early twentieth-century physics than Einstein's laconic advice to his readers that 'it will be helpful in understanding the phenomena which we are about to describe, to bear in mind the difference between sea waves and a shower of bullets.' Such a simple image with which to communicate such a dauntingly complex idea: that is the essence of effective science writing, of which there are many other examples gathered in this book, such as Lucretius's dust-motes dancing in a sunlit room, offered up as an image of atomic motion

(p. 20), or Marcus Chown's surreal description of a troop of ants marching across an indented trampoline – a particularly neat illustration of why the 'force' of gravity is an illusion created by the curvature of space (p. 348).

My intention in compiling this anthology, however, was not just to show-case some accessible translations of key scientific ideas, but to situate the birth of those ideas in their cultural and historical contexts. To this end, I have sought out more than a hundred pieces of writing that either reflect the particular situation in which a moment of scientific understanding took place, or that reveal something of the personalities of the scientists involved. James Watson's account of the events leading up to the discovery of the structure of DNA, for example, is inseparable from the character of his written voice, and it is clear both from his own recollections, as well as of those who were there at the time, that the egotism and insensitivity that mark every page of *The Double Helix* were important contributing factors in Watson's scientific success – typified by the insults he directed at the crystallographer Rosalind Franklin, whose results he had purloined at a critical stage of their research:

> By choice she did not emphasize her feminine qualities. Though her features were strong, she was not unattractive and might have been quite stunning had she taken even a mild interest in clothes. This she did not. There was never lipstick to contrast with her straight black hair, while at the age of thirty-one her dresses showed all the imagination of English blue-stocking adolescents . . . The thought could not be avoided that the best home for a feminist was in another person's lab (see p. 377).

Watson's book caused an uproar when it appeared in 1968, although it wasn't the sexism so much as the cynicism that shocked so many of its readers, its breezy portrayal of laboratory research as an all-out race between the winners and the losers breaking the unwritten code of 'gentlemanly' conduct that was still an integral part of how Science (with a capital 's') presented itself to the world. But in telling the story in the way that he did, by exposing the intrigues and infighting that attended one of the greatest achievements in the history of science, Watson transformed the way that scientists were perceived by the wider public, whose preconceptions of Nobel Prize-winning molecular biologists are unlikely to have included 'a pair of loudmouthed young men who devoted more time to talking and drinking than to experiment', as Crick and Watson were described by a contemporary of theirs, who wondered, as many have wondered since, how such an unlikely double-act could have been destined for scientific immortality.

*

'Put yourself in the story' is the advice most often given to non-fiction writers, and a glance at the extracts assembled here show that scientists throughout history have done exactly that, from Francis Bacon's exuberant account of his sonic adventures in the world of echoes, to Edward O. Wilson's amiable confession that he feeds bits of coffeecake to the ants on his kitchen floor ('they also like tuna and whipped cream', he notes). Science writing has always tended towards the first-person narrative voice, partly because, in its earliest days, it was usually communicated via personal correspondence, and partly because the device works to simplify the task of describing what was done, what was seen, and what was discovered by an observer. It also enables scientists to reflect upon their immediate thoughts and feelings, a notion that some non-scientists might find a little surprising, given the widespread belief that scientific objectivity is incompatible with everyday human emotion, as though scientists somehow disengage their intuitive faculties on the way to the laboratory. But one needs only to alight at random on any of the pieces I have collected here to see that there is nothing cold or emotionally detached about scientific work. Take Marie Curie's description of her and her husband's enchantment at the sight of the radium they had isolated from the tons of muddy pitchblende they had shovelled in and out of that stinking shed in Paris:

> I shall never be able to express the joy of the untroubled quietness of this atmosphere of research and the excitement of actual progress with the confident hope of still better results. The feeling of discouragement that sometimes came after some unsuccessful toil did not last long and gave way to renewed activity. We had happy moments devoted to a quiet discussion of our work, walking around our shed.
>
> One of our joys was to go into our workroom at night; we then perceived on all sides the feebly luminous silhouettes of the bottles or capsules containing our products. It was really a lovely sight and one always new to us. The glowing tubes looked like faint, fairy lights.

If such candour strikes the reader today as hopelessly 'unscientific', that must have something to do with the way in which science has been routinely mistaught in schools – particularly in the discouragement of any form of self-expression in favour of some half-baked parody of Olympian objectivity. I can still recall being marked down by a chemistry teacher for handing in a piece of descriptive writing that would have earned high praise in an English lesson (though I have often wondered what that same teacher would have made of Samuel Taylor Coleridge's overheated account of one of Humphry Davy's chemical demonstrations: 'ether burns bright indeed in the atmosphere, but o! how brightly whitely vividly beautiful in Oxygen gas'); and the day that our chemistry classwork headings were changed from 'What we did', 'What we saw', and 'What we learned' to 'Method', 'Observations', and 'Conclusions' was the day that my interest in the subject began to wane, hastened by the realization that 'writing up' was not the same thing as 'writing'.

But, of course, it is, or at least, it should be, and what struck me most often in the course of assembling this anthology was the fact that science is an inherently narrative enterprise, a heightened encounter with the workings of the world that requires a commensurately heightened language with which to describe and account for it. And when elements of that descriptive language turn out to be unavailable, as has often been the case, then they have to be invented there and then: Curie herself came up with the term 'radioactivity' to describe the stream of particles released by the disintegration of unstable atomic nuclei, and in the pages that follow there are many such examples of the creation of an expressive scientific discourse with which to describe and organize the world, from Robert Hooke coining the word 'cell' while looking at a piece of magnified cork through his home-made compound microscope – the honeycomb structure apparently reminded him of the rows of cells in a monastery – or Richard Owen proposing the name 'dinosaur' to describe the long-extinct 'terrible lizards' that so haunted the Victorian popular imagination, to Wallace Broecker innocently unleashing the phrase 'global warming', a term coined in response to a sudden reversal of the mid-twentieth-century cooling phase, the 'climatic surprise' that served to place environmental science at the heart of an ongoing policy dispute that looks likely to simmer on for years. Even the word 'scientist' had to be invented, as it was in 1833, only making its first appearance halfway through this book! (See p. 216.)

So is good science synonymous with good science writing? Well, yes and no: as Samuel Johnson noted in the preface to his *Dictionary*, 'language is only the instrument of science, and words are but the signs of ideas', and though arguments about language are as vital to the sciences as they are to the humanities – see, for example, John R. Baker's impassioned essay on the state of scientific English on p. 129 – I have tried to avoid giving valuable space to mere displays of eloquence, preferring to treat this anthology as a kind of natural history of ideas, a selective, broadly chronological, overview of the ways in which the world has been rendered explicable through reason and discovery.

But at the same time the process of selection has been largely determined by the interest and readability of my chosen extracts, there being little point, it seems to me, in reproducing reams of technical material that only a few specialists could understand. That is why the likes of Copernicus and Newton are not here in their own words – Copernicus may have written in an unintentionally impenetrable style, but Newton apparently made his work deliberately difficult in order to discourage 'little Smatterers in Mathematicks' from getting in touch with him and wasting his time – so they are represented in the words of others, gifted contemporaries and followers keen to elucidate the new materialist philosophies of nature that were beginning to transform the way the world was understood. It is difficult to imagine from today's perspective the profound cultural shock of realizing that the earth is just another planet in orbit around the sun, but we can get at least a glimpse of the strangeness and excitement generated by such novel ideas through reading contemporary accounts, such as the one by Thomas Digges that appears on pp. 63–6:

> But in this our age one rare wit … hath by long study, painfull practise, and rare invention delivered a new Theory or model of the world, shewing that the Earth resteth not in the Centre of the whole world, but only in the Centre of this our mortall world or Globe of Elements which, environed and enclosed in the Moon's Orb, and together with the whole Globe of mortality is carried yearly round about the Sunne, which like a king in the midst of all reigneth and giveth laws of motion to ye rest, spherically dispersing his glorious beams of light through all this sacred Celestiall Temple.

This is wonderful writing – literally so, in that it is filled with wonder at the newly revealed patterns of the universe – but it also sheds fascinating light on the means by which Copernican ideas began to spread around the world towards the end of the sixteenth century. And most of the other pieces in this collection share these same paired qualities of readability and documentary value, twin factors that I soon identified as being my minimum criteria when deciding what to include. For though it went without saying that anything badly written would be rejected from the start – this book, after all, is intended to be a pleasure to read – when it came to making more difficult choices about what to keep and what to let go, I always gave priority to pieces that revealed some aspect of their immediate historical circumstances. James Lind's account of his clinical trials on board HMS *Salisbury* in 1746 and 1747 is a case in point, for not only is it a superbly crafted case history, well written and clearly explained, it also affords us a surgeon's-eye view of everyday life on an eighteenth-century warship, complete with barrels of baked biscuits and a scurvy-ridden crew (p. 175). Naturally, such details make for good storytelling as well as good science, but at the same time they offer far more than picturesque adornment, for they also serve to illustrate the range of everyday situations – ships, private houses, the back rooms of pubs – in which scientific knowledge was created in the days before the rise of industrial laboratories and university science departments.

This gradual restriction of the spaces of science is one of the many themes that unfold in the background to this collection, which is one reason that I chose to arrange it as a chronological sequence, to allow historical fluctuations to reveal themselves as part of the wider narrative. Note, for example, the rise and fall of a generation of early nineteenth-century women scientists that included Caroline Herschel (pp. 185–92), Jane Marcet (pp. 199–203) and Mary Somerville (pp. 213–15), who were recognized in their lifetimes as being at the forefront of European science. Yet by the end of the nineteenth century the scientific professions had more or less closed their doors to such women, who were barred from graduating from most universities (the University of London was a rare exception) and routinely turned down for membership of the top scientific institutions. It was not until later in the following century that equality of access to a scientific career became anything like a reality.

But at the same time I was keen to make connections across the centuries, as though tracing the lineage of certain recurrent preoccupations, so in a number of places I have confounded chronology in order to suggest thematic correspondences between, say, Aristotle's account of the saltness of the sea and that of Rachel Carson more than two thousand years later (pp. 15–18), or between Richard Dawkins's take on the evolution of the eye and Lucretius's impressive pre-emptive assault on the argument from design (p. 403). Though I should stress that such pairings have little value as historical explanations – they are not much more than illuminating echoes – they nevertheless offer a modest corrective to the old-fashioned view of the history of science as a steady accumulation of knowledge and certainty, a sort of stately progression from ignorance to enlightenment, rather than a series of ad-hoc adventures that may or may not lead to anything. For if there's one inescapable conclusion to be drawn from the contents of this collection, it's that science is only sometimes about the discovery of laws and patterns in nature, and even less about 'eureka' moments of insight and revelation; most of the time it's about the raw human dramas of politics, passion, intrigue and ingenuity, all of which, and much else besides, can be found among the one hundred and one extraordinary stories that follow.

BABYLONIAN AWAKENING

Science was first pursued in an organized way some four or five thousand years ago in Sumeria and Babylonia, the two most culturally advanced regions of ancient Mesopotamia (present-day Iraq). Although such ancient cultures might seem unimaginably remote from us today, we still inhabit a world of Babylonian quantification, in the form of the sexagesimal (base-60) numerical system that we've incorporated into our modern conventions of time-keeping (60 minutes to the hour) and geometry (360 degrees in a circle). In some ways the Babylonian base-60 system was more user-friendly than our modern decimal system, 60 being wholly divisible by 1, 2, 3, 4, 5, 6 and 10; while 10, by contrast, is wholly divisible only by 1, 2 and 5.

The Babylonians were much preoccupied with astro-meteorological observations that they recorded on clay tablets, of which several hundred from the period survive, containing such gnomic statements as 'When a dark halo surrounds the moon, the month will bring clouds and rain', or 'When a cloud grows dark in heaven, a wind will blow'. While a selection of these would have made fascinating reading, I decided to begin this anthology instead with a modern imagining of the Babylonian worldview, written by the Hungarian-born novelist Arthur Koestler (1905–83). Though Koestler was something of a mystical thinker, he wrote about the advent of science with insight and empathy, and his attempt, in this extract, to convey the strangeness and otherness of a long-vanished mindset is a valuable corrective to our tendency to look for modern parallels in the mists of early human history:

The world of the Babylonians, Egyptians and Hebrews was an oyster, with water underneath, and more water overhead, supported by the solid firmament. It was of moderate dimensions, and as safely closed in on all sides as a cot in the nursery or a babe in the womb. The Babylonians' oyster was round, the earth was a hollow mountain, placed in its centre, floating on the waters of the deep; above it was a solid dome, covered by the upper waters. The upper waters seeped through the dome as rain, and the lower waters rose in fountains and springs. Sun, moon and stars progressed in a slow dance across the dome, entering the scene through the doors in the East and vanishing through doors in the West.

The universe of the Egyptians was a more rectangular oyster or box; the earth was its floor, the sky was either a cow whose feet rested on the four corners of the earth, or a woman supporting herself on her elbows and knees; later, a vaulted metal lid. Around the inner walls of the box, on a kind of elevated gallery, flowed a river on which the sun and moon gods sailed their barques, entering and vanishing through various stage doors. The fixed stars were lamps, suspended from the vault, or carried by other gods. The planets sailed their own boats along canals originating in the Milky Way, the celestial twin of the Nile. Towards the fifteenth of each month, the moon god was attacked by a ferocious sow, and devoured in a fortnight of agony; then he was re-born again. Sometimes the sow swallowed him whole, causing a lunar eclipse. But these tragedies were, like those in a dream, both real and not; inside his box or womb, the dreamer felt fairly safe.

This feeling of safety was derived from the discovery that, in spite of the tumultuous private lives of the sun and moon gods, their appearances and movements remained utterly dependable and predictable. They brought night and day, the seasons and the rain, harvest and sowing time, in regular cycles. The mother leaning over the cradle is an unpredictable goddess; but her feeding breast can be depended on to appear when needed. The dreaming mind may go through wild adventures, it may travel through Olympus and Tartarus, but the pulse of the dreamer has a regular beat that can be counted. The first to learn counting the pulse of the stars were the Babylonians.

Some six thousand years ago, when the human mind was still half asleep, Chaldean priests were standing on watch-towers, scanning the stars, making maps and time-tables of their motions. Clay tablets dating from the reign of Sargon of Akkad, around 3800 BC, show an already old-established astronomical tradition. The time-tables became calendars which regulated organized activity, from the growing of crops to religious ceremonies. Their observations became amazingly precise: they computed the length of the year with a deviation of less than 0.001 per cent from the correct value, and their figures relating to the motions of sun and moon have only three times the margin of error of nineteenth-century astronomers armed with mammoth telescopes. In this respect, theirs was an Exact Science; their observations were verifiable, and enabled them to make precise predictions of astronomical events; though based on mythological assumptions, theory 'worked'. Thus at the very beginning of this long journey, Science emerges in the shape of Janus, the double-faced god, guardian of doors and gates: the face in front alert and observant, while the other, dreamy and glassy-eyed, stares in the opposite direction.

The most fascinating objects in the sky – from both points of view – were the planets, or vagabond stars. Only seven of these existed among the thousands of lights suspended from the firmament. They were the Sun, the Moon, Nebo – Mercury, Ishtar – Venus, Nergal – Mars, Marduk – Jupiter, and Ninib – Saturn. All other stars remained stationary, fixed in the pattern of the firmament, revolving once a day round the earth-mountain, but never changing their places in the pattern. The seven vagabond stars revolved with them, but at the same time they had a motion of their own, like flies wandering over the surface of a spinning globe. Yet they did not wander all across the sky: their movements were confined to a narrow lane, or belt, which was looped around the firmament at an angle of about twenty-three degrees to the equator. This belt – the Zodiac – was divided into twelve sections, and each section was named after a constellation of fixed stars in the neighbourhood. The Zodiac was the lovers' lane in the skies, along which the planets ambled. The passing of a planet through one of the sections had a double significance: it yielded figures for the observer's time-table, and symbolic messages of the mythological drama played out behind the scenes. Astrology and Astronomy remain to this day complementary fields of vision of Janus sapiens.

Source: Arthur Koestler, *The Sleepwalkers* (London: Hutchinson, 1959), pp. 19–21.

ANCIENT EGYPTIAN MATHEMATICS

Like the Babylonians, the ancient Egyptians were impressive mathematicians who passed on their knowledge through prototype 'textbooks' such as the Rhind Mathematical Papyrus of c. 1650 BC (fig. 1). The papyrus, which is named after the Scottish Egyptologist Alexander Henry Rhind (1833–63), who bought it in Luxor in 1858, contains eighty-seven mathematical problems, the majority of which apply to practical matters such as dividing a bushel of barley among ten farmers so that each farmer gets $1/8$ of a bushel more than the one before, or working out the area of a circular field with a particular diameter. Some of the problems have a slightly more theoretical bias, and seem designed to enhance the student's ability to manipulate the number system. The first six problems, for example, ask how to divide n loaves between ten men, where $n = 1$ in the first problem, $n = 2$ in the second, $n = 6$ in the third, and so on up to $n = 9$ in the sixth.

The following gently irreverent account of the Egyptian mathematical context is by the celebrated Egyptologist Thomas Eric Peet (1882–1934), who taught for many years at the University of Liverpool:

The outstanding feature of Egyptian mathematics is its intensely practical character. This is not peculiar to mathematics, for it is typical of all the sciences in Egypt. As Plato alone of the Greeks seems to have realized, the Egyptians were essentially a 'nation of shopkeepers,' and interest in or speculation concerning a subject for its own sake was totally foreign to their minds.

To realize this we have only to take a glance through the problems of the Rhind Papyrus. Here everything is expressed in concrete terms. The Egyptian does not speak or think of 8 as an abstract number, he thinks of 8 loaves or 8 sheep. He does not work out the slope of the sides of a pyramid because it interests him to know it, but because he needs a practical working rule to give to the mason who is to dress the stones. If he resolves $2/13$ into $1/8 + 1/52 + 1/104$ it is not because this fact in itself appeals in any way to his curiosity, but simply because sooner or later he will come across the fraction $2/13$ in a sum, and since he has no machinery for dealing with fractions whose numerators are greater than unity he will then urgently need the resolution above stated.

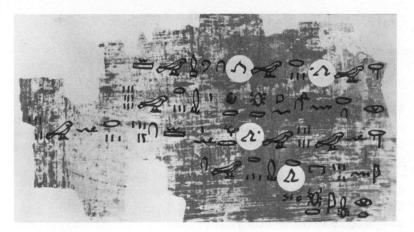

1. Problem no. 28 from the Rhind Mathematical Papyrus, c. 1650 BC. In ancient Egyptian notation, a pair of legs walking left symbolized addition (+), while legs walking right symbolized subtraction (−). According to Peet, this problem 'is incomplete and elliptically worded. Written in full it would run as follows: A number: two-thirds of it is added to itself and one-third of the total is subtracted. The result is 10. Find the number. Answer: 9.' N.B., while the Babylonians wrote from left to right, the Egyptians wrote from right to left.

Perhaps it is in keeping with this attitude that there is in our papyrus practically no instance of the use of a general formula, each case being worked out on its own merits, and cases which to us seem analogous being sometimes dealt with by totally different methods.

In these facts we may see the cause why Egyptian mathematics stagnated, as they undoubtedly did. By the XIIth Dynasty the mathematician was already able to work out any problem which he was liable to meet in ordinary life. He could measure a field or a granary, and divide wages or booty in fixed proportions, and after all what more was needed? For further progress one of two conditions must be realized. Either altered conditions of civilization must present fresh problems for solution, or a genius must arise thirsting for knowledge for its own sake. Egypt could never produce such a genius, and it remained for Greece to do it . . .

System of Notation

The Egyptian system as we find it at the beginning of the Dynastic Period, and as it continued throughout history, was decimal. A unit was represented by a vertical stroke I, two by two strokes, and so on up to 9. Ten was represented by ∩, 20 by two such signs, and so on up to 90. For 100 a new unit ९ appears, and this repeated the requisite number of times served from 200 up to 900. For 1,000 ↑ was used, for 10,000 ⎰, for 100,000 ⟍, and for 1,000,000 ⚊. Thus 143,257 would be written

In the cursive ink-written script known as hieratic many of these numbers took on ligatured and contracted forms, the four strokes, for example, being shortened into a horizontal line. Hieratic forms for the numerals already existed as early as the First Dynasty, and ran through Egyptian history until replaced by the demotic in the Persian period.

Despite the fact that in historical times the system is definitely decimal it contains faint traces of having originally been quinary. The evidence for this is too intricate to be discussed here, but the main points of the latest pronouncement on the subject, that of Jéquier, are as follows. The numbers from 1 to 5 have names resembling the African (Hamitic) names, and are part of the African inheritance of the Egyptians. The numbers from 6–10 have names offering some analogies with the Semitic names and are a later acquisition. The tens from 10 to 40 have special names which correspond neither to those of the Egyptian 1–5 nor to those of the tens in either Hamitic or Semitic languages. The tens from 50 to 90 are formed from the numbers 5–9, of which they are perhaps plural forms. These results must not be regarded as final, and will doubtless meet with considerable criticism. What would appear almost certain, however, is that there are remnants in the Egyptian system of a primitive quinary system based on finger numbering (the number 5 was represented by the figure of a hand), complicated by a later extension to a decimal system formed by the addition of the second hand. Exactly what portions of this system are due to African and Semitic origins respectively is still a matter of almost complete conjecture.

The defects of this system are obvious. In the first place it was cumbrous, for in order to write such a number as 879 no fewer than 24 signs had to be made. This was to a certain extent neutralized in hieratic, where almost every unit, ten, hundred, and thousand developed a contracted form. The other defect of the system was the absence of anything in the nature of value by position, a disadvantage which it shared with the Greek notation and which was only circumvented by the Arab mathematicians, who are said to have derived positional notation from the Hindus and passed it on to us.

As against these defects the system had one virtue which the Greek could not claim: it lent itself admirably to multiplication and division by 10, for in order to multiply 98 by 10 it was only necessary to turn the 8 units into ten-signs and the 9 ten-signs into hundred-signs. The result of this was that multiplication and division by 10 played a large role in the elementary processes of Egyptian reckoning.

Source: T. Eric Peet, *The Rhind Mathematical Papyrus, British Museum 10057 and 10058: Introduction, Transcription, Translation and Commentary* (London: Hodder & Stoughton/Liverpool University Press, 1923), pp. 10–11.

ZENO'S PARADOX AND OTHER PRESOCRATIC FRAGMENTS

Though the Presocratic philosophers were never a cohesive group or school, and their active dates span several centuries, the term is nevertheless useful for referring to those early Mediterranean thinkers, such as Thales and Democritus, who (perhaps under the influence of the Babylonians and Egyptians) began to enquire into the processes of nature. Their thoughts survive mostly in fragment form, buried in the works of later Greek and Roman scholars, but there is enough material with which to form of a picture of a series of restless minds attempting to grapple with the questions that would go on to form the future disciplines of science. So here is Anaximander attempting to explain lunar eclipses and peals of thunder; Xenophanes describing fossilized fish; Empedocles proposing a theory of chemical affinity; and Diogenes asserting that the roundness of the earth is the result of its own rotation.

Though most of these thinkers are now largely forgotten, some of their ideas remain familiar today, such as Leucippus's and Democritus's atomist theory, which states that everything in the universe, including us, is made up of tiny indivisible particles in a state of continual random motion; or Zeno's famous paradox of Achilles and the tortoise, which claims that the faster of two runners could never overtake the slower if the slower is given a head start. Although Zeno's paradoxes were not meant to be taken literally, they were designed to show that the ways in which we imagine that we understand the world are not as straightforward as we think. As Aristotle pointed out, Zeno was clearly wrong – Achilles will naturally overtake the tortoise – but his arguments nevertheless 'give trouble to those who try to solve the problems they raise':

Thales of Miletus (c. 624–547 BC)

Thales is traditionally the first to have revealed the investigation of nature to the Greeks: he had many predecessors, as Theophrastus thinks, but so far surpassed them as to blot out all who came before him. He is said to have left nothing in the form of writings except the so-called 'Nautical Star-Guide'.

(Simplicius, *Commentary on the Physics*)

A witty and attractive Thracian servant-girl is said to have mocked Thales for falling into a well while he was observing the stars and gazing upwards; declaring that he was eager to know the things in the sky, but that what was behind him and just by his feet escaped his notice. (Plato, *Theaetetus*, 174A)

Water is, according to Thales, the most powerful of the elements. He thinks it was the first of them, and that all the others sprang from it. We Stoics, too, are also of the same opinion.
(Seneca, *Quaestiones Naturales*, Bk. III)

Thales said that the sun is eclipsed when the moon is in front of it, and he indicated the day on which it is eclipsed, which some call the thirtieth and others the new moon.
(Anon., *Commentary on the Odyssey*,
Oxyrhynchus Papyrus 3710)

We are indebted to old Thales for many discoveries and for this theorem in particular: he is said to have been the first to have recognized and stated that in every isosceles triangle the angles at the base are equal, and to have called the equal angles 'similar' in the archaic style. (Proclus, *Commentary on Euclid*)

Anaximander of Miletus (c. 610–540 BC)

Anaximander was a pupil of Thales . . . He said that a certain nature, the limitless, is the principle of the things which exist. From it come the heavens and the worlds in them. It is eternal and ageless, and it contains all the worlds. He also calls it time, since the generation and the destruction of the things which exist are determinate.

The earth is aloft, not supported by anything but resting where it is because of its equal distance from everything. Its shape is rounded, circular, like a stone pillar. Of its surfaces, we stand on one while the other is opposite. The heavenly bodies are a circle of fire, separated off from the fire in the world and enclosed by air. There are vents – tubular channels – at which the heavenly bodies appear; hence eclipses occur when the vents are blocked, and the moon appears now waxing and now waning according to the blocking or opening of the channels. The circle of the sun is twenty-seven times greater than the earth and the circle of the moon eighteen times greater. The sun is highest, the circles of the fixed stars lowest.

(Hippolytus, *Refutation of All Heresies* I)

Anaximander refers all the phenomena of thunder to air. Peals of thunder are, he says, the sounds of blows on a cloud. He explains the inequality of the peals by the inequality of the blows. To the question, why it thunders in a clear sky also, he answers that even in absence of cloud the atmosphere is shaken and rent by the bursting forth of air. But why is there thunder sometimes and yet no lightning? The rarity and feebleness of the air render it incapable of producing flame, while yet sufficient to produce sound. Lightning, according to him, then, is really a disturbance where the atmosphere is merely parted and rushes thither and thither, displaying a faint fire that will not issue from its place. As for the thunderbolt, it is the career of the more active and denser air.   (Seneca, *Quaestiones Naturales*, Bk. II)

Xenophanes of Colophon (c. 570–475 bc)

He said that the sea is salty because of the many mixtures flowing along in it. But Metrodorus says that it is on account of its filtering through the earth that it becomes salty. Further, Xenophanes thinks that a mixture of the land with the sea comes about, but that in time the land becomes freed from the moisture, and he asserts that there are proofs for these ideas: that shells are found inland and in mountains, and he says that in quarries in Syracuse imprints of fish and seals were found; and in Paros the imprint of coral in the deep of the marble and on Malta slabs of rock containing all sorts of sea creatures. He says that these things came about when long ago everything was covered with mud, and then the imprint dried in the clay . . . And he says that all men are destroyed when the earth is carried down into the sea and turns into mud; then they begin to be born again. And this is how all the worlds begin.

(Hippolytus, *Refutation of All Heresies* I)

Zeno's Paradox

The infamous paradoxes devised by Zeno of Elea (*fl.* 450 BC) offer a neat illustration of Einstein's observation that 'all the essential ideas in science were born in a dramatic conflict between reality and our attempts at understanding.' They are pure thought experiments, an impertinent challenge to the logicians to prove them false: because merely knowing the paradoxes to be false was easy – proving them to be so was next to impossible:

Zeno's reasoning, however, is fallacious, when he says that if everything when it occupies an equal space is at rest, and if that which is in locomotion is always occupying such a space at any moment, the flying arrow is therefore motionless. This is false, for time is not composed of indivisible moments any more than any other magnitude is composed of indivisibles.

Zeno's arguments about motion, which cause so much disquietude to those who try to solve the problems that they present, are four in number. The first asserts the non-existence of motion on the ground that that which is in locomotion must arrive at the half-way stage before it arrives at the goal. This we have discussed above.

The second is the so-called 'Achilles', and it amounts to this, that in a race the quickest runner can never overtake the slowest, since the pursuer must first reach the point whence the pursued started, so that the slower must always hold a lead. This argument is the same in principle as that which depends on bisection, though it differs from it in that the spaces with which we successively have to deal are not divided into halves. The result of the argument is that the slower is not overtaken: but it proceeds along the same lines as the bisection-argument (for in both a division of the space in a certain way leads to the result that the goal is not reached, though the 'Achilles' goes further in that it affirms that even the quickest runner in legendary tradition must fail in his pursuit of the slowest), so that the solution must be the same. And the axiom that that which holds a lead is never overtaken is false: it is not overtaken, it is true, while it holds a lead: but it is overtaken nevertheless if it is granted that it traverses the finite distance prescribed. These then are two of his arguments.

The third is that already given above, to the effect that the flying arrow is at rest, which result follows from the assumption that time is composed of 'moments': if this assumption is not granted, the conclusion will not follow.   (Aristotle, *Physics* Bk. VI)

Empedocles of Acragas (c. 495–435 BC)

Empedocles said – as Aristotle stated in *On Generation and Corruption* – that in all sublunary things (water, oil, etc.) channels and solid parts are mingled. He called the channels hollow and the solid parts dense. Where the solid parts and the channels, i.e. the hollow and the dense parts, are commensurate in such a way as to pass through one another, he said that mixing and blending take place (for example water and wine), but where they are incommensurate, he said they do not mix (for example oil and water); for he says that water 'has an affinity with wine, but with oil it will not . . .' Applying this to all bodies, he attempted to explain the sterility of mules.

(Michael of Ephesus, *Commentary on the Generation of Animals*)

Anaxagoras of Clazomenae (c. 500–428 BC)

Pericles gained much from his association with Anaxagoras, and in particular he is thought to have risen above that superstition which amazement at the celestial phenomena produces in those who are ignorant of their causes and who, because of their inexperience, are fascinated and confused about things divine – a state of mind which is changed by a scientific account, which creates a sure piety based on good hopes in place of a fearful and feverish superstition.

(Plutarch, *Pericles*)

Anaxagoras says all the phenomena correspond to the descent of some force from there to the lower regions. So when the fire encounters cold clouds it emits a sound; when it cleaves them there is a flash; less violence in the fires produces lightning; greater, thunderbolts.

(Seneca, *Quaestiones Naturales*, Bk. II)

Leucippus (fl. early 5th century BC)

Leucippus of Elea or Miletus (he is variously said to be from both cities), although he was committed to Parmenides' theory, did not pursue the same path about existing things as Parmenides and Xenophanes, but the opposite, as it were. For whereas they held that the totality was one, motionless, ungenerated, and limited, and rules out seeking what-is-not, he theorized that the elements were atoms innumerable and in motion, and that there was unceasing coming to be and change in existing things. Moreover, what-is exists no more than what-is-not, and they are both equally causes of generated things ...

For, supposing that the substance of the atoms is solid and full, he said that it exists and that it is carried about in what is empty, which he called non-existent and which he says exists no less than does what is existent.

(Simplicius, *Commentary on Aristotle's Physics*)

Democritus of Abdera (c. 460–370 BC)

Democritus thinks that the nature of eternal things consists in small substances, limitless in quantity, and for them he posits a place, distinct from them, and limitless in extent. He calls places by the names 'empty', 'nothing' and 'limitless'; and each of the substances he calls 'thing', 'solid' and 'existent'. He thinks that the substances are so small that they escape our senses, and that they possess all sorts of forms and all sorts of shapes and differences in size. From them, as from elements, he produces and compounds the visible and perceptible masses. The atoms struggle and are carried about in the empty because of their dissimilarities and the other differences mentioned, and as they are carried about they collide and intertwine in a way which makes them touch and be near one another but which does not produce any truly single nature whatever from them; for it is utterly foolish to think that two or more things might ever become one. He explains that the substances remain together for a certain time because the bodies entangle with and grasp hold of one another; for some of them are scalene, some hooked, some concave, some convex, and others have innumerable other differences. So he thinks that they hold on to one another and remain together up to the time when some stronger necessity reaches them from their surroundings and shakes them and scatters them apart ...

(Simplicius, *Commentary on the Heavens*)

Diogenes of Apollonia (*fl.* 450 BC)

Air is the element. There are unlimited worlds and unlimited void.
The air by being condensed and rarefied is generative of the worlds.
Nothing comes to be from or perishes into what is not. The earth is
round and is supported in the center (of the cosmos) and has undergone its process of formation through the rotation resulting from the
hot and the solidification caused by the cold.

(Diogenes Laertius, *Lives of the Philosophers* IX)

This is the beginning of his book: 'It is my opinion that the author, at
the beginning of any account, should make his principle or starting-
point indisputable, and his explanation simple and dignified'.

(Diogenes Laertius, *Lives of the Philosophers* IX)

Sources: I have made fair use of a range of sources for the fragments, including: Aristotle,
Physica, trans. R. P. Hardie and R. K. Gaye (Oxford: Clarendon Press, 1908); John Clarke, *Physical
Science in the Time of Nero: Being a Translation of the Quaestiones Naturales of Seneca* (London:
Macmillan, 1910); G. S. Kirk, J. E. Raven and M. Schofield (eds), *The Presocratic Philosophers:
A Critical History with a Selection of Texts*, 2nd edn (Cambridge: Cambridge University Press,
1983); *Xenophanes of Colophon: Fragments*, trans. J. H. Lesher (Toronto: University of Toronto
Press, 1992); Patricia Curd and Richard D. McKirahan, Jr (eds), *A Presocratics Reader: Selected
Fragments and Testimonia* (Indianapolis: Hackett, 1995); Jonathan Barnes (ed.), *Early Greek Phil-
osophy*, 2nd edn (London: Penguin, 2001); Daniel W. Graham (ed.), *The Texts of Early Greek
Philosophy: The Complete Fragments and Selected Testimonies of the Major Presocratics* (Cambridge:
Cambridge University Press, 2010).

THE SALTNESS OF THE SEA

'Why is the sea salty?' was a simple question that nevertheless puzzled ancient philosophers, who offered a variety of competing answers, from Democritus's suggestion that the oceans have shrunk and thereby become more concentrated, to Empedocles's crude bodily analogy of the sea as the sweat of the earth.

Aristotle

In this survey of the subject from his *Meteorologica* (c. 340 BC), the influential Greek philosopher and naturalist Aristotle (c. 384–322 BC) set out the parameters of the debate so far, offering some salty observations of his own on the Presocratics' use of metaphorical language in connection with their search for natural knowledge. As will be seen in later episodes (on p. 127, for example), efforts to constrain the style of scientific communication have regularly been made over the centuries:

We must now explain why the sea is salt, and ask whether it eternally exists as identically the same body, or whether it did not exist at all once and some day will exist no longer, but will dry up as some people think.

Every one admits this, that if the whole world originated the sea did too; for they make them come into being at the same time. It follows that if the universe is eternal the same must be true of the sea. Any one who thinks like Democritus that the sea is diminishing and will disappear in the end reminds us of Aesop's tales. His story was that Charybdis had twice sucked in the sea: the first time she made the mountains visible; the second time the islands; and when she sucks it in for the last time she will dry it up entirely. Such a tale is appropriate enough to Aesop in a rage with the ferryman, but not to serious inquirers . . .

To return to the saltness of the sea: those who create the sea once for all, or indeed generate it at all, cannot account for its saltness. It makes no difference whether the sea is the residue of all the moisture that is about the earth and has been drawn up by the sun, or whether all the flavour existing in the whole mass of sweet water is due to the admixture of a certain kind of earth. Since the total volume of the sea is the same once the water that evaporated has returned, it follows that it must either have been salt at first too, or, if not at first, then not now either. If it was salt from the very beginning, then we want to know why that was so; and why, if salt water was drawn up then, that is not the case now.

Again, if it is maintained that an admixture of earth makes the sea salt (for they say that earth has many flavours and is washed down by the rivers and so makes the sea salt by its admixture), it is strange that rivers should not be salt too. How can the admixture of this earth have such a striking effect in a great quantity of water and not in each river singly? For the sea, differing in nothing from rivers but in being salt, is evidently simply the totality of river water, and the rivers are the vehicle in which that earth is carried to their common destination.

It is equally absurd to suppose that anything has been explained by calling the sea 'the sweat of the earth', like Empedocles. Metaphors are poetical and so that expression of his may satisfy the requirements of a poem, but as a scientific theory it is unsatisfactory. Even in the case of the body it is a question how the sweet liquid drunk becomes salt sweat whether it is merely by the departure of some element in it which is sweetest, or by the admixture of something, as when water is strained through ashes. Actually the saltness seems to be due to the same cause as in the case of the residual liquid that gathers in the bladder. That, too, becomes bitter and salt though the liquid we drink and that contained in our food is sweet. If then the bitterness is due in these cases (as with the water strained through lye) to the presence of a certain sort of stuff that is carried along by the urine (as indeed we actually find a salt deposit settling in chamber-pots) and is secreted from the flesh in sweat (as if the departing moisture were washing the stuff out of the body), then no doubt the admixture of something earthy with the water is what makes the sea salt.

Rachel Carson

Although best known today as the author of *Silent Spring* (1962), her influential polemic against the use of chemical pesticides, Rachel Carson (1907–64) worked for many years as a marine biologist, and began her writing career with a trilogy of beautifully crafted books about the sea. The following extract, in which Carson outlines the mineral content of the oceans, is from the second in the series, *The Sea Around Us*, that appeared in 1951:

The ocean is the earth's greatest storehouse of minerals. In a single cubic mile of sea water there are, on the average, 166 million tons of dissolved salts, and in all the ocean waters of the earth there are about 50 quadrillion tons. And it is in the nature of things for this quantity to be gradually increasing over the millennia, for although the earth is constantly shifting her component materials from place to place, the heaviest movements are forever seaward.

It has been assumed that the first seas were only faintly saline and that their saltiness has been growing over the eons of time. For the primary source of the ocean's salt is the rocky mantle of the continents. When those first rains came – the centuries-long rains that fell from the heavy clouds enveloping the young earth – they began the processes of wearing away the rocks and carrying their contained minerals to the sea. The annual flow of water seaward is believed to be about 6500 cubic miles, this inflow of river water adding to the ocean several billion tons of salts.

It is a curious fact that there is little similarity between the chemical composition of river water and that of sea water. The various elements are present in entirely different proportions. The rivers bring in four times as much calcium as chloride, for example, yet in the ocean the proportions are strongly reversed – 46 times as much chloride as calcium. An important reason for the difference is that immense amounts of calcium salts are constantly being withdrawn from the sea water by marine animals and are used for building shells and skeletons – for the microscopic shells that house the foraminifera, for the massive structures of the coral reefs, and for the shells of oysters and clams and other mollusks. Another reason is the precipitation of calcium from sea water. There is a striking difference, too, in the silicon content of river and sea water – about 500 per cent greater in rivers than in the sea. The silica is required by diatoms to make their shells, and so the immense quantities brought in by rivers are largely utilized by these ubiquitous plants of the sea. Often there are exceptionally heavy growths of diatoms off the mouths of rivers. Because of the enormous total chemical requirements of all the fauna and flora of the sea, only a small part of the salts annually brought in by rivers goes to increasing the quantity of dissolved minerals in the water. The inequalities of chemical make-up are further reduced by reactions that are set in motion immediately the fresh water is discharged into the sea, and by the enormous disparities of volume between the incoming fresh water and the ocean.

There are other agencies by which minerals are added to the sea – from obscure sources buried deep within the earth. From every volcano chlorine and other gases escape into the atmosphere and are carried down in rain onto the surface of land and sea. Volcanic ash and rock bring up other materials. And all the submarine volcanoes, discharging through unseen craters directly into the sea, pour in boron, chlorine, sulphur, and iodine.

All this is a one-way flow of minerals to the sea. Only to a very limited extent is there any return of salts to the land. We attempt to recover some of them directly by chemical extraction and mining, and indirectly by harvesting the sea's plants and animals. There is another way, in the long, recurring cycles of the earth, by which the sea itself gives back to the land what it has received. This happens when the ocean waters rise over the lands, deposit their sediments, and at last withdraw, leaving over the continent another layer of sedimentary rocks. These contain some of the water and salts of the sea. But it is only a temporary loan of minerals to the land and the return payment begins at once by way of the old, familiar channels – rain, erosion, run-off to the rivers, transport to the sea.

Sources: Aristotle, *Meteorologica*, trans. E. W. Webster (Oxford: Clarendon Press, 1931), 356^a–357^b [Bk. II, iii]; Rachel Carson, *The Sea Around Us* (New York: Oxford University Press, 1951), pp. 188–90.

ATOMS AND INFINITY

As was seen earlier (p. 13), the ancient Greek materialist philosophers Leucippus, Democritus and Epicurus (c. 341–270 BC) argued that everything in the world, including us, was made of minute, indivisible particles called atoms (from the Greek *atomos*, meaning 'that which cannot be divided'). But it was a Roman poet, Titus Lucretius Carus (c. 100–55 BC), who first popularized Greek atomist theory in an extraordinary 7,400-line didactic poem known as *De rerum natura* ('On the Nature of Things'), from the first English translation of which I have chosen two representative extracts. The translator, Lucy Hutchinson (1620–81), was the author of *Order and Disorder* (1679), the first epic poem to be written by an Englishwoman, and as the following selections show, her poetic gifts were given full reign in this exuberant scholarly rendition of Lucretius's great work.

The first passage is concerned with demonstrating the infinity of the universe, and asks the reader to conduct a curiously modern thought experiment: imagine the edge of the universe, and then imagine a spear or dart thrown past that edge – into what? Into yet more universe, of course:

> We' have proovd materiall bodies solid be,
> And uncorrupted moove eternally,
> Now letts enquire, whither they have a bound,
> Or are unlimited; Againe we 'ave found
> Vacuitie an open empty space,
> Where each thing mooves, and hath its proper place;
> Let's then consider too, whither this be
> An universall vast profunditie,
> Or else enclosd with bounds; the world hath none;
> For whatere yeilds to circumscription,
> It selfe hath an extreame, that must admitt
> Or something else beyond, that limitts it.
> And when our sence arrives at that extreame,
> Which doth the utmost terme of allthings seeme,
> It matters not, where we that region site,
> Which must unbounded be, and infinite;
> Since nothing can those utmost limitts passe,
> Soe that though parts may be assignd to place,

Th'immense universe yet must we leave
Unlimited, and free; but now conceive
All space could be comprizd in bounds, and soe
Some man might to those uttmost regions goe,
Should he there hurl his dart, with nimble force,
Would it flie on, in the designed course,
Or opposition find? One of these two
Must be confest; and granting either, you
Barre all retreate, and both ways are compelld,
To allow the world, within no limitt held;
For whither aniething the shaft oppose
Or stop its flight, and it with bounds enclose,
Or whither it be carried forth, it will
Not finde an utmost terme, for we shall still
Persue it wheresoere it flies, and when
Tis brought to the extreamest region, then
Enquire where you at length the shaft will place,
Which wheresoere it flies, will still find space.

This second extract contains one of Lucretius's most celebrated images, of
dust-motes dancing in a sunlit room, just as atoms dance through the immensity of space and time:

Now that in this vast deepe it seeme not strange
Materiall bodies ceaselessly should range,
Thinke 'tis a place, which hath it selfe no bounde,
Where no fixt seat can be for bodies found;
Whose fathomlesse unlimited extent
Allreadie proov'd, needs not new argument.
And since tis soe, in this unbounded space,
The principles rest not in any space:
But dayly chang'd in motion still abide;
Some in loose order joynd, extending wide;
Some that touch neere, with stricter unions closd,
Into a narrower compasse are disposd;
And interwoven soe, they cannot spread.
By these are the more solid bodies bred,
As stones, hard iron, and some few more. But they
Which from each others touch doe start away,

And in a wider scope themselves dilate,
They the transparent ayre and light create.
Besides all these, many loose bodies be
Which nere receiv'd into societie,
Alone still wander through this vast extent,
Whose image dayly objects represent.
For if you marke, when the high sun conveys
Into an obscure roome his piercing rayes,
Even where the light flowes in with glorious streames,
Armies of attoms sport in those bright beames,
And meeting in perpetuall skirmishies,
Here joyne, there part, their motions never cease;
From whose vicissitudes we may comprize,
What motions the first bodies exercise,
In the unbounded world; thus small things may
Illustrate greate, and guide us in the way
Which to cleare knowledge leads; Againe when we
Those mooving attoms in the sunbeames see,
The perplext agitations in there declare,
Such secret tumults in the matter are;
For these troopes smitten with undiscerned force
Are oft driv'n back, and often change their course,
Here mount, there sinke, on every side reverst,
All by th' impulsive matter thus disperst.
For principles first moove themselves, then those
Whose bodies fewest substances compose,
Who next them plac'd, their mooving power provokes,
By the impulsion of its secret strokes.
These moovd by them, moove the next rank, from whence
Motion proceeds, untill it meete our sence;
Which sees the attoms in the sunbeams strive,
But not the force, whence they that power derive.

J. J. Thomson

The British physicist Joseph John ('J. J.') Thomson (1856–1940) won the 1906 Nobel Prize in Physics for his discovery of the electron, an elementary sub-atomic particle whose existence had first been postulated in the mid-nineteenth century. He attempted to name these particles 'corpuscles', but later physicists preferred to use the more classical term 'electron'. In this extract from a public lecture given in Oxford in 1914, Thomson looks back at the early Greek atomists, before making a quietly brilliant insight about 'physical facts': that facts are not the dull, pedantic part of science, but the fun part, the impulsive part that gives the greatest kick to scientific creativity. For 'nature is far more wonderful and unconventional than anything we can evolve from our inner consciousness':

The theory that matter in spite of its apparent continuity is in reality made up of a great number of very small particles, is as old as the science of Physics itself, and was enunciated almost as soon as men began to reason about physical phenomena. It would, however, be misleading to suppose that there is any very close connexion between the modern Atomic Theory and the views of Democritus and Lucretius. The old theory was in intention and effect metaphysical rather than physical, theological rather than scientific. The physics of two thousand years ago was far too scanty and uncertain to afford any support or test for such a theory; indeed, if I were called upon to prove to you that Democritus was right when he held that matter was discontinuous, and Aristotle wrong when he said it was not so, I should have to appeal to facts not one of which was known either to Democritus or Aristotle. The great and invaluable service which the Greek atomists have rendered to science is that they were the first to attempt on mechanical principles to explain complicated physical phenomena as the result of combinations of simpler ones; they pointed out the goal which science is still struggling to reach. For two thousand years the Atomic Theory itself made no progress, because, though in form a physical theory, it had no real connexion with physical phenomena, no facts were known by which it could be tested, and it was too vague to suggest for itself effects which could be put to the text of experiment. It was sterile because it was divorced from experience. It affords a striking proof that a theory can only grow by the co-operation of thought and facts, and that all that is valuable in a physical theory is not only tested, but in most cases suggested, by the study of physical phenomena. In the interplay between mind and matter in scientific discovery, the parts played by the two are, I think, widely different from those usually assigned to them in popular estimation. There is a widespread belief that the mind itself is desperately speculative, that it is only kept from wild imaginings by the control of its stolid and prosaic partner, the physical facts. The true state of affairs is, I think, that it is the mind which acts as the brake in this combination, that the impulsive partner is the facts, and that these spur on the mind to take leaps which it would shudder at when not under the influence of this stimulus. Nature is far more wonderful and unconventional than anything we can evolve from our inner consciousness. The most far-reaching generalizations which may influence philosophy as well as revolutionize physics, may be suggested, nay, forced on the mind by the discovery of some trivial phenomenon. To take an example, an improvement in the method of exhausting air from closed vessels enabled experimenters to send an electric discharge through gas more highly rarefied than had previously been possible. When they did this they observed that the glass of the vessel shone with a peculiar phosphorescent light: the study of this light led to the discovery of cathode rays, cathode rays led on to Rontgen rays, and the study of those rays started ideas which have entirely changed our conceptions of matter.

Sources: Lucretius, *De rerum natura*, trans. Lucy Hutchinson (British Library Add. MS 19333, ff. 23ᵛ–24ᵛ [Bk. I, 955–90]; ff. 29ᵛ–30ᵛ [Bk. II, 90–137]); J. J. Thomson, *The Atomic Theory: The Romanes Lecture 1914* (Oxford: Clarendon, 1914), pp. 3–4.

WHIRLWINDS

A whirlwind is a mini-tornado that forms in an updraught of air as it passes over discontinuously heated ground. If the moving air is funnelled through some kind of constriction it can begin to rotate, picking up dust as it does so, growing upwards into a sometimes quite vigorous twister. True tornadoes, on the other hand, form below the base of a supercell stormcloud when powerful downdraughts drag the cloud's rotating heart (its mesocyclone) towards the ground, creating a visible funnel that begins to throw up soil and debris as soon as it makes contact. As the following two extracts show, the causes of such phenomena have been known for centuries.

Seneca

Lucius Annaeus Seneca (c. 4 BC – AD 65) was a Roman statesman and philosopher, who based his last book, *Natural Questions*, on a lifetime of reading about the natural sciences. Seneca evidently understood that air and water are subject to similar kinds of turbulent flow, and knew that the most powerful storms were driven by the rotation of the earth. In the following extract, on the phenomenon of the whirlwind (Latin: *turbo*), he employs a deft analogy with the way water swirls around an obstruction to create a vortex, as well as what appears to be an accurate description of the collapse of a tornadic stormcloud's rain-free base:

At this point, if you have no objection, one may raise the question why a whirlwind occurs. In rivers, when their course has been without any obstacle for a long distance, the channel is a straight, uniform one. But when they meet some boulder that juts from the bank, the stream is driven back and whirls the waters in a circle without a way of escape, so that in their revolution they are constantly sucked in toward the centre to form a whirlpool. In like manner the wind pours out in full force as long as no obstacle stands in the way. But when it is reflected from some jutting projection, or is massed in a quarter which combines to form a thin downward channel, then it revolves upon its own axis, and produces an eddy similar to that in which, as we have just said, the water revolves. This revolving wind, which always traverses the same spot and is roused to fury by the mere giddy whirling, is a whirlwind. If it is a very fierce one, and revolves longer than ordinary, it ignites and causes what the Greeks call a fire-wind (*prester*), which is just a fiery whirlwind. The bursting of such winds from the clouds produces almost all the disasters by which herds are carried off and ships lifted, bodily, right out of the water. Further, some winds produce different ones by dispersing the air and driving it before them in other directions than that toward which they themselves have bent their course.

It occurs to me at the moment to mention a parallel to wind that may be drawn from drops of moisture. The single drops may begin to incline downwards and be on the verge of giving way, but yet do not manage to fall. When, however, several have united and the mass has imparted strength, then they are said to flow and to move. So, as long as there are slight movements of the atmosphere disturbed at several points, they do not produce wind. The latter begins only when all those movements are united and concentrated in a single effort. Air differs from wind in degree alone. A more violent air is a wind; air in turn is gently flowing atmosphere.

. . . I said a moment ago that the whirlwind's eddy could not long endure, nor could it mount higher than the moon, or as far as the place of stars. Of course, the whirlwind is caused by the mutual struggle of several winds, and the contest cannot be kept up for any long time. When the wandering uncertain air assumes a rotatory form, in the last instance the force of all the winds yields to the single strongest one. No hurricane lasts long. The more strength squalls have, the shorter their duration. When winds reach their maximum, they quickly abate all their violence. By that headlong speed they must needs hasten to their own destruction. So no one has ever seen a whirlwind last a whole day, or even an hour. Its velocity is astonishing, its brevity no less astonishing. Moreover, on the earth and near it, its rotation is swifter and more violent; the higher it is, the less condensed and compact is it, and that is the reason of its more rapid dissipation. Add the fact, too, that even if it reached the highest region where the stars' path lies, it would most certainly be broken up by the motion which causes the universe to revolve. For what can compare in rapidity with the revolution of the world? Thereby the strength of all the winds combined in one would be shattered, aye, and the strong solid chain that binds the earth, not to say a wisp of whirling air.

John P. Finley

The first book devoted to a meteorological treatment of full-scale Midwest tornadoes appeared in 1887, written in a heightened 'frontier' style by John Park Finley (1854–1943), a lieutenant in the US Army Signal Corps:

The populous region of the United States is forever doomed to the devastation of the tornado. As certain as that night follows day is the coming of the funnel-shaped cloud. So long as the sun shines upon the vast regions in the Mississippi and Missouri valleys, there will forever occur those atmospheric conditions which terminate in the destructive violence of the tornado. Nature's laws are unerring in their certainty of procedure, the earth must travel in its orbit about the sun and the seasons must recur in regular sequence as the result of this wonderful periodicity of movement. The earth must revolve upon its axis, and daylight and darkness, heat and cold, must succeed each other with infallible precision. Without these great and regular mutations dependent upon the solar system, atmospheric phenomena would cease altogether. Granting that the solar system must continue intact, we have but to watch and protect ourselves as best we may against the fury of the elements. Ignorance of our surroundings is a most unfortunate plea for those who stubbornly fail to heed the warnings of science . . .

What is a tornado? In defining this storm it would seem almost a necessity to rehearse its long line of striking characteristics, but this in the common acceptation of the term would not strictly be a definition. For the sake of brevity, we will state that the tornado is that form of atmospheric disturbance which takes the outward, visible fashion or figure of a funnel-shaped cloud, revolving about a vertical axis from right to left with an inconceivably rapid movement and an immensity of power almost beyond calculation.

Conditions of Formation. – These may be divided into classes. First, those within the reach of and which may be known or investigated by an isolated observer. Second, those conditions only to be witnessed and analyzed by the intelligent and practiced eye of the

2. A dramatic engraving of an 1883 tornado from the first book devoted to the subject, John P. Finley's *Tornadoes: What they are and how to observe them* (1887).

student of the weather map. To the single observer, located mayhap at his farm home, the workshop, or the store, there are important atmospheric conditions which he may carefully watch and study with profit, viz.: the gradual setting in and prolonged movement of the air from the north and south points; the gradual but continued *fall* of the thermometer with a prevalence of the northerly currents, or a *rise* with the predominance of the southerly. If the northerly currents are the prevailing air-movements at your place of observation, the atmospheric disturbance is forming to the southward, but if the prevailing air-currents are from the south the storm is forming to the northward of your location. Carefully study cloud development, color as well as form, also manner and direction of approach. The approach of the cirrus cloud (perhaps at a height of six to eight miles) from the southwest is very significant, and is the first evidence of the gradual but certain advance of the upper southwest current, which eventually plays so important a part in the development of the tornado-cloud. Clouds are but the embodiment of air-currents, yet they are full of meaning. A study of the upper currents of the atmosphere would be impossible without their manifestations, and that, too, in a variety of forms. Without cloud formation, the face of the sky would become a blank, and intelligent reasoning thereof a superhuman task.

Wind direction, temperature, and clouds are the proper subjects of observation and thought by the isolated observer. The barometer is of little if any importance in this line of inquiry. If you cannot compare your barometric observations with those taken at near or distant points and at the same moment of actual time, they are of no practical moment, even though your instrument is a standard one and your corrections for temperature and elevation carefully applied. The storm you are watching for (the tornado) is an extremely local affair, whereas the barometer indicates general changes, affecting a large extent of country. Your instrument, if a standard, does not lack possession of the delicate sensitiveness requisite for all the purposes of its construction, but if it were placed in the immediate track of the tornado-cloud, it would not indicate its presence until the crash of the storm was upon the instrument, when of course it would be too late . . .

As these conditions continue to prevail there is a growing contrast of temperature to the north and south of the major axis, owing to the long-continued movement of the atmosphere from opposite directions, such movement eventually affecting the disposition of air in the warmer regions of the extreme south and likewise the colder regions of the extreme north. The contrast of temperature now naturally increases with marked rapidity, and the formation of clouds commences in earnest. Huge masses of dark and portentous appearance bank up in the northwest and southwest with amazing rapidity, and soon the scene becomes one of awful grandeur. The struggle for mastery in the opposing currents is thus indicated by the gathering cloud formations. The condensation of vapor from the extremely humid southerly currents by contact with the augmenting cold of their struggling opponents continues. It increases rapidly. Finally, when resistance to the unstable equilibrium can no longer be maintained (controlled by the rate of temperature change and rapidity of condensation), the opposing forces are, as it were, broken asunder, followed by the upward rush of huge volumes of air. The outward indication of this event is first shown in the whirling, dashing clouds over the broken surface of the heavy bank of condensed vapor, forming the background. A scene not easily depicted or realized by one who has not witnessed it, but never to be effaced from the memory of the actual observer. There is an awful terror in the majesty of the power here represented, and in the unnatural movement of the clouds, which affects animals as well as human beings. The next stage in the further development of this atmospheric disturbance is the gradual descent of the funnel-shaped cloud from a point apparently just beneath the position of the enactment of the first scene. The tornado is now before us, not fully developed, but soon to acquire that condition when the terrible violence of its power will make the earth tremble, animals terror-stricken, and men's hearts quake with fear.

Sources: *Physical Science in the Time of Nero: Being a Translation of the Quaestiones Naturales of Seneca* (London: Macmillan, 1910), pp. 204–5; 280–81 (Bk.V, 13.1–4; Bk.VII, 9.2–4); John P. Finley, *Tornadoes: What they are and how to observe them; with practical suggestions for the protection of life and property* (New York: The Insurance Monitor, 1887), pp. 8–25.

THE PENALTY OF RUST

The Roman naturalist Pliny the Elder (c. AD 23–79) completed his multi-volume *Natural History* in AD 77, two years before he met his death during the great eruption of Vesuvius – the terms Plinian and ultra-Plinian are still used by volcanologists to describe the more violent kinds of eruption. The following extract from book thirty-four of his vast compendium offers an insight into the advanced state of Roman metallurgy, in which the understanding of the effects of rust and verdigris (copper carbonate) extended to their deliberate use as decorative sculptural devices; it also features an early expression of disquiet at the military exploitation of science and technology in the shape of the 'winged missile', a sentiment that anticipates the later unease that was felt by scientists such as Erwin Chargaff in the wake of the bombings of Japan (see p. 360):

Verdigris is also applied to many purposes, and is prepared in numerous ways. Sometimes it is detached already formed, from the mineral from which copper is smelted: and sometimes it is made by piercing holes in white copper, and suspending it over strong vinegar in casks, which are closed with covers; it being much superior if scales of copper are used for the purpose. Some persons plunge vessels themselves, made of white copper, into earthen pots filled with vinegar, and scrape them at the end of ten days. Others, again, cover the vessels with husks of grapes, and scrape them in the same way, at the end of ten days. Others sprinkle vinegar upon copper filings, and stir them frequently with a spatula in the course of the day, until they are completely dissolved. Others prefer triturating these filings with vinegar in a brazen mortar: but the most expeditious method of all is to add to the vinegar shavings of coronet copper. Rhodian verdigris, more particularly, is adulterated with pounded marble; some persons use pumice-stone or gum.

The adulteration, however, which is the most difficult to detect, is made with copperas; the other sophistications being detected by the crackling of the substance when bitten with the teeth. The best mode of testing it is by using an iron fire-shovel; for when thus subjected to the fire, if pure, the verdigris retains its colour, but if mixed with copperas, it becomes red. The fraud may also be detected by using a leaf of papyrus, which has been steeped in an infusion of nut-galls; for it becomes black immediately upon the genuine verdigris being applied. It may also be detected by the eye; the green colour being unpleasant to the sight. But whether it is pure or adulterated, the best method is first to wash and dry it, and then to burn it in a new earthen vessel, turning it over until it is reduced to an ash; after which it is pounded and put by for use. Some persons calcine it in raw earthen vessels, until the earthenware becomes thoroughly baked: others again add to it male frankincense. Verdigris is washed, too, in the same manner as cadmia.

It affords a most useful ingredient for eye-salves, and from its mordent action is highly beneficial for watery humours of the eyes. It is necessary, however, to wash the part with warm water, applied with a fine sponge, until its mordency is no longer felt . . .

Next to copper we must give an account of the metal known as iron, at the same time the most useful and the most fatal instrument in the hand of mankind. For by the aid of iron we lay open the ground, we plant trees, we prepare our vineyard-trees, and we force our vines each year to resume their youthful state, by cutting away their decayed branches. It is by the aid of iron that we construct houses, cleave rocks, and perform so many other useful offices of life. But it is with iron also that wars, murders, and robberies are effected, and this, not only hand to hand, but from a distance even, by the aid of missiles and winged weapons, now launched from engines, now hurled by the human arm, and now furnished with feathery wings. This last I regard as the most criminal artifice that has been devised by the human mind; for, as if to bring death upon man with still greater rapidity, we have given wings to iron and taught it to fly. Let us therefore acquit Nature of a charge that here belongs to man himself.

Indeed there have been some instances in which it has been proved that iron might be solely used for innocent purposes. In the treaty which Porsena granted to the Roman people, after the expulsion of the kings, we find it expressly stipulated, that iron shall be only employed for the cultivation of the fields; and our oldest authors inform us, that in those days it was considered unsafe to write with an iron pen. There is an edict extant, published in the third consulship of Pompeius Magnus, during the tumults that ensued upon the death of Clodius, prohibiting any weapon from being retained in the City.

Still, however, human industry has not failed to employ iron for perpetuating the honours of more civilized life. The artist Aristonidas, wishing to express the fury of Athamas subsiding into repentance, after he had thrown his son Learchus from the rock, blended copper and iron, in order that the blush of shame might be more exactly expressed, by the rust of the iron making its appearance through the shining substance of the copper; a statue which still exists at Rhodes. There is also, in the same city, a Hercules of iron, executed by Alcon, the endurance displayed in his labours by the god having suggested the idea. We see too, at Rome, cups of iron consecrated in the Temple of Mars the Avenger. Nature, in conformity with her usual benevolence, has limited the power of iron, by inflicting upon it the penalty of rust; and has thus displayed her usual foresight in rendering nothing in existence more perishable, than the substance which brings the greatest dangers upon perishable mortality . . . Human blood revenges itself upon iron; for if the metal has been once touched by this blood it is much more apt to become rusty.

As Pliny's words imply, corrosion was one of nature's conundrums, especially when the advent of precision balances in the mid-eighteenth century revealed that a rusted piece of iron weighs more than it did in its pristine state. This is because the metal has oxidized – it has taken on oxygen molecules from the air – though there is more to corrosion than the simple appearance of rust. Firstly, a corroded metal surface has been attacked by certain gases in the atmosphere, as well as by moisture, and these two agents interact in chemically complicated ways. In fact a piece of corroding metal behaves a little like a battery, with small localized electric currents running around the surface. Corrosion is almost a form of electrolysis, as William Alexander and Arthur Street pointed out in their near-definitive survey *Metals in the Service of Man* (a book that first appeared in 1944, and was regularly updated in new editions over the course of the following four decades). 'In a battery an electric current is produced by suspending two metals in a chemical solution', they wrote. 'When the circuit is completed one metal, known as the anode, dissolves, while an electric current flows through the solution from this corroding metal to the other, called the cathode.' This, they argued, is closely analogous to the situation of a piece of metal in contact with moisture in the air: 'the presence of particles of an impurity, or contact with some other metal, allows a difference of voltage to be set up, thus causing a minute electric current to flow. The moisture, containing air or some dissolved chemical substance, conducts the electricity and local attack is begun.'

Source: *The Natural History of Pliny*, trans. John Bostock and H. T. Riley, 6 vols (London: Henry G. Bohn, 1855), VI, pp. 195–7; 205–9 (Bk. XXXIV).

THE FERRIS WHEEL UNIVERSE

Aristotle had maintained that the planets move in stately circles at a uniform speed around the earth, but by the beginning of the Christian era this view could no longer be reconciled with astronomical observations. The dilemma was seemingly solved by the Greek-Egyptian mathematician Claudius Ptolemy (c. AD 100–178), who argued that the planets move at variable speeds, performing small looping circles, known as epicycles, within their wider circular orbits of the earth. As new observations served to complicate the picture, the epicycles – and the mathematics needed to describe them – became ever more complex and ingenious. Ptolemy's calculations were collected in a thirteen-volume compendium entitled the *Syntaxis mathematica*, better known by its Arabic name, the *Almagest* ('the greatest'), much of which is dense, difficult and obscurely expressed, though I am fond of Ptolemy's assertion, from the beginning of Book I, that the earth does not rotate, otherwise objects would fling off its surface like mud from a spinning wheel, and 'the animals and other weights would be left hanging in the air, and the Earth would very quickly fall out of the heavens.'

The Ptolemaic universe, in which the earth remained fixed at the centre while the stars and planets performed ever more elaborate quadrilles around it, dominated cosmology for more than 1,400 years, and remained the orthodox view of the heavens even after Nicholas Copernicus published his *De Revolutionibus Orbium Coelestium* in 1543 (see p. 63). It was not until Johannes Kepler devised his laws of planetary motion in the early decades of the seventeenth century that Ptolemy's infamous epicycles were finally abandoned. Arthur Koestler – whose grand evocation of the Babylonian worldview opened this anthology – seemed perturbed by the hold that Ptolemy's work seemed to have over centuries of astronomical thought, describing it as 'the work of a pedant with much patience and little originality, doggedly piling "orb in orb"'. No wonder he called it 'the Ferris Wheel universe', a clockwork artefact in which vast numbers of invisible wheels revolved in intricate sympathy, turning independently from the main structure like the cabins on a Ferris wheel, describing a series of epicycles that draw an amazing variety of 'curves, garlands, ovals, and even straight lines!' As Alphonso X of Castile (1221–84) apparently said, when he was taught the fundamentals of the Ptolemaic system, 'If the Lord Almighty had consulted me before embarking upon the Creation, I should have recommended something simpler.'

What follows is my own experimental attempt at translating a short passage from Book I of the *Almagest*, cribbed partly from the Latin edition, partly from a nineteenth-century French edition (with heavy use of a pile of dictionaries!), and partly from the available English translations. My aim has been to capture the strange, iridescent poetry of the Ptolemaic scheme in all its (literally) eccentric glory. In this passage, Ptolemy outlines the case for believing that the earth is at the centre of the universe, and that it does not move from its fixed position:

Undecimus

consideratione patuit:erit τ locus augis facilime cognitus. Ptolemeus eni distantiam tertie habitudinis ab auge numerauit.51.gradus:τ.14.minuta. Erat aut locus huius tertie habitudinis verus in.14.gra.τ.14.minu.capri corni.quare cotra signorum consequentiam a.14.mi.14.gra.capricorni si nu merauerimus.51.gradus τ.14.minuta:ad finem.23.gradus.46.minu.scor pionis perueniemus: In quo etiam Ptolemeus augi locum in principio re gni Antonij deputauit.

Propositio   xv.

IN qua vo parte zodiaci saturni locus medius sit in aliqua trium habitudinu:quantucp ab auge epi cycli media distet inuestigare. Locus augis iam notus est ex precedenti. Media vo vniuf cuiusqz trium habitudinu ab auge distantia superius inuenta est:quare medius locus erit notus. Cp si super puncto.g.tertie habitudinis epicyclum.b.t.k.descripserimus:erit arcus.b.t.k.distantie planete ab auge epicycli media in tertia habitudine no ignotus. Est eni angulus.g.5.l.cogni tus ex.12.buius. Sed τ angulus.g.e.l.vere distate tertie habitudinis ab au ge per.13.notus.quare residuus intrinsecus.e.g.5.cognitus:τ arcus.t.k.nu meratus. Cuem si a semicirculo.b.t.vemperis: relinquetur arcus.b.k.qui querebatur notus.

Propositio   xvi.

Centrici τ epicycli duab9 semidiametris liga pro portionibus elaborare. Certissima quadam ad hoc propositum opus est considera tione. Ptolemeus noster in anno secundo Antonij : sexto die mensis Mesir : sexti sz transacto:ante medietatem noctis.4. horis equalibus Saturn loci instrumeto suo ad Aldebaran rectificato τ ad lunam relatione:deprehendit in.9.g.τ.4.m.aquarij : vbi sz medium celi instrumeto indice esset in Alexandria vltimus gradus arietis. et sol cursu suo medio in.28.partibus τ.4s.minutis sagittarij. Estimauit aut inter cornu septentrionale τ saturnum tunc fm visum quidem cadere.30.m. ad successionem signorum. Sed locus visuo lune tunc fm numeratione Pto lemei fuit in.8.gradu τ.54.minu.aquarij.vnde certuo fuit locus saturni. Et quia tantuorquod intercedit huic psiderationi τ habitudini tertie superius memorate notum erat : notus fuit medius motuo logitudinis saturni in hoc tempore. Qui tamersi nondum rectificatuo habebatur:tamen non poterit sen sibilem in hoc opere errorem ingerere. Erat etiam medius locus saturni in hac habitudine tertia notus:quare τ in hac csideratione motuo medius sa turni non ignorabitur. Simili pacto distantia lune ab auge epicycli media in hac consideratione innotuit. Post hec itaqz recitata pingamus circu lum eccentricum epicycli delatorem.a.b.g.super centro.d. In cuius viame tro.a.g.punctuo.a.sit aug.g.oppositum augis.3.centrum equantie.τ.e.cen trum mundi. Sitqz in eius circuferentia punctuo.b.centrum epicycli.b.t.k. et locus planete in eodem punctuo.k.productis lincie.e.b.t.et.d.b.et.3.b.b. erit.b.aug media epicycli.et.t.aug vera. Iteqz vue lince.e.k.et.b.k.producã tur:vueqz perpendiculares.d.m.et.e.l.super lincam.b.l.aliaqz perpendicu

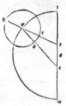

3. Ptolemy's Earth-centred cosmology dominated Western astronomy for 1,400 years until displaced by the calculations of Copernicus and Kepler. This page from a fifteenth-century Latin translation of Ptolemy's *Almagest* shows how epicycles were used to explain Saturn's apparent retrograde motions during its orbit of the Earth.

We know that the earth is spherical for the following obvious reasons: we can see that the sun, moon and other stars do not rise or set at the same time for everyone on earth, but always do so earlier for those who live in the east, and later for those who live in the west. And we also find that an eclipse, especially a lunar eclipse, which takes place at a particular hour, will be recorded as occurring at a variety of times, which is always later in the east than in the west (relative to our noon). So it is reasonable to conclude that the earth is spherical, because its curving surface cuts off the view of the heavens for every set of observers in turn . . . the same thing also happens at sea, when mountains seem to rise from the waters as we approach them, due to the curvature of the water's surface . . .

With that settled, let us turn to the *position* of the earth, where we find that these and other appearances can only be accounted for by its being in the middle of the universe, as though at the centre of a vast sphere. For if it wasn't, the earth would have to be either off its axis but equidistant from the poles, on its axis but nearer one of the poles, or else neither on its axis nor equidistant from the poles.

None of these alternatives stand up to scrutiny; for if you succeeded in relocating the earth along its axis, you would never see another equinox again, since the horizon would always bisect the heavens unequally . . .

So, if the earth was not at the centre of the universe, all the heavenly phenomena that we see, such as the seasonal lengthening and shortening of the days, would be thrown into complete disarray. Moreover, lunar eclipses would no longer occur only when the moon is diametrically opposite the sun, since the earth would often come between them even when they were not diametrically opposed, at intervals of less than a semi-circle . . .

Following on from these objections, it can be shown that the earth cannot move in any of these alternative directions, nor alter its place at the centre of the universe, because the implied outcomes would be the same. So it's a waste of time to look for explanations of the apparent motion of objects towards the centre, since it is it obvious from actual observations that the earth occupies the centre, and that all massive objects are naturally carried towards it. You need only consider the following fact: in all parts of the earth, which has already been shown to be spherical and in the centre of the universe, the direction and path of the movement of all massive bodies is always at right angles to the tangent plane (as drawn through the falling body's point of collision with the earth). It is clear that, if such objects were not stopped at the earth's surface, they would go all the way to the very centre, since the line to the centre is always perpendicular to the plane tangent of the sphere's surface at the point of intersection.

Anyone who thinks it a paradox that a body as heavy as the earth appears not to be supported by anything and yet stays where it is, is falling into the trap of judging on the basis of their own small experience instead of taking into account the nature of the wider universe. Because I don't believe they would still find it strange if they bore in mind that the mass of the earth, when compared with the magnitude of all that surrounds it, is but a tiny dot in comparison. Looked at in those terms, it is clear that a smaller object will always be overwhelmed and pressed in from all directions by a vast and uniform medium such as space. For there is no such thing as 'above' or 'below' in the great expanse of the universe, just as there is no 'up' and 'down' inside a sphere. Instead, light and airy bodies are scattered towards the outer circumference – in the overhead direction which we call 'up' – while dense and weighty bodies, by contrast, are carried towards the centre, in the direction we call 'down', that is, towards our feet and beyond, towards the centre of the earth. Naturally, these heavy bodies gravitate towards the centre due to the uniform pressure and resistance that is present throughout the universe. Since the earth's mass is great compared to that of the smaller bodies that fall towards it, it stays motionless even when struck from all sides. If the earth had even one motion in common with other massive objects, it is obvious that, because of its size, it would fall through the universe faster than any of them: animals and other weights would be left hanging in the air, and the earth would very soon fall out of the heavens. But you only have to think about it to see how nonsensical that is.

Source: Ptolemy, *Almagest*, Bk. I, chapters 4–7.

THE COLOUR OF SHAME

A native of Basra, in modern-day Iraq, Abū 'Alī al-Ḥasan ibn al-Haytham (c. 965–1040), known in the West as Alhazen, pioneered the study of light and optics, and was the first to show that light travels in straight lines. His chief work, *Kitāb al-Manāẓir* (*The Book of Optics*, c. 1021) – the first major treatment of the subject since Ptolemy's *Optica* of nine hundred years before – was widely read in European translations, as well as in its original Arabic, and according to the art historian Ernst Gombrich, it 'taught the medieval West the distinction between sense, knowledge, and inference, all of which come into play in perception.' Alhazen's emphasis on the importance of conducting first-hand experimentation would go on to influence Roger Bacon (p. 56), Johannes Kepler, and Francis Bacon (p. 80), whose later formulations of scientific method owed a great deal to Alhazen's example.

The first of this pair of extracts from the *Kitāb al-Manāẓir*, specially translated for this anthology, forms part of a longer argument against the belief that colours are optical illusions and therefore have no basis in reality. Note the way that Alhazen makes use of an everyday phenomenon in order to argue his point:

Something that proves that colours really do exist in the object and are not merely optical effects is the redness that suffuses someone's face whenever he feels shame or embarrassment, as well as the whitish yellow that appears when he experiences fear. His face can be a regular colour one minute, without much red in it, but if he suddenly feels shame, a crimson glow will cover his face, there for all to see. There has been no alteration in the light conditions, and the shamed person has not moved further away, therefore nothing has changed with respect to the observer's immediate perception of light and colour. The red that can be seen is the red of shame, it has no other cause. Since shame comes from within, it has nothing to do with the light conditions or the observer's eyesight; its colour originates from inside, not outside the body, so it is a physical entity, and not a trick of the light.

The same is true of the whiteness of fear, which can appear in someone's face if he is suddenly terrified – he may have been a healthy colour before, but fearful feelings will quickly transform the colour of his face.

So the altered colours of shame and fear are proof that visible colours exist as objects in coloured bodies, and are not simply external effects of the light. Colour is a reality in itself, despite what some people claim to the contrary. It exists as a form within the coloured body. There can be no doubt about this; there might be uncertainties about some of the other physical properties of colours, but not about the fact that they exist within the body and do not result from external phenomena.

This second extract also uses everyday experiences and experiments to reinforce a point about the nature of light, in this case the after-effects of bright or coloured lights on the eyes:

If you look directly at a powerful light, such as the sun, your eyes hurt and you have to look away. The same is true of bright light reflected in a mirror – it's painful and you have to either close your eyes or avert your gaze.

But if you look for a long time at a brightly lit object, like a sunlit window, and then turn to look into the shadows, you can't see anything for a while, although as your eyes slowly get used to the gloom you can begin to make out objects. The same thing happens if you look at a roaring fire, then look away into the darkness.

Moreover, if you stare at a white object in the daylight, then look away into the shadows, you carry on 'seeing' that object, even when you close your eyes. After a while, the impression of the object fades, and normal vision returns. The same thing happens if you look at the daytime sky through a window, and then turn to look indoors, you carry on 'seeing' the shape of the window in the form of a brightness in front of your eyes, even if you close your eyes,

What this demonstrates is that light has an effect on sight.

Recall how if you stare at a green, sunlit meadow and then turn to look into a shady area, that area appears to be coloured dark green. If you look at some white or pale objects, they, too, appear tinged with green. And then even if you close your eyes tight, you continue to 'see' the greenness of the meadow, until your vision slowly returns to normal. The same is true if you stare at sunlit flowers, say purple or blue ones, and then turn to look into a shaded area, the objects in that area will appear purple or blue.

What this demonstrates is that brightly lit colours also have an effect on sight.

Consider this: we can see the stars at night, but not during the day. This is because the air is brightly lit during the day but not at night. We see the stars because the sky is dark at night. By day, the sky is too brightly lit for them to be visible.

Imagine you are at a fire-side at night, with the light of the fire illuminating the ground, at the edges of which there are a number of small objects with detailed surfaces: if the fire is in your eyes, you can hardly see these objects or their detailed surfaces, but if you shield your eyes so that the fire is no longer in your eyes, you can see them perfectly well.

This shows that light shining directly into the eyes prevents the eyes from seeing near objects in detail.

It is the same with handwriting on a sheet of paper: if the page reflects the light directly you cannot read the writing, but if you incline the page, altering its position in relation to the light source, you will be able to read it.

Or a low, glowing fire which is visible in the dark, but in bright sunlight would disappear; which is to say, you would see the smoke but not the fire . . .

Similarly, certain sea creatures have luminescent skins and shells that appear to glow in the dark as if on fire, but when seen in the daylight they do not appear to glow. Fireflies are also seen to flash at night but not in the day. At night you see their travelling glow but by day you can see these insects flying about but not their famous 'fire'.

What these examples show is that strong light makes certain visible objects disappear, while an absence of light renders them visible once more.

Likewise, a glass of deep-coloured liquid looks black and opaque in dim light, but in strong light its colour and transparency become visible to the eye. The same is true of coloured gemstones, which look dark and opaque in poor light, but when held up to the light of the sun their colours glow as the light passes through them . . .

All this shows that our sight perceives the colours of bodies in accordance with the kind of light that shines upon them.

Source: Ibn al-Haytham, *Kitāb al-Manāẓir*, Bk. I, chapters 3 and 4.

IMPOSTERS

When faced with competing scientific claims, how do we know who to trust? Scientific credibility has always depended on the identification of trustworthy agents, as well as on the corresponding exposure of untrustworthy ones, as the following pair of extracts show. The first piece is from a medical treatise by the Persian physician Rhazes (Muhammad ibn Zakariyyā al-Rāzī, c. 860–925); the second from an investigation of spiritualism by the Victorian chemist William Crookes (1832–1919).

Medieval Charlatans

This first passage, from a nineteenth-century translation of Rhazes's short treatise *Kitāb al-Manṣūr (Al-Manṣūr's Book of Medicine)*, is an exposé of the various tricks that were performed upon patients by the many medical mountebanks who touted their services in early tenth-century Persia. Rhazes advised his readers always to check the credentials, experience, morality and lifestyle of any medic who offered to treat them, while despairing of the credulousness of those who persisted in paying for obviously fraudulent cures:

There are so many little Arts used by Mountebanks and pretenders to physic, that an entire treatise, had I a mind to write one, would not contain them: but their impudence, and daring boldness is equal to the guilt and inward conviction they have of tormenting and putting persons to pain in their last hours, for no reason at all. Now some of them profess to cure the falling-sickness, and thereupon make an issue in the hinder part of the head, in form of a cross, and pretend to take something out of the opening, which they held all the while in their hands. Others give out, that they can draw snakes or lizards out of their patients' noses, which they seem to perform by putting up a pointed iron probe, with which they wound the nostril, until the blood comes: then they draw out the little artificial animal composed of liver, etc. Some are confident, they can take out the white specks in the eye. Before they apply the instrument to that part, they put in a piece of fine rag into the eye, and taking it out with the

instrument, pretend it is drawn immediately from the eye. Some
again undertake to suck water out of the ear, which they fill with a
tube from their mouth, and hold the other end to the ear; and so
spurting the water out of their mouths, pretend it came from the ear.
Others pretend to get out worms, which grow in the ear, or roots in
the teeth. Others can extract frogs from the under part of the tongue;
and by lancing make an incision, into which they clap the frog, and
so take it out. What shall I say of bones inserted into wounds and
ulcers, which, after remaining there for some time, they take out
again? Some, when they have taken out a stone from the bladder,
persuade their patients, that still there's another left; they do this for
the reason, to have it believed, that they have taken out another.
Sometimes they probe the bladder, being altogether ignorant and
uncertain, whether there be a stone or no. But if they don't find it,
they pretend at least to take out one they have in readiness before,
and show that to them. Sometimes they make an incision in the anus
for the piles, and by repeating the operation often bring it to a fistula
or an ulcer, when there was neither before. Some say they take
phlegm, of a substance like unto glass, out of the penis or other part
of the body, by the conveyance of a pipe, which they hold with water
in their mouths. Some pretend that they can contract and collect all
the floating humours of the body to one place by rubbing it with winter cherries; which causes a burning or inflammation; and then they
expect to be rewarded, as if they cured the distemper; and after they
have suppled the place with oil, the pain presently goes off. Some
make their patients believe they have swallowed glass; so, taking a
feather, which they force down the throat, they throw them into a
vomiting, which brings up the stuff they themselves had put in with
that very feather. Many things of this nature do they get out, which
these impostors with great dexterity have put in, tending many times
to the endangering the health of their patients, and often ending in
the death of them. Such counterfeits could not pass with discerning
men, but that they did not dream of any fallacies, and made no doubt
of the skill of those whom they employed: till at last when they suspect, or rather look more narrowly into their operations, the cheat is
discovered. Therefore no wise men ought to trust their lives in their
hands, nor take any more of their medicines, which have proved so
fatal to many.

Nineteenth-Century Spiritualists

Modern scientists are often accused of failing to approach neighbouring areas such as astrology or spiritualism with a suitably open mind. The following offers a counter-example to the charge, though, perhaps inevitably, William Crookes's spiritualist researches in the 1860s and 1870s put a lasting dent in his scientific reputation. As Crookes pointed out, 'we have by no means exhausted all human knowledge, or fathomed the depths of all the physical forces', and he began his inquiries 'with no preconceived notions whatever'. But he was soon faced with the unwillingness of his subjects to have their claims and methods tested in controlled conditions, and what began as a fair-minded investigation into spiritualist phenomena ended as an irritable dismissal. As Crookes concluded, 'I consider it the duty of scientific men who have learnt exact modes of working, to examine phenomena which attract the attention of the public, in order to confirm their genuineness, or to explain if possible the delusions of the honest and to expose the tricks of deceivers':

The modes of reasoning of scientific men appear to be generally mis-understood by spiritualists with whom I have conversed, and the reluctance of the trained scientific mind to investigate this subject is frequently ascribed to unworthy motives. I think, therefore, it will be of service if I here illustrate the modes of thought current amongst those who investigate science, and say what kind of experimental proof science has a right to demand before admitting a new depart-ment of knowledge into her ranks. We must not mix up the exact and the inexact. The supremacy of accuracy must be absolute.

The first requisite is to be sure of facts; then to ascertain condi-tions; next, laws. Accuracy and knowledge of detail stand foremost amongst the great aims of modern scientific men. No observations are of much use to the student of science unless they are truthful, and made under test conditions; and here I find the great mass of spiritu-alistic evidence to fail. In a subject which, perhaps, more than any other, lends itself to trickery and deception, the precautions against fraud appear to have been, in most cases, totally insufficient, owing, it would seem, to an erroneous idea that to ask for such safeguards was to imply a suspicion of the honesty of some one present. We may use our own unaided senses, but when we ask for instrumental means to increase their sharpness, certainty, and trustworthiness under circumstances of excitement and difficulty, and when one's natural senses are liable to be thrown off their balance, offence is taken.

In the countless number of recorded observations I have read, there appear to be few instances of meetings held for the express purpose of getting the phenomena under test conditions, in the presence of persons properly qualified by scientific training to weigh and adjust the value of the evidence which might present itself. The only good series of test experiments I have met with were tried by the Count de Gasparin, and he, whilst admitting the genuineness of the phenomena, came to the conclusion that they were not due to supernatural agency.

The pseudo-scientific spiritualist professes to know everything: no calculations trouble his serenity, no hard experiments, no long laborious readings; no weary attempts to make clear in words that which has rejoiced the heart and elevated the mind. He talks glibly of all sciences and arts, overwhelming the inquirer with terms like 'electro-biologize,' 'psychologize,' 'animal magnetism,' &c. – a mere play upon words, showing ignorance rather than understanding. Popular science such as this is little able to guide discovery rushing onwards to an unknown future; and the real workers of science must be extremely careful not to allow the reins to get into unfit and incompetent hands.

In investigations which so completely baffle the ordinary observer, the thorough scientific man has a great advantage. He has followed science from the beginning through a long line of learning, and he knows, therefore, in what direction it is leading; he knows that there are dangers on one side, uncertainties on another, and almost absolute certainty on a third: he sees to a certain extent in advance. But, where every step is towards the marvellous and unexpected, precautions and tests should be multiplied rather than diminished. Investigators must work; although their work may be very small in quantity if only compensation be made by its intrinsic excellence. But, even in this realm of marvels, – this wonder-land towards which scientific inquiry is sending out its pioneers, – can anything be more astonishing than the delicacy of the instrumental aids which the workers bring with them to supplement the observations of their natural senses?

The spiritualist tells of bodies weighing 50 or 100 lbs. being lifted up into the air without the intervention of any known force; but the scientific chemist is accustomed to use a balance which will render sensible a weight so small that it would take ten thousand of them to weigh one grain; he is, therefore, justified in asking that a power professing to be guided by intelligence, which will toss a heavy body up to the ceiling, shall also cause his delicately-poised balance to move under test conditions.

The spiritualist tells of tapping sounds which are produced in different parts of a room when two or more persons sit quietly round a table. The scientific experimenter is entitled to ask that these taps shall be produced on the stretched membrane of his phonautograph.

The spiritualist tells of rooms and houses being shaken, even to injury, by superhuman power. The man of science merely asks for a pendulum to be set vibrating when it is in a glass case and supported on solid masonry.

The spiritualist tells of heavy articles of furniture moving from one room to another without human agency. But the man of science has made instruments which will divide an inch into a million parts; and he is justified in doubting the accuracy of the former observations, if the same force is powerless to move the index of his instrument one poor degree.

The spiritualist tells of flowers with the fresh dew on them, of fruit, and living objects being carried through closed windows, and even solid brick-walls. The scientific investigator naturally asks that an additional weight (if it be only the 1000th part of a grain) be deposited on one pan of his balance when the case is locked. And the chemist asks for the 1000th of a grain of arsenic to be carried through the sides of a glass tube in which pure water is hermetically sealed.

The spiritualist tells of manifestations of power, which would be equivalent to many thousands of 'foot-pounds,' taking place without known agency. The man of science, believing firmly in the conservation of force and that it is never produced without a corresponding exhaustion of something to replace it, asks for some such exhibitions of power to be manifested in his laboratory, where he can weigh, measure, and submit it to proper tests.

For these reasons and with these feelings I began an inquiry suggested to me by eminent men exercising great influence on the thought of the country. At first, like other men who thought little of the matter and saw little, I believed that the whole affair was a superstition, or at least an unexplained trick. Even at this moment I meet with cases which I cannot *prove* to be anything else; and in some cases I am sure that it is a delusion of the senses.

I by no means promise to enter fully into this subject; it seems very difficult to obtain opportunities, and numerous failures certainly may dishearten anyone. The persons in whose presence these phenomena take place are few in number, and opportunities for experimenting with previously arranged apparatus are rarer still. I should feel it to be a great satisfaction if I could bring out light in any direction, and I may safely say I care not in what direction. With this end in view, I appeal to any of my readers who may possess a key to these strange phenomena, to further the progress of the truth by assisting me in my investigations. That the subject has to do with strange physiological conditions is clear, and these in a sense may be called 'spiritual' when they produce certain results in our minds. At present the phenomena I have observed baffle explanation; so do the phenomena of thought, which are also spiritual, and no philosopher has yet understood. No man however denies them.

The explanations given to me, both orally and in most of the books I have read, are shrouded in such an affected ponderosity of style, such an attempt at disguising poverty of ideas in grandiloquent language, that I feel it impossible, after driving off the frothy diluent, to discern a crystalline residue of meaning. I confess that the reasoning of some spiritualists would almost seem to justify Faraday's severe statement – that many dogs have the power of coming to much more logical conclusions. Their speculations utterly ignore all theories of force being only a form of molecular motion, and they speak of Force, Matter, Spirit, as three distinct entities, each capable of existing without the others; although they sometimes admit that they are mutually convertible . . .

It has been my wish to show that science is gradually making its followers the representatives of care and accuracy. It is a fine quality, that of uttering undeniable truth. Let, then, that position not be lowered, but let words suit facts with an accuracy equal to that with which the facts themselves can be ascertained; and in a subject encrusted with credulity and superstition, let it be shown that there *is* a class of facts to be found upon which reliance can be placed, so far, that we may be certain they will never change. In common affairs a mistake may have but a short life, but in the study of nature an imperfect observation may cause infinite trouble to thousands. The increased employment of scientific methods will promote exact observation and greater love of truth among inquirers, and will produce a race of observers who will drive the worthless residuum of spiritualism hence into the unknown limbo of magic and necromancy.

Sources: Rhazes, *A Treatise on the Small-Pox and Measles*, trans. William Alexander Greenhill (London: Sydenham Society, 1848), pp. 80–82; William Crookes, 'Spiritualism Viewed by the Light of Modern Science', *Quarterly Journal of Science* 7 (July 1870), pp. 316–21.

THE GATES OF KNOWLEDGE

Abu Alī al-Husain ibn Adballah ibn Sīnā (c. 980–1037), known in the west as Avicenna, was a Persian physician and polymath who made important contributions to a wide range of subjects including chemistry, mathematics and geology (he was particularly interested in the various processes of mountain formation), though his greatest impact was in the field of medicine. He wrote numerous medical treatises based principally on his own clinical experience, the best known of which, the *Qanun* (*The Canon of Medicine*, completed c. 1025), became a standard text in universities throughout Europe and the Middle East. While the book offered a summary of existing medical knowledge, it also contained a number of Avicenna's original ideas, including the concept of a 'syndrome', a collection of clinically readable signs and symptoms that could be treated as a whole; indeed, it was through Avicenna's work that a number of innovative medical practices, such as the quarantining of infectious patients, were first introduced to the West.

As the extracts I have chosen illustrate, the range of theoretical and practical material was impressive, taking us from a brief guide to avoiding seasickness (see the last entry below) to a philosophical restatement of Galen's famous tripartite distinction between health, sickness, and a vague kind of in-between condition – 'a habit of body', neither ill nor well – that most of us put up with as our ongoing physical state:

The Practice of Medicine

The practice of medicine is not the work which the physician carries out, but is that branch of medical knowledge which, when acquired, enables one to form an opinion upon which to base the proper plan of treatment . . . the theory guides to an opinion, and the opinion is the basis of treatment. Once the purpose of each aspect of medicine is understood, you can become skilled in both even if you are never called upon to use your knowledge.

Another thing is that there is no need to assert that there are *three* states of the human body – sickness, health, and a state which is neither health nor disease. The first two cover everything. Careful consideration of the subject will make it clear to the physician either that the threefold grouping is unnecessary or the group which we reject is unnecessary. The first two states really cover everything. Careful consideration will convince the physician that the third state is dual – on the one hand an infirmity, and on the other a habit of body or a condition which cannot be called strict health although the actions and functions of the body are normal. One must not risk defining health in an arbitrary fashion and include in it a condition which does not belong to it. However, we do not propose to argue this matter out because a disputation of that kind does not really further medicine.

The Difference Between Disease and Symptom

Disease is an abnormal state of the body which primarily and independently produces a disturbance in the normal functions of the body. It may be an abnormality of temperament or form (structure).

Symptom is a manifestation of some abnormal state in the body. It may be harmful as the colic pain or harmless as the flushing of cheeks in peri pneumonia.

The Difference Between Symptoms and Signs

A symptom may not only be a symptom but also a sign. A symptom refers to that which relates to its own intrinsic character or in relation to that to which it belongs. A sign is that which guides the physician to a knowledge of the real essential nature of the disease.

Leaving It to Nature

When you do not know the nature of a malady, leave it to nature; do not strive to hasten matters. For either nature will bring about the cure or it will itself reveal clearly what the malady really is.

The Psychological Factors

Remember, too, that among the advantageous contributory factors in treatment is the help afforded by anything which exalts the sensitive and vital drives: for instance, joyfulness. In consequence, one sets out to please one's patient, and even tranquilize him by anything which can reasonably gratify him. Sometimes one may advantageously arouse his sense of shame, making him blush, and so leading the sick person to avoid what is harmful to him.

On Sea-Sickness

Those who travel by sea often suffer from scotoma and vertigo, and the motion brings on nausea and vomiting, especially during the first few days of the voyage, after which it subsides. It is not wise to allow nausea and vomiting to continue longer than is required for getting rid of superfluities.

Measures to prevent sea-sickness – it is justifiable to endeavor to prevent sea-sickness. Thus, take fruit such as of quinces, apples, and pomegranates. Parsley seed made into a drink will prevent nausea as long as one lies quite still; and if one cannot lie still, it soothes the sense of nausea. Absinthe has the same effect.

Among the things which prevent sea-sickness are: nourishing the mouth of the stomach with tonic acetous substances, and such things as prevent vapors from rising into the head. Namely, lentils in vinegar (or dried and boiled with a little pennyroyal, or boiled till soft and then triturated and dried and kept in an earthen vessel); juice of sour grapes; a little pennyroyal; thyme; bread broken up in weak and fragrant wine, or in cold water. Thyme is sometimes added to that.

The nostrils should also be smeared over on the inside with white lead (cosmetic) ointment.

Persistent sea-sickness – avoid all food. Take a little vinegar and honey with water in which thyme has been infused, or pennyroyal water with some fine polenta; or take some weak fragrant wine, with fine polenta. Take antibilious remedies.

Simple precautionary measures: (1) Counteract the disagreeable smell of the ship by sniffing at quinces, thyme or pennyroyal. (2) Do not look at the sea. (3) Beware of the drinking water. (4) Note the diet already mentioned. (5) Have remedies against vermin. Mercury, oil, long birthwort or wearing wool smeared with oil or mercury ensures against lice.

Among the most interesting of Avicenna's other surviving texts is a robustly immodest account of his own life and work, in which he claimed to have read and mastered the whole of medicine by the age of sixteen ('medicine is not one of the difficult sciences'), and that soon the most distinguished doctors were seeking him out for his opinion. After that, he wrote, 'the gates of knowledge began opening for me', and he turned his attentions to the study of law, philosophy, logic and metaphysics, mastering them all in turn, sleeping for only a few hours a night, and keeping himself awake at his desk with regular cups of wine. And even when asleep, he recalled, 'I would see those very problems in my dream; and many questions became clear to me in my sleep. I continued in this until all of the sciences were deeply rooted within me and I understood them as far as is humanly possible. Everything which I knew at that time is just as I know it now; I have not added anything to it to this day.'

Sources: Avicenna, The Canon of Medicine, adapted by Laleh Bakhtiar (Chicago: Great Books of the Islamic World, 1999), pp. 10–11; 171–2; 454–5; The Life of Ibn Sina, trans. William E. Gohlman (Albany: State University of New York Press, 1974), p. 25.

DREAM POOL ESSAYS

The writings of the eleventh-century court official Shên Kua (c. 1031–95) offer unparalleled insights into early Chinese scientific culture, especially his *Méng Xī Bǐ Tán* ('Dream Pool Essays', c. 1086), a sprawling collection of more than five hundred thoughtful entries on a polymathic range of subjects, including mathematics, astronomy, geology and botany. Here are five representative examples, beginning with an invocation of the traditional Five Element theory (Fire, Earth, Metal, Water, Wood) that so shaped the early Chinese view of nature. Medieval Chinese technology was extraordinarily advanced – in fact the last of these five excerpts contains, according to the scientist and historian Joseph Needham (1900–95), the world's first description of a magnetic compass, an achievement that predated its European debut by several hundred years:

The Five Elements

In the Chhien-shan district of Hsin-chou there is a bitter spring which forms a rivulet at the bottom of a gorge. When its water is heated it becomes *tan fan* (lit. gall-alum; probably impure copper sulphate). When this is heated it gives copper. If this 'alum' is heated for a long time in an iron pan the pan is changed to copper. Thus water can be converted into copper – an extraordinary change of substance, really unfathomable. According to the *Huang Ti Nei Ching, Su Wên* (the medical classic) 'there are five elements in the sky, and five elements in the earth. The *chi* of Earth, when in the sky, is moisture. Earth produces metal and stone (as ores in the mountains), but water can also produce metal and stone.' These instances then are proofs [that the principles of the *Su Wên* are right]. It is like water dripping in caverns and (slowly) forming stalactites, or like the formation of crystals from well and spring water at the spring and autumn equinoctial seasons, or like selenite (*yin ching shih*) deposits from strong brines; all show the transformations from moisture. So also the *chi* of Wood, when in the sky, is wind, and both Wood and wind can generate Fire. Such is the nature of the Five Elements.

The Pole Star

Before Han times it was believed that the pole star was in the centre of the sky, so it was called Chi hsing ('Summit star'). Tsu Kêng-Chih found out with the help of the sighting-tube that the point in the sky which really does not move was a little more than 1° away from the summit-star. In the Hsi-Ning reign-period (+ 1068–1077) I accepted the order of the emperor to take charge of the Bureau of the Calendar. I then tried to find the true pole by means of the tube. On the very first night I noticed that the star which could be seen through the tube moved after a while outside the field of view. I realised, therefore, that the tube was too small, so I increased the size of the tube by stages. After three months' trials I adjusted it so that the star would go round and round within the field of view without disappearing. In this way I found that the pole-star was distant from the true pole somewhat more than 3°. We used to make diagrams of the field, plotting the positions of the star from the time when it entered the field of view, observing after nightfall, at midnight, and early in the morning before dawn. Two hundred of such diagrams showed that the 'pole-star' was really a circumpolar star. And this I stated in my detailed report to the emperor.

Meteorite Fall

In the 1st year of the Chih-Phing reign period (+ 1064), there was a tremendous noise like thunder at Chhang-chou about noon. A fiery star as big as the moon appeared in the south east. In a moment there was a further thunderclap while the star moved to the south-west, and then with more thunder it fell in the garden of the Hsü family in the I-hsing district. Fire was seen reflected in the sky far and near, and fences in the garden round about were all burnt. When they had been extinguished, a bowl-shaped hole was seen in the ground, with the meteorite glowing within it for a long time. Even when the glow ceased it was too hot to be approached. Finally the earth was dug up, and a round stone as big as a fist, still hot, was found, with one side elongated. Its colour and weight were like iron. The governor, Chêng Shen, sent it to the Chin Shan temple at Jun-chou, where it is still kept in a box and shown to visitors.

Petrified Bamboo

In recent years there was a landslide on the bank of a large river in Yung-Ning Kuan near Yenchow. The bank collapsed, opening a space of several dozens of feet, and under the ground a forest of bamboo shoots was thus revealed. It contained several hundred bamboos with their roots and trunks all complete, and all turned to stone. A high official happened to pass by, and took away several, saying that he would present them to the emperor. Now bamboos do not grow in Yenchow. These were several dozens of feet below the present surface of the ground, and we do not know in what dynasty they could possibly have grown. Perhaps in very ancient times the climate was different so that the place was low, damp, gloomy, and suitable for bamboos. On Chin-Hua Shan in Wuchow there are stone pine-cones, and stones formed from peach-kernels, stone bulrush roots, stone fishes, crabs, and so on, but as these are all (modern) native products of that place, people are not very surprised at them. But these petrified bamboos appeared under the ground so deep, though they are not produced in that place today. This is a very strange thing.

The First Description of a Compass

Magicians rub the point of a needle with the lodestone; then it is able to point to the south. But it always inclines slightly to the east, and does not point directly at the south. [It may be made to] float on the surface of water, but it is then rather unsteady. It may be balanced on the finger-nail, or on the rim of a cup, where it can be made to turn more easily, but these supports being hard and smooth, it is liable to fall off. It is best to suspend it by a single cocoon fibre of new silk attached to the centre of the needle by a piece of wax the size of a mustard-seed – then, hanging in a windless place, it will always point to the south. Among such needles there are some which, after being rubbed, point to the north. I have needles of both kinds by me. The south-pointing property of the lodestone is like the habit of cypress-trees of always pointing to the west. No one can explain the principles of these things.

Source: Joseph Needham *et al.* (eds), *Science and Civilisation in China*, 7 vols (Cambridge: Cambridge University Press, 1954–2008), V, pt 4 (1980), p. 201; III (1959), p. 262, pp. 433–4, p. 614; IV, pt 1 (1962), pp. 249–50.

EMPIRICAL RAINBOWS

'Who was the first real scientist?' is an essentially unanswerable question, though that has never prevented it from being asked. Even the most cursory attempt to answer it would yield dozens of possible candidates, some of whom we have already met: Thales, Democritus, Aristotle, Ptolemy, Shên Kua, as well as, of course, the ever reliable 'anon'. One school of thought, how-ever, claims that the work of Roger Bacon (c. 1214–92), an English Franciscan friar and philosopher, constitutes the first expression of a self-consciously empirical approach to nature that was distinct from the speculative outlook of the ancients. 'Experimental Science', as Bacon described this new way of looking in his seven-volume encyclopaedia, the *Opus Majus* (completed c. 1267), was certainly an impressive precursor of the modern scientific method that would be introduced by his namesake Francis Bacon some four hundred years later (see p. 80); but, in pointing out that we can know nothing truly until we have experienced it for our-selves, Bacon was also reiterating the views of earlier scholars such as Alhazen and Rhazes (see p. 41), whose works he had studied in Paris and Oxford, and from whom (as can be seen from the following), he derived many of his experimental notions:

Since this Experimental Science is wholly unknown to the rank and file of students, I am therefore unable to convince people of its utility unless at the same time I disclose its excellence and its proper signi-fication. This science alone, therefore, knows how to test perfectly what can be done by nature, what by the effort of art, what by trick-ery, what the incantations, conjurations, invocations, deprecations, sacrifices, that belong to magic, mean and dream of, and what is in them, so that all falsity may be removed and the truth alone of art and nature may be retained. This science alone teaches us how to view the mad acts of magicians, that they may not be ratified but shunned, just as logic considers sophistical reasoning.

This science has three leading characteristics with respect to other sciences. The first is that it investigates by experiment the notable conclusions of all those sciences. For the other sciences know how to discover their principles by experiments, but their conclu-sions are reached by reasoning drawn from the principles discovered.

But if they should have a particular and complete experience of their own conclusions, they must have it with the aid of this noble science. For it is true that mathematics has general experiments as regards its conclusions in its figures and calculations, which also are applied to all science and to this kind of experiment, because no science can be known without mathematics. But if we give our attention to particular and complete experiments and such as are attested wholly by the proper method, we must employ the principles of this science which is called experimental. I give as an example the rainbow and phenomena connected with it, of which nature are the circle around the sun and the stars, the streak also lying at the side of the sun or of a star, which is apparent to the eye in a straight line, and is called by Aristotle in the third book of the Meteorologics a perpendicular, but by Seneca a streak, and the circle is called a corona, phenomena which frequently have the colors of the rainbow. The natural philosopher discusses these phenomena, and the writer on Perspective has much to add pertaining to the mode of vision that is necessary in this case. But neither Aristotle nor Avicenna in their Natural Histories has given us a knowledge of phenomena of this kind, nor has Seneca, who composed a special book on them. But Experimental Science attests them.

Let the experimenter first, then, examine visible objects, in order that he may find colors arranged as in the phenomena mentioned above and also the same figure. For let him take hexagonal stones from Ireland or from India, which are called rainbows in Solinus on the Wonders of the World, and let him hold these in a solar ray falling through the window, so that he may find all the colors of the rainbow, arranged as in it, in the shadow near the ray. And further let the same experimenter turn to a somewhat dark place and apply the stone to one of his eyes which is almost closed, and he will see the colors of the rainbow clearly arranged just as in the bow. And since many employing these stones think that the phenomena is due to the special virtue of those stones and to their hexagonal shape, therefore let the experimenter proceed further, and he will find this same peculiarity in crystalline stones correctly shaped, and in other transparent stones. Moreover, he will find this not only in white stones like the Irish crystals, but also in black ones, as is evident in the dark crystal and in all stones of similar transparency. He will find it besides in crystals of a shape differing from the hexagonal, provided they

have a roughened surface, like the Irish crystals, neither altogether
smooth, nor rougher than they are. Nature produces some that have
surfaces like the Irish crystals. For a difference in the corrugation
causes a difference in the colors. And further let him observe rowers,
and in the drops falling from the raised oars he finds the same colors
when the solar rays penetrate drops of this kind. The same phenomenon is seen in water falling from the wheels of a mill; and likewise
when one sees on a summer's morning the drops of dew on the grass
in meadow or field, he will observe the colors. Likewise when it is
raining, if he stands in a dark place and the rays beyond it passing
through the falling rain, the colors will appear in the shadow near
by; and frequently at night colors appear around a candle. Moreover,
if a man in summer, when he rises from sleep and has his eyes only
partly open, suddenly looks at a hole through which a ray of the
sun enters, he will see colors. Moreover, if seated beyond the sun
he draws his cap beyond his eyes, he will see colors; and similarly
if he closes an eye the same thing happens under the shade of the
eyebrows; and again the same phenomenon appears through a glass
vessel filled with water and placed in the sun's rays. Or similarly if
one having water in his mouth sprinkles it vigorously into the rays
and stands at the side of the rays. So, too, if rays in the required
position pass through an oil lamp hanging in the air so that the light
falls on the surface of the oil, colors will be produced. Thus in an
infinite number of ways colors of this kind appear, which the diligent
experimenter knows how to discover.

The 'streak' mentioned by Bacon in connection with the writings of Seneca
was a sun-dog, or parhelion, caused by the refraction of sunlight by ice crystals present in high cirriform clouds. Sun-dogs tend to occur when the sun is
low in the sky, and exhibit a range of rainbow colours, with the red end of
the spectrum appearing towards the inner edges, nearest the sun. Seneca's
description of the phenomenon was admirably succinct, and worth quoting
in full:

I must now go on to speak of *Streaks* (watergalls, sun-dogs), which
are as bright and varied as the rainbow, and commonly received by
us as equally indicative of rain. No great labour need be spent in
explaining them, for they are just incomplete rainbows. They have
the variegated appearance of the bow, but none of its curve. They lie

in a straight line. They are formed near the sun, as a rule, in a moist cloud that has begun to break up. Thus, they have the same colour as is found in the rainbow, but there is a difference in the shape, due to the corresponding difference in the clouds over which they stretch.

Sources: *The Opus Majus of Roger Bacon*, trans. Robert Belle Burke, 2 vols (Philadelphia: University of Pennsylvania Press, 1928), II, pp. 587–9; *Physical Science in the Time of Nero: Being a Translation of the Quaestiones Naturales of Seneca* (London: Macmillan, 1910), pp. 33–4 (Bk. I, 9).

THE ART OF NUMBRING

Schoolchildren have always complained about arithmetic, even in the six-teenth century. The following dialogue between a master and his student, from a mathematical textbook by Robert Recorde (c. 1510–58), represents an early attempt at convincing maths-resistant youngsters of the advantages of cultivating numbers. Recorde, a Welsh-born physician and mathemati-cian, is best known today for having introduced the equals sign, 'a paire of paralleles of one length, thus: =, because noe 2 thynges can be more equalle', as he described it in his wonderfully titled treatise, *The Whetstone of Witt* (1557). This sub-Shakespearean exchange, from an earlier volume, begins with a challenging thought experiment: imagine a world without numbers – an experiment that I remember my own maths teacher deploying, more than five hundred years after Recorde first devised it as a means of over-coming his pupils' fear of figures:

Master. If number were so vile a thing as you did esteem it, then need it not to be used so much in mens communication. Exclude number, and answer to this question: How many years old are you?

Scholar. Mum.

Master. How many dayes in a week? How many weeks in a year? What lands hath your Father? How many men doth he keep? How long is it since you came from him to me?

Scholar. Mum.

Master. So that if number want, you answer all by Mummes: How many miles to London?

Scholar. A poak full of plums.

Master. Why, thus you may see, what rule number beareth, and that if number bee lacking it maketh men dumb, so that to most questions they must answer Mum.

Scholar. This is the cause, sir, that I judged it so vile, because it is so common in talking every while: Nor plenty is not dainty, as the common saying is.

Master. No, nor store is no sore, perceive you this? The more common that the thing is, being needfully required, the better is the thing, and the more to be desired. But in numbring, as some of it is light and plain, so the most part is difficult, and not easie to attain.

The easier part serveth all men in common, and the other requireth some learning. Wherefore as without numbring a man can do almost nothing, so with the help of it, you may attain to all things.

Scholar. Yes, sir, why then it were best to learn the Art of numbring, first of all other learning, and then a man need learn no more, if all other come with it.

Master. Nay not so: but if it be first learned, then shall a man be able (I mean) to learn, perceive, and attain to other Sciences; which without it he could never get.

Scholar. I perceive by your former words, that Astronomy and Geometry depend much on the help of numbring: but that other Sciences, as Musick, Physick, Law, Grammer, and such like, have any help of Arithmetick, I perceive not.

Master. I may perceive your great Clerklinesse by the ordering of your Sciences: but I will let that passe now, because it toucheth not the matter that I intend, and I will shew you how Arithmetick doth profit in all these somewhat grosly, according to your small understanding, omitting other reasons more substantiall.

First (as you reckon them) Musick hath not onely great help of Arithmetick, but is made, and hath his perfectnesse of it: for all Musick standeth by number and proportion: And in Physick, beside the calculation of criticall dayes, with other things, which I omit, how can any man judge the pulse rightly, that is ignorant of the proportion of numbers?

And so for the Law, it is plain, that the man that is ignorant of Arithmetick, is neither meet to be a Judge, neither an Advocate, nor yet a Proctor. For how can hee well understand another mans cause, appertaining to distribution of goods, or other debts, or of summes of money, if he be ignorant of Arithmetick? This oftentimes causeth right to bee hindered, when the Judge either delighteth not to hear of a matter that hee perceiveth not, or cannot judge for lack of understanding: this commeth by ignorance of Arithmetick.

Now, as for Grammer, me thinketh you would not doubt in what it needeth number, sith you have learned that Nouns of all sorts, Pronouns, Verbs, and Participles are distinct diversly by numbers: besides the variety of Nouns of Numbers, and Adverbs. And if you take away number from Grammer, then is all the quantity of Syllables lost. And many other ways doth number help Grammer.

Whereby were all kindes of Meeters found and made? Was it not by number?

But how needfull Arithmetick is to all parts of Philosophy, they may soon see, that do read either Aristotle, Plato, or any other Philosophers writings. For all their examples almost, and their probations, depend of Arithmetick. It is the saying of Aristotle, that hee that is ignorant of Arithmetick, is meet for no Science. And Plato his Master wrote a little sentence over his Schoolhouse door, Let none enter in hither (quoth he) that is ignorant of Geometry. Seeing hee would have all his Scholars expert in Geometry, much rather he would the same in Arithmetick, without which Geometry cannot stand.

And how needfull Arithmetick is to Divinity, it appeareth, seeing so many Doctors gather so great mysteries out of number, and so much do write of it. And if I should go about to write all the commodities of Arithmetick in civill acts, as in governance of Common-weales in times of peace, and in due provision & order of Armies, in time of war, for numbering the Host, summing of their wages, provision of victuals, viewing of Artillery, with other Armour; beside the cunningest point of all, for casting of ground, for encamping of men, with such other like: And how many wayes also Arithmetick is conducible for all private Weales, of Lords and all Possessioners, of Merchants, and all other occupiers, and generally for all estates of men, besides Auditors, Treasurers, Receivers, Stewards, Bailiffes, and such like, whose Offices without Arithmetick are nothing: If I should (I say) particularly repeat all such commodities of the noble Science of Arithmetick, it were enough to make a very great book.

Source: Robert Recorde, *The Grounde of Artes, Teachyng the Worke and Practise of Arithmetike, much necessary for all States of Men* (London: Reynold Wolfe, 1543), pp. 1–9.

COPERNICAN REVOLUTIONS

When the Polish astronomer Nicholas Copernicus (1473–1543) published his *De Revolutionibus Orbium Coelestium* (*On the Revolutions of the Celestial Orbs*) in 1543, few people took much notice. For even though the book turned out to be literally revolutionary – it overturned fourteen centuries of Ptolemaic doctrine by placing the sun at the centre of our solar system, with the earth spinning round it with the rest of the planets – it was a difficult, technical work that required specialist knowledge from its readers. It was not until many years later that Copernican ideas began to circulate widely, helped along by popular expositions, of which the following was one of the first. Written by Thomas Digges (1546–95), an English mathematician and astronomer, it appeared in the 1576 edition of his late father Leonard Digges's bestselling almanac, *A Prognostication Everlasting*. Digges senior, like most of his contemporaries, had been a confirmed Ptolemaic geocentrist, but his son Thomas was instead a disciple of the new heliocentric system, and it was through his simplified English translation, appended to a cheap, popular yearbook, that the Anglophone world was introduced to the implications of the new cosmology. The work had a particular influence on the language and ideas of William Shakespeare – note Digges's use of the phrase 'thin air', predating the famous line in *The Tempest* by more than thirty years:

Having of late (gentle Reader) corrected and reformed sundry faults that by negligence in printing have crept into my fathers *General Prognostication*: Among other things I found a description or Model of the world and situation of Spheres Celestiall and Elementary according to the doctrine of *Ptolemy*, whereunto all Universities (led thereto chiefly by the authority of *Aristotle*) sithens have consented. But in this our age one rare wit (seeing the continuall errors that from time to time more & more have been discovered, besides the infinite absurdities in their Theories, which they have been forced to admit that would not confess any mobility in the ball of the earth) hath by long study, painfull practise, and rare invention delivered a new Theory or model of the world, shewing that the Earth resteth not in the Centre of the whole world, but only in the Centre of this our mortall world or Globe of Elements which, environed and enclosed in the Moon's Orb, and together with the whole Globe of mortality is carried yearly round about the Sunne, which like a king in the midst of all reigneth and giveth laws of motion to ye rest, spherically dispersing his glorious beams of light through all this sacred Celestiall Temple. And the earth itself to be one of the Planets, having its peculiar and straying courses, turning every 24 hours round upon his own Centre, whereby the Sun and great Globe of fixed stars seem to sway about and turn, albeit indeed they remain fixed. So many ways is the sense of mortall men abused, but reason and deep discourse of wit having opened these things to *Copernicus*, & the same being with demonstrations Mathematicall most apparently by him to the world delivered. I thought it convenient together with the old Theory also to publish this, to the end such noble English minds (as delight to reach above baser sort of men) might not be altogether defrauded of so noble a part of Philosophy. And to the end it might manifestly appear that *Copernicus* meant not, as some have fondly excused him, to deliver these grounds of the Earth's mobility only as Mathematicall principles, feigned and not as Philosophicall truly averred, I have also from him delivered both the Philosophicall reasons by *Aristotle* and others produced to maintain the Earth's stability, and also their solutions and insufficiency, wherein I cannot a little commend the modesty of that grave Philosopher *Aristotle*, who seeing (no doubt) the insufficiency of his own reasons in seeking to confute the Earth's motion, with these words: *De his explicatum est ea qua potuimus facultate* ['this is the best explanation of these things we can offer']; howbeit his disciples have not with like sobriety maintained the same. Thus much for my own part in this case I will only say: There is no doubt but of a true ground truer effects may be produced than of principles that are false, and of true principles falsehood or absurdity cannot be inferred. If therefore the Earth be situate immovable in the Centre of the world, why find we not Theories upon that ground to produce effects as true and certain as these of *Copernicus*? Why cast we not away those *Circulos Æquantes* and motions irregular, seeing our own Philosopher *Aristotle* him self the light of our Universities hath taught us: *Simplicis corporis simplicem oportet esse motum* ['the motion of simple bodies ought to be simple']. But if contrary it be found impossible (the Earth's stability being granted) but that we must necessarily fall into these absurdities, and cannot by any means avoid them: Why shall we so much dote in the appearance of our senses, which many ways may be abused, and not suffer our selves to be directed by the rule of Reason, which the great God hath given us as a Lampe to lighten the darkness of our understanding and the perfect guide to lead us to the golden branch of Verity amid the forest of errors.

Behold a noble Question to be of the Philosophers and Mathematicians of our Universities argued not with childish Invectives, but with grave reasons Philosophicall and irreprovable Demonstrations Mathematicall. And let us not in matters of reason be led away with authority and opinions of men, but with the Stellified Poet let us say:

> *Non quid Aristoteles vel quiuis dicat eorum*
> *Dicta nihil moror, à vero cum forte recedunt,*
> *Magni sæpi viri mendacia magna loquuntur,*
> *Nec quisquam est adeo sagax, quin sæpius erret.*
> [Whatever Aristotle or others may say,
> I will not esteem statements which so depart from truth;
> Earnest, large-souled, and illustrious men often fall into error;
> Great authors misdirect many minds when they themselves fall short.]
> Marcellus Palingenius Stellatus, *Zodiacus Vitae* (1540)

The Globe of Elements enclosed in the Orb of the Moon I call the Globe of Mortality because it is the peculiar Empire of death. For above the Moone they fear not his force . . .

In the midst of this Globe of Mortality hangeth this dark star or ball of earth and water balanced and sustained in the midst of the thin air only with that property which the wonderfull workman hath given at the Creation to the Centre of this Globe with his magnetical force vehemently to draw and hale unto it self all such other Elementall things as retain the like nature. This ball every 24 hours by naturall, uniform and wonderfull sly and smooth motion rolleth round making with his Period our naturall day, whereby it seems to us that the huge infinite immoveable Globe should sway and turn about.

The Moon's Orb that environeth and containeth this dark star and the other mortall, changeable, corruptible Elements and Elementary things is also turned round every 29 days 31 minutes 50 seconds, 8 thirds, 9 fourths, and 20 fifths, and this Period may most aptly be called the Month . . .

Herein good Reader, I have waded farther than the vulgar sort

Demonstratively and Practically, & God sparing life I mean though not as a Judge to decide, yet at the Mathematicall bar in this case to plead in such sort, as it shall manifestly appear to the World whether it be possible upon the Earth's stability to deliver any true or probable Theory, & then refer the pronouncing of sentence to the grave Senate of indifferent discreet Mathematicall Readers.

Farewell and respect my travails as thou shalt see them tend to the advancement of truth and discovering the Monstrous loathsome shape of error.

Source: Thomas Digges, *A Perfect Description of the Celestial Orbs*, in Leonard Digges, *A Prognostication Everlasting* (London: Thomas Marsh, 1576), ff. 42–4.

SUPERNOVA

The Danish nobleman Tycho Brahe (1546–1601) was the last major astronomer to work without the aid of a telescope. His fame rests on a single observation, made on 11 November 1572, of a bright new celestial object that seemed suddenly to have appeared in the constellation of Cassiopeia. Observing it over the next few months, he concluded that it was neither a comet nor a planet, but an entirely new star – a *Nova Stella* – 'one that has never previously been seen before our time, in any age since the beginning of the world', he wrote. By the time this new star had faded from view, in March 1574, it had played its part in finishing off the discredited Aristotelian doctrine that the heavenly spheres are permanent, fixed and immutable. Here, instead, was a dramatic example of celestial change, bright enough to have been seen with the naked eye, and completely irrefutable.

It was, of course, a supernova – an incandescent stellar explosion – that had taken place in the Milky Way, some 7,500 light years from Earth. The word *nova* (Latin for 'new') went on to be adopted by astronomers from the title of Tycho's book, *De Nova Stella*:

Last year [1572], in the month of November, on the eleventh day of that month, in the evening, after sunset, when, according to my habit, I was contemplating the stars in a clear sky, I noticed that a new and unusual star, surpassing the other stars in brilliancy, was shining almost directly above my head; and since I had, almost from boyhood, known all the stars of the heavens perfectly (there is no great difficulty in attaining that knowledge), it was quite evident to me that there had never before been any star in that place in the sky, even the smallest, to say nothing of a star so conspicuously bright as this. I was so astonished at this sight that I was not ashamed to doubt the trustworthiness of my own eyes. But when I observed that others, too, on having the place pointed out to them, could see that there was really a star there, I had no further doubts. A miracle indeed, either the greatest of all that have occurred in the whole range of nature since the beginning of the world, or one certainly that is to be classed with those attested by the Holy Oracles, the staying of the Sun in its course in answer to the prayers of Joshua, and the darkening of the Sun's face at the time of the Crucifixion. For all

philosophers agree, and facts clearly prove it to be the case, that in
the ethereal region of the celestial world no change, in the way either
of generation or of corruption, takes place; but that the heavens and
the celestial bodies in the heavens are without increase or diminu-
tion, and that they undergo no alteration, either in number or in size
or in light or in any other respect; that they always remain the same,
like unto themselves in all respects, no years wearing them away. Furthermore, the observations of all the founders of the science, made
some thousands of years ago, testify that all the stars have always
retained the same number, position, order, motion, and size as they
are found, by careful observation on the part of those who take
delight in heavenly phenomena, to preserve even in our own day ...

It is a difficult matter, and one that requires a subtle mind, to try
to determine the distances of the stars from us, because they are so
incredibly far removed from the earth; nor can it be done in any way
more conveniently and with greater certainty than by the measure of
the parallax, if a star have one. For if a star that is near the horizon
is seen in a different place than when it is at its highest point and near
the vertex, it is necessarily found in some orbit with respect to which
the Earth has a sensible size. How far distant the said orbit is, the size
of the parallax compared with the semi-diameter of the Earth will
make clear. If, however, a star that is as near to the horizon as to the
vertex, is seen at the same point of the Primum Mobile, there is no
doubt that it is situated either in the eighth sphere or not far below
it, in an orbit with respect to which the whole Earth is as a point.

In order, therefore, that I might find out in this way whether this
star was in the region of the Element or among the celestial orbits,
and what its distance was from the Earth itself, I tried to determine
whether it had a parallax, and, if so, how great a one; and this I did
in the following way: I observed the distance between this star and
Schedir of Cassiopeia (for the latter and the new star were both
nearly on the meridian), when the star was at its nearest point to the
vertex, being only 6 degrees removed from the zenith itself (and for
that reason, though it were near the Earth, would produce no parallax in that place, the visual position of the star and the real position
then uniting in one point, since the line from the center of the Earth
and that from the surface nearly coincide). I made the same observation when the star was farthest from the zenith and at its nearest
point to the horizon, and in each case I found that the distance from

the above-mentioned fixed star was exactly the same, without the variation of a minute: namely 7 degrees and 55 minutes. Then I went through the same process, making numerous observations with other stars. Whence I conclude that this new star has no diversity of aspect, even when it is near the horizon. For otherwise in its least altitude it would have been farther away from the above-mentioned star in the breast of Cassiopeia than when in its greatest altitude. Therefore, we shall find it necessary to place this star, not in the region of the Element, below the Moon, but far above, in an orbit with respect to which the Earth has no sensible size. For if it were in the highest region of the air, below the hollow region if the Lunar sphere, it would, when nearest the horizon, have produced on the circle a sensible variation of altitude from that which it held when near the vertex . . .

Therefore, this new star is neither in the region of the Element, below the Moon, nor among the orbits of the seven wandering stars, but it is in the eighth sphere, among the other fixed stars, which was what we had to prove. Hence it follows that it is not some peculiar kind of comet or some other kind of fiery meteor become visible. For none of these are generated in the heavens themselves, but they are below the Moon, in the upper region of the air, as all philosophers testify; unless one would believe with Albategnius that comets are produced, not in the air, but in the heavens. For he believes that he has observed a comet above the Moon, in the sphere of Venus. That this can be the case, is not yet clear to me. But, please God, sometime, if a comet shows itself in our age, I will investigate the truth of the matter. Even should we assume that it can happen (which I, in company with other philosophers, can hardly admit), still it does not follow that this star is a kind of comet: first, by reason of its very form, which is the same as the form of the real stars and different from the form of all the comets hitherto seen, and then because, in such a length of time, it advances neither latitudinally nor longitudinally by any motion of its own, as comets have been observed to do. For, although these sometimes seem to remain in one place several days, still, when the observation is made carefully by exact instruments, they are seen not to keep the same position for so very long or so very exactly. I conclude, therefore, that this star is not some kind of comet or a fiery meteor, whether these be generated beneath the Moon or above the Moon, but that it is a star shining in the firmament itself – one that has never previously been seen before our time, in any age since the beginning of the world.

Source: *A Source Book in Astronomy*, ed. Harlow Shapley and Helen E. Howarth (New York: McGraw-Hill, 1929), pp. 13–19.

THE MOONS OF JUPITER

In 1609 the forty-five-year-old mathematician Galileo Galilei (1564–1642) heard about an optical instrument that had just been invented by a Dutch spectacle maker named Hans Lippershey; it was a simple refracting telescope, a spyglass made from a pair of lenses fixed at either end of a metal tube. Galileo decided to make one for himself, and his improvements to the original design increased its powers of magnification from three to thirty times. This rudimentary instrument revolutionized astronomy by challenging the prevailing view of the heavens; the moon, for example, was not the smooth sphere that Aristotle had claimed, but was visibly scarred with ridges and craters, while Jupiter turned out to have moons of its own, a discovery that put a serious dent in the belief that everything in the universe orbited the earth.

The following extracts from a nineteenth-century translation of Galileo's *Sidereus Nuncius* ('The Starry Messenger'), the book that made him famous, describe how he came to build his astronomical telescope, before moving on to the momentous discovery of the Galilean moons: Io, Europa, Ganymede and Callisto, as they came to be known (Galileo's attempt to name them 'the Medicean Stars', in honour of his patron Cosimo II de' Medici, having proved unsuccessful):

About ten months ago a report reached my ears that a Dutchman had constructed a telescope, by the aid of which visible objects, although at a great distance from the eye of the observer, were seen distinctly as if near; and some proofs of its most wonderful performances were reported, which some gave credence to, but others contradicted. A few days after, I received confirmation of the report in a letter written from Paris by a noble Frenchman, Jacques Badovere, which finally determined me to give myself up first to inquire into the principle of the telescope, and then to consider the means by which I might compass the invention of a similar instrument, which after a little while I succeeded in doing, through deep study of the theory of Refraction; and I prepared a tube, at first of lead, in the ends of which I fitted two glass lenses, both plane on one side, but on the other side one spherically convex, and the other concave. Then bringing my eye to the concave lens I saw objects satisfactorily large and near, for they

appeared one-third of the distance off and nine times larger than
when they are seen with the natural eye alone. I shortly afterwards
constructed another telescope with more nicety, which magnified
objects more than sixty times. At length, by sparing neither labour
nor expense, I succeeded in constructing for myself an instrument
so superior that objects seen through it appear magnified nearly a
thousand times, and more than thirty times nearer than if viewed by
the natural powers of sight alone.

It would be altogether a waste of time to enumerate the number
and importance of the benefits which this instrument may be
expected to confer, when used by land or sea. But without paying
attention to its use for terrestrial objects, I betook myself to observations of the heavenly bodies.

Galileo first examined the moon, making a series of detailed drawings and
observations, before moving on to examine some of the familiar constellations such as Orion and the Pleiades. But an extraordinary discovery, made
on 7 January 1610, would transform the entire scientific landscape:

There remains the matter, which seems to me to deserve to be
considered the most important in this work, namely, that I should
disclose and publish to the world the occasion of discovering and
observing four Planets, never seen from the very beginning of the
world to our own times, their positions, and the observations made
during the last two months about their movements and their changes
of magnitude; and I summon all astronomers to apply themselves to
examine and determine their periodic times, which it has not been
permitted me to achieve up to this day, owing to the restriction of
my time. I give them warning, however, again, so that they may not
approach such an inquiry to no purpose, that they will want a very
accurate telescope, and such as I have described in the beginning of
this account.

On the 7th day of January in the present year, 1610, in the first
hour of the following night, when I was viewing the constellations of
the heavens through a telescope, the planet Jupiter presented itself to
my view, and as I had prepared for myself a very excellent instrument, I noticed a circumstance which I had never been able to notice
before, owing to want of power in my other telescope, namely, that
three little stars, small but very bright, were near the planet; and

although I believed them to belong to the number of the fixed stars, yet they made me somewhat wonder, because they seemed to be arranged exactly in a straight line, parallel to the ecliptic, and to be brighter than the rest of the stars, equal to them in magnitude. The position of them with reference to one another and to Jupiter was as follows:

East   *   *   O   *   *West*

On the east side there were two stars, and a single one towards the west. The star which was furthest towards the east, and the western star, appeared rather larger than the third.

I scarcely troubled at all about the distances between them and Jupiter, for, as I have already said, at first I believed them to be fixed stars; but when, on January 8th, led by some fatality, I turned again to look at the same part of the heavens, I found a very different state of things, for there were three little stars all west of Jupiter, and nearer together than on the previous night, and they were separated from one another by equal intervals, as the accompanying figure shows.

East   O   *   *   *   *West*

At this point, although I had not turned my thoughts at all upon the approximation of the stars to one another, yet my surprise began to be excited, how Jupiter could one day be found to the east of all the aforesaid fixed stars when the day before it had been west of two of them; and forthwith I became afraid lest the planet might have moved differently from the calculation of astronomers, and so had passed those stars by its own proper motion. I, therefore, waited for the next night with the most intense longing, but I was disappointed of my hope, for the sky was covered with clouds in every direction.

But on January 10th the stars appeared in the following position with regard to Jupiter, the third, as I thought, being

East   *   *   O   *West*

hidden behind by the planet. They were situated just as before, exactly in the same straight line with Jupiter, and along the Zodiac . . .

When I had seen these phenomena, as I knew that corresponding changes of position could not by any means belong to Jupiter, and as, moreover, I perceived that the stars which I saw had always been the same, for there were no others either in front or behind, within a great distance, along the Zodiac – at length, changing from doubt into surprise, I discovered that the interchange of position which I saw belonged not to Jupiter, but to the stars to which my attention had been drawn, and I thought therefore that they ought to be observed henceforward with more attention and precision.

Accordingly, on January 11th I saw an arrangement of the following kind:

*East   *   *   O   West*

namely, only two stars to the east of Jupiter, the nearer of which was distant from Jupiter three times as far as from the star further to the east; and the star furthest to the east was nearly twice as large as the other one; whereas on the previous night they had appeared nearly of equal magnitude. I, therefore, concluded, and decided unhesitatingly, that there are three stars in the heavens moving about Jupiter, as Venus and Mercury round the Sun; which at length was established as clear as daylight by numerous other subsequent observations. These observations also established that there are not only three, but four, erratic sidereal bodies performing their revolutions round Jupiter . . .

These are my observations upon the four Medicean planets, recently discovered for the first time by me; and although it is not yet permitted me to deduce by calculation from these observations the orbits of these bodies, yet I may be allowed to make some statements, based upon them, well worthy of attention.

And, in the first place, since they are sometimes behind, sometimes before Jupiter, at like distances, and withdraw from this planet towards the east and towards the west only within very narrow limits of divergence, and since they accompany this planet alike when its motion is retrograde and direct, it can be a matter of doubt to no one that they perform their revolutions about this planet, while at the same time they all accomplish together orbits of twelve years' length about the centre of the world. Moreover, they revolve in unequal circles, which is evidently the conclusion to be drawn from the fact that I have never been permitted to see two satellites in conjunction when their distance from Jupiter was great, whereas near Jupiter two, three, and sometimes all four, have been found closely packed together. Moreover, it may be detected that the revolutions of the satellites which describe the smallest circles round Jupiter are the most rapid, for the satellites nearest to Jupiter are often to be seen in the east, when the day before they have appeared in the west, and contrariwise. Also, the satellite moving in the greatest orbit seems to me, after carefully weighing the occasions of its returning to positions previously noticed, to have a periodic time of half a month. Besides, we have a notable and splendid argument to remove the scruples of those who can tolerate the revolution of the planets round the Sun in the Copernican system, yet are so disturbed by the motion of one Moon about the Earth, while both accomplish an orbit of a year's length about the Sun, that they consider that this theory of the universe must be upset as impossible: for now we have not one planet only revolving about another, while both traverse a vast orbit about the Sun, but our sense of sight presents to us four satellites circling about Jupiter, like the Moon about the Earth, while the whole system travels over a mighty orbit about the Sun in the space of twelve years.

Source: *The Sidereal Messenger of Galileo Galilei, and a part of the Preface to Kepler's Dioptrics,* trans. Edward Stafford Carlos (London: Rivingtons, 1880), pp. 10–11; 44–70.

THE CIRCULATION OF BLOOD

Now we move from a great discovery of the outer world to a great discovery of the inner. William Harvey's confirmation of the circulation of blood around the body is one of the best-known achievements in the history of science. Harvey (1578–1657), Physician Extraordinary to King James I, had developed an interest in the phenomenon of blood while a medical student at the University of Padua – Galileo's old stamping ground – where his experiments began to raise nagging doubts about the long-established Galenic doctrine that the liver was the daily source of fresh venous blood, which the body then consumed. Once Harvey had calculated that over 540 pounds of blood passes through a human heart every twenty-four hours, it was clear that the liver couldn't possibly maintain such a vast level of production. Earlier writers, including the Syrian-born physician Ibn al-Nafis (1213–88) and the Spanish theologian Michael Servetus (c. 1511–53), had proposed that blood circulated between the heart and the lungs – a view that was confirmed by later anatomists – but it was Harvey who proved that a fixed quantity of blood flows around the entire body, pumped away from the heart in arteries and towards the heart in veins.

Yet the strength of established opinion meant that Harvey's ideas took decades to be accepted, and in the meantime many of his patients deserted him for his traditional blood-letting rivals. As the immunologist Sir Peter Medawar rightly pointed out, 'it is difficult nowadays to put oneself in the position of those who at one time believed that the movement of blood is essentially an ebb and flow – an advance and retreat. Thus it is easy to under-estimate the importance of the discovery of the circulation of the blood, a discovery which by its sheer magnitude and by the style in which it was made earned William Harvey a place on the team of All-Time Greats captained by Sir Isaac Newton, which counts among its members Galileo, Darwin, and those few others who in their time changed the direction of the flow of thought.'

The following extract has been taken from an early twentieth-century translation of Harvey's revolutionary treatise, De Motu Cordis ('Movement of the Heart and Blood in Animals'), which was published in Latin in 1628. Note how the clarity of his expository style is aided by the use of contemporary mechanistic imagery:

I am convinced that the motion of the heart is as follows: First, the auricle contracts, and this forces the abundant blood it contains as the cistern and reservoir of the veins, into the ventricle. This being filled, the heart raises itself, makes its fibers tense, contracts, and beats. By this beat it at once ejects into the arteries the blood received from the auricle; the right ventricle sending its blood to the lungs through the vessel called the *vena arteriosa*, but which in structure and function is an artery; the left ventricle sending its blood to the aorta, and to the rest of the body through the arteries.

These two motions, one of the auricles, the other of the ventricles, are consecutive, with a rhythm between them, so that only one movement may be apparent, especially in warm-blooded animals where it happens rapidly. This is like a piece of machinery in which one wheel moves another, though all seem to move simultaneously, or like the mechanism in fire-arms, where touching the trigger brings down the flint, lights a spark, which falls in the powder and explodes it, firing the ball, which reaches the mark. All these events because of their quickness seem to occur simultaneously in the twinkling of an eye. Likewise in swallowing: lifting the tongue and pressing the mouth forces the food to the throat, the larynx and the epiglottis are closed by their own muscles, the gullet rises and opens its mouth like a sac, and receiving the bolus forces it down by its transverse and longitudinal muscles. All these diverse movements, carried out by different organs, are done so smoothly and regularly that they seem to be a single movement and action, which we call swallowing.

So it happens in the movement and action of the heart, which is a sort of deglutition or transference of blood from the veins to the arteries. If anyone with these points in mind will carefully watch the cardiac action in a living animal, he will see, not only what I have said, that the heart contracts in a continuous movement with the auricles, but also a peculiar side-wise turning toward the right ventricle as if it twists slightly on itself in performing its work. It is easy to see that when a horse drinks that water is drawn in and passed to the stomach with each gulp, the movement making a sound, and the pulsation may be heard and felt. So it is with each movement of the heart when a portion of the blood is transferred from the veins to the arteries, that a pulse is made which may be heard in the chest.

The motion of the heart, then, is of this general type. The chief

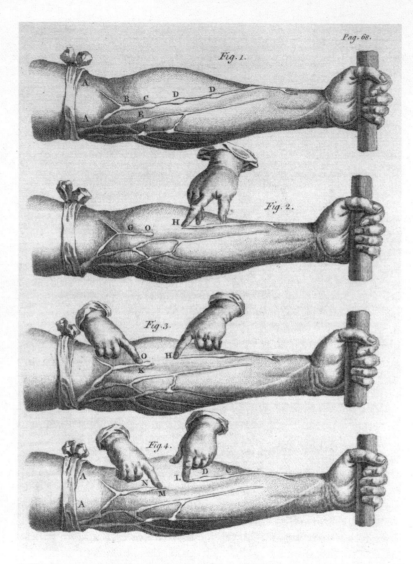

4. This engraving from Harvey's *De Motu Cordis* (1628) shows swellings in the veins of an arm caused by the slight loosening of a ligature tied above the elbow. The swelling indicated that blood was once more flowing back into the lower arm.

function of the heart is the transmission and pumping of the blood through the arteries to the extremities of the body. Thus the pulse which we feel in the arteries is nothing else than the impact of blood from the heart.

Whether or not the heart, besides transferring, distributing and giving motion to the blood, adds anything else to it, as heat, spirits, or perfection, may be discussed later and determined on other grounds. It is enough now to have shown that during the heart beat the blood is transferred through the ventricles from the veins to the arteries, and distributed to the whole body.

Source: William Harvey, *Anatomical Studies on the Motion of the Heart and Blood in Animals*, trans. Chauncey D. Leake (Springfield, Ill.: Charles C. Thomas, 1929), pp. 47–9.

A QUIRE OF ECHOES

Francis Bacon (1561–1626) was an essayist, lawyer, and politician, as well as one of the most influential figures in the history of experimental science. He is best known today for coining the phrase 'knowledge is power', though I prefer his less triumphalist slogan, from *The Advancement of Learning* (1605), that 'if a man will begin with certainties, he shall end in doubts; but if he will be content to begin with doubts, he shall end in certainties': an admirable expression of the evolving scientific outlook. In 1620 Bacon formally proposed a new empirical scientific method based on the patient accumulation of observational data; any hypotheses arising from the data could then be tested by rigorous experimentation. Forty years later this 'Baconian' method was adopted by the founders of the Royal Society of London, and it remains to this day a central, if idealized, component of modern scientific methodology (though, as we have seen already, it owed much to the example of earlier scholars such as Alhazen).

Bacon was a prolific writer, and selecting an illustrative example of his work proved more of a challenge than I had expected. In the end I chose one of my favourite pieces from *Sylva Sylvarum; or, a Natural History* (1627), a posthumously published volume filled with diverse observations about natural curiosities, drawn mostly from Bacon's own experiments in the field. Although the book contains no great scientific insights, it nevertheless reveals the strength of early modern curiosity about the workings of the natural world. The following passage describes the author's first-hand attempts to understand the phenomenon of multiple echoes:

There be three kinds of reflexions of sounds; a reflexion concurrent; a reflexion iterant, which we call echo; and a super-reflexion, or an echo of an echo . . . the latter two we will now speak of.

The reflexion of species visible, by mirrors, you may command; because passing in right lines, they may be guided to any point: but the reflexion of sounds is hard to master; because the sound filling great spaces in arched lines, cannot be so guided: and therefore we see there hath not been practised any means to make artificial echoes . . . Nevertheless it hath been tried, that one leaning over a well of twenty-five fathom deep, and speaking, though but softly, (yet not so soft as a whisper,) the water returned a good audible echo. It would be tried, whether speaking in caves, where there is no issue save where you speak, will not yield echoes, as wells do.

The echo cometh as the original sound doth, in a round orb of air: it were good to try the creating of the echo where the body repercussing maketh an angle: as against the return of a wall, &c. Also we see that in mirrors there is the like angle of incidence, from the object to the glass, and from the glass to the eye. And if you strike a ball side-long, not full upon the surface, the rebound will be as much the contrary way. Whether there be any such resilience in echoes, (that is, whether a man shall hear better if he stand aside the body repercussing, than if he stand where he speaketh, or anywhere in a right line between,) may be tried. Trial likewise would be made, by standing nearer the place of repercussing than he that speaketh; and again by standing further off than he that speaketh; and so knowledge would be taken, whether echoes, as well as original sounds, be not strongest near hand.

There may be places where you shall hear a number of echoes one after another: and it is when there is variety of hills or woods, some nearer, some further off: so that the return from the further, being last created, will be likewise last heard.

As the voice goeth round, as well towards the back as towards the front of him that speaketh; so likewise doth the echo: for you have many back-echoes to the place where you stand.

To make an echo that will report three, or four, or five words distinctly, it is requisite that the body repercussing be a good distance off: for if it be near, and yet not so near as to make a concurrent echo, it choppeth with you upon the sudden. It is requisite likewise that the air be not much pent: for air at a great distance pent, worketh the same effect with air at large in a small distance. And therefore in the trial of speaking in the well, though the well was deep, the voice came back suddenly, and would bear the report of but two words.

For echoes upon echoes, there is a rare instance thereof in a place which I will now exactly describe. It is some three or four miles from Paris, near a town called Pont-Charenton; and some bird-bolt shot or more from the river of Seine. The room is a chapel or small church. The walls all standing, both at the sides and at the ends. Two rows of pillars, after the manner of aisles of churches, also standing; the roof all open, not so much as any embowment near any of the wall left . . . Speaking at the one end, I did hear it return the voice thirteen several times; and I have heard of others, that it would return sixteen times: for I was there about three of the clock in the afternoon; and it is best (as all other echoes are) in the evening. It is manifest that it is not echoes from several places, but a tossing of the voice, as a ball, to and fro; like to reflexions in looking-glasses; where if you place one glass before and another behind, you shall see the glass behind with the image, within the glass before; and again, the glass before in that; and divers such super-reflexions, till the *species speciei* at last die. For it is every return weaker and more shady. In like manner the voice in that chapel createth *speciem speciei*, and maketh succeeding super-reflexions; for it melteth by degrees, and every reflexion is weaker than the former: so that if you speak three words, it will (perhaps) some three times report you the whole three words; and then the two latter words for some times; and then the last word alone for some times; still fading and growing weaker . . .

The like echo upon echo, but only with two reports, hath been observed to be, if you stand between a house and a hill, and lure towards the hill. For the house will give a back echo; one taking it from the other, and the latter the weaker.

There are certain letters that an echo will hardly express: as S for one, especially being principal in a word. I remember well, that when I went to the echo at Pont-Charenton, there was an old Parisian, who took it to be the work of spirits, and of good spirits. For (said he) call *Satan* and the echo will not deliver back the devil's name; but will say, *va t'en*; which is as much in French as *apage* or *avoid*. And thereby I did hap to find that an echo would not return S, being but a hissing and an interior sound.

Echoes are some more sudden, and chop again as soon as the voice is delivered; as hath been partly said: others are more deliberate, that is, give more space between the voice and the echo; which is caused by the local nearness or distance: some will report a longer train of words, and some a shorter; some more loud, (full as loud as the original, and sometimes more loud,) and some weaker and fainter.

Where echoes come from several parts at the same distance, they must needs make (as it were) a quire of echoes, and so make the report greater, and even a continued echo; which you shall find in some hills that stand encompassed, theatre-like.

Source: Francis Bacon, *Sylva Sylvarum: or, a Natural History*, in *The Works of Francis Bacon*, ed. James Spedding, Robert Leslie Ellis and Douglas Denon Heath, 10 vols (London: Longmans, 1870), II, pp. 425–8.

THE AIR-PUMP

Joseph Wright of Derby's celebrated painting *An Experiment on a Bird in the Air-Pump* (1768) leaves the viewer with an unanswered question: will the bird survive? Anyone familiar with the work of Robert Boyle (1627–91) will suspect that the answer might be 'no'. An aristocratic Anglo-Irish physicist and chemist, Boyle was a leading light in the founding of the Royal Society of London in 1660, the year that saw the publication of his *New Experiments Physico-Mechanicall, Touching the Spring of the Air*. This was the book that gave us Boyle's Law, which states that at a constant temperature, the volume of an ideal gas varies inversely with its pressure. The book described a series of controlled experiments that Boyle conducted using the newly invented air-pump, a customized version of which he commissioned from the twenty-five-year-old technical whiz-kid Robert Hooke (see following entry). The air-pump was a closed glass vessel with a sealable lid from which air could be extracted using a simple pump and bellows, in order to produce something approaching a vacuum. Boyle soon discovered that an absence of air extinguished burning candles, and killed all manner of unfortunate animals, as he recounts in the following extracts:

Experiment 40

It may well seem worth trying, whether or no in our exhausted Glass the want of an ambient body, of the wonted thickness of air, would disable even light and little animals, as bees and other winged insects, to fly. But though we easily foresaw how difficult it would be to make such an Experiment; yet not to omit our endeavors, we procur'd a large flesh-fly, which we convey'd into a small receiver. We also another time shut into a great receiver a Humming bee, that appear'd strong and lively, though we had rather have made the trial with a butter-fly, if the cold season would have permitted us to find any. The fly, after some exsuctions of the air, dropp'd down from the side of the Glass whereon she was walking: But, that the experiment with the bee might be more instructive, we convey'd in with her a bundle of flowers, which remain'd suspended by a string near the upper part of the receiver: And having provok'd the bee, we excited her to flie up and down the capacity of the vessel, till at length, as we desir'd, she lighted upon the flowers; whereupon we presently began to draw out the air, and observ'd, that though for some time the bee seem'd to take no notice of it, yet within a while after she did not flie, but fall down from the flowers, without appearing to make any use of her wings to help her self. But whether this fall of the bee, and the other insect, proceeded from the mediums being too thin for them to flie in, or barely from the weakness, and as it were swooning of the animals themselves, you will easily gather from the following experiment.

Experiment 41

To satisfie ourselves in some measure about the account upon which Respiration is so necessary to the animals, that nature hath furnish'd with lungs, we took (being then unable to procure any other lively bird, small enough to be put into the receiver) a lark, one of whose wings had been broken by a shot, of a man that we had sent to provide us some birds for our experiment; but notwithstanding this hurt, the lark was very lively, and did, being put into the receiver, divers times spring up in it to a good height. The vessel being hastily, but carefully clos'd, the pump was diligently plied, and the bird for a while appear'd lively enough; but upon a greater exsuction of the air, she began manifestly to droop and appear sick, and very soon after was taken with as violent and irregular convulsions, as are wont to be observ'd in poultry, when their heads are wrung off: for the bird threw herself over and over two or three times, and dyed with her breast upward, her head downwards, and her neck awry. And though upon the appearing of these convulsions, we turn'd the stop-cock, and let in the air upon her, yet it came too late . . . Soon after we got a hen-sparrow, which being caught with bird-lime was not at all hurt; when we put her into the receiver, almost to the top of which she would briskly raise herself, the experiment being tried with this bird, as it was with the former, she seem'd to be dead within seven minutes, one of which were imploy'd in cementing on the cover: But upon the speedy turning of the key, the fresh air flowing in began slowly to revive her, so that after some pantings she open'd her eyes, and regain'd her feet, and in about a $1/4$ of an hour, after threatening to make an escape at the top of the Glass, which had been upstopp'd to let in the fresh air upon her: But the receiver being clos'd the second time, she was kill'd with violent convulsions, within five minutes from the beginning of the pumping.

A while after we put in a mouse, newly taken, in such a trap as had rather affrighted than hurt him; whilst he was leaping up very high in the receiver, we fasten'd the cover to it, expecting that an animal used to live in narrow holes with very little fresh air, would endure the want of it better than the lately mention'd birds: But though, for a while after the pump was set awork, he continued leaping up as before; yet 'twas not long ere he began to appear sick and giddy, and to stagger, after which he fell down as dead, but without such violent convulsions as the birds died with. Whereupon, hastily turning the key, we let in some fresh air upon him, by which he recover'd, after a while, his senses and his feet, but seem'd to continue weak and sick: But at length, growing able to skip as formerly, the pump was plyed again for eight minutes, about the middle of which space, if not before, a very little air by a mischance got in at the stop-cock; and about two minutes after that, the mouse divers times leap'd up lively enough, though after about two minutes more he fell down quite dead, yet with convulsions far milder than those therewith the two birds expired. This alacrity so little before his death, and his not dying sooner than at the end of the eighth minute, seem'd ascribable to the air (how little soever) that slipt into the receiver . . .

I forgot to mention, that having caus'd these three creatures to be open'd, I could, in such small bodies, discover little of what we sought for, and what we might possibly have found in larger animals; for though the lungs of the birds appear'd very red, and as it were inflam'd, yet that colour being usual enough in the lungs of such winged creatures, deserves not so much our notice, as it does that in almost all the destructive experiments made in our engine, the animals appear'd to die with violently convulsive motions: From which, whether physicians can gather any things towards the discovery of the nature of convulsive distempers, I leave to them to consider.

Having proceeded thus far, though (as we have partly intimated already) there appear'd not much cause to doubt, but that the death of the fore-mention'd animals proceeded rather from the want of air, than that the air was over-clogg'd by the steams of their bodies, exquisitely pent up in the Glass; yet I, that love not to believe any thing upon conjectures, when by a not over-difficult experiment I can try whether it be true or no, thought it the safest way to obviate objections, and remove scruples, by shutting up another mouse as close as I could in the receiver, wherein it liv'd above three quarters of an hour; and might probably have done so much longer, had not a *Virtuoso* of quality, who in the mean time chanc'd to make me a visit, desir'd to see whether or no the mouse could be kill'd by the exsuction of the ambient air, whereupon we thought fit to open, for a little while, an intercourse betwixt the air in the receiver and that without it, that the mouse might thereby (if it were needful for him) be refresh'd, and yet we did this without uncementing the cover at the top, that it might not be objected, that perhaps the vessel was more closely stopp'd for the exsuction of the air than before.

The experiment had this event, that after the mouse had liv'd ten minutes . . . he dyed with convulsive fits, wherein he made two or three bounds into the air, before he fell down dead.

Nor was I content with this, but for Your Lordships further satisfaction, and my own, I caus'd a mouse, that was very hungry, to be shut in all night, with a bed of paper for him to rest upon: And to be sure that the receiver was well clos'd, I caus'd some air to be drawn out of it, whereby, perceiving that there was no sensible leak, I presently readmitted the air at the stop-cock, lest the want of it should harm the little animal; and then I caus'd the engine to be kept all night by the fire side, to keep him from being destroy'd by the immoderate cold of the frosty night. And this care succeeded so well, that the next morning I found that the mouse not only was alive, but had devour'd a good part of the cheese that had been put in with him. And having thus kept him alive full twelve hours, or better, we did, by sucking out part of the air, bring him to droop, and to appear swell'd; and by letting in the air again, we soon reduc'd him to his former liveliness.

Robert Boyle, *New Experiments Physico-Mechanicall, Touching the Spring of the Air, and its Effects (Made for the most part, in a New Pneumatical Engine)* (Oxford: Tho. Robinson, 1660), pp. 326–34.

THE DISCOVERY OF PLANT CELLS

In January 1665, Robert Hooke (1635–1703), the Curator of Experiments at the Royal Society of London, published a magnificently illustrated book, *Micrographia*, 'the most ingenious book that ever I read in my life', according to Samuel Pepys, whose diary records that he stayed up reading it until two o'clock in the morning. Featuring dramatic fold-out engravings based on detailed drawings prepared by Hooke as he peered into his home-built microscope, *Micrographia* instantly directed scientific attention to the world of the very small. Describing this strange new world, however, required the coining of a new vocabulary, and in the course of his account of the microscopic structure of cork, Hooke came up with the term 'cells', since the box-like arrangement of the tissues reminded him of the rows of cells in a monastery. The following extract – from the chapter on cork – conveys Hooke's growing excitement at the sheer *smallness* of the cellular structures he could see, 'a thing almost incredible, did not our *Microscope* assure us of it by ocular demonstration':

I took a good clear piece of Cork, and with a Pen-knife sharpen'd as keen as a Razor, I cut a piece of it off, and thereby left the surface of it exceeding smooth, then examining it very diligently with a *Microscope*, me thought I could perceive it to appear a little porous; but I could not so plainly distinguish them, as to be sure they were pores, much less what Figure they were of: But judging from the lightness and yielding quality of the Cork, that certainly the texture could not be so curious, but that possibly, if I could use some further diligence, I might find it to be discernable with a *Microscope*, I with the same sharp Pen-knife, cut off from the former smooth surface an exceeding thin piece of it, and placing it on a black object Plate, because it was it self a white body, and casting the light on it with a deep *plano-convex Glass*, I could exceeding plainly perceive it to be all perforated and porous, much like a Honey-comb, but that the pores of it were not regular; yet it was not unlike a Honey-comb in these particulars.

First, in that it had a very little solid substance, in comparison of the empty cavity that was contain'd between, as does more manifestly appear by the Figure A and B of the XI. *scheme*, for the

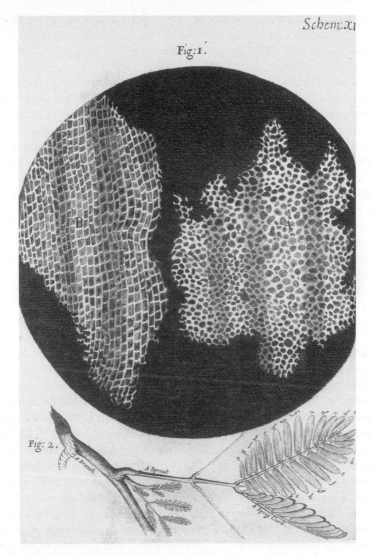

5. The publication of Robert Hooke's *Micrographia* in 1665 served to direct scientific attention to the world of the very small. This engraved drawing of highly magnified cork wood as seen through a compound microscope was the first image of 'cells' in a biological specimen; the term was coined by Hooke, who claimed that the box-like arrangements reminded him of the rows of cells in a monastery.

Interstitia, or walls (as I may so call them) or partitions of those pores were neer as thin in proportion to their pores, as those thin films of Wax in a Honey-comb (which enclose and constitute the sexangular cells) are to theirs.

Next, in that these pores, or cells, were not very deep, but consisted of a great many little Boxes, separated out of one continued long pore, by certain Diaphragms, as is visible by the Figure B, which represents a sight of these pores split the long-ways.

I no sooner discern'd these (which were indeed the first *microscopical* pores I ever saw, and perhaps, that were ever seen, for I had not met with any Writer or Person, that had made any mention of them before this) but me thought I had with the discovery of them, presently hinted to me the true and intelligible reason of all the *Phænomena* of Cork; As,

First, if I enquir'd why it was so exceeding light a body? my *Microscope* could presently inform me that here was the same reason evident that there is found for the lightness of froth, an empty Honey-comb, Wool, a Spunge, a Pumice-stone, or the like; namely, a very small quantity of a solid body, extended into exceeding large dimensions . . .

But, to return to our Observation. I told several lines of these pores, and found that there were usually about threescore of these small Cells placed end-ways in the eighteenth part of an Inch in length, whence I concluded there must be neer eleven hundred of them, or somewhat more than a thousand in the length of an Inch, and therefore in a square Inch above a Million, or 1166400, and in a Cubick Inch, above twelve hundred Millions, or 1259712000, a thing almost incredible, did not our *Microscope* assure us of it by ocular demonstration; nay, did it not discover to us the pores of a body, which were they *diaphragm'd*, like those of Cork, would afford us in one Cubick Inch, more than ten times as many little Cells, as is evident in several charr'd Vegetables; so prodigiously curious are the works of Nature, that even these conspicuous pores of bodies, which seem to be the channels or pipes through which the *Succus nutritius*, or natural juices of Vegetables are convey'd, and seem to correspond to the veins, arteries, and other Vessels in sensible creatures, that these pores I say, which seem to be the Vessels of nutrition to the vastest body in the World, are yet so exceeding small, that the *Atoms* which *Epicurus* fancy'd would go neer to prove too bigg to enter them, much more to constitute a fluid body in them. And how infinitely smaller then must be the Vessels of a Mite, or the pores of one of those little Vegetables I have discovered to grow on the back-side of a Rose-leaf, and shall anon more fully describe, whose bulk is many millions of times less than the bulk of the small shrub it grows on; and even that shrub, many millions of times less in bulk than several trees (that have heretofore grown in *England*, and are this day flourishing in other hotter Climates, as we are very credibly inform'd) if at least the pores of this small Vegetable should keep any such proportion to the body of it, as we have found these pores of other Vegetables to do to their bulk. But of these pores I have said more elsewhere.

Hooke's microscopical researches were the cause of much excitement among his scientific contemporaries, and in 1676 the Dutch naturalist Antoni van Leeuwenhoek (1632–1723) wrote to the Royal Society to announce his discovery of 'little animals' seen through the microscope: ''twas wonderful to see: and I judge that some of these little creatures were above a thousand times smaller than the smallest ones I have ever yet seen, upon the rind of cheese, in wheaten flour, mould and the like': bacteria had just been discovered.

However, not everyone was convinced of the benefits to science bestowed by the new technology. The writer Margaret Cavendish, Duchess of Newcastle (1623–73), was an outspoken critic of the new instrumental approach to natural knowledge, and in 1666 she published a lengthy and controversial attack on the fashion for optical instruments:

Although I am not able to give a solid judgment of the Art of *Micrography*, and the several dioptrical instruments belonging thereto, by reason I have neither studied nor practised that Art; yet of this I am confident, that this same Art, with all its Instruments, is not able to discover the interior natural motions of any part or creature of Nature; nay, the question is, whether it can represent yet the exterior shapes and motions so exactly, as naturally they are; for Art doth more easily alter than inform: As for example; Art makes Cylinders, Concave and Convex-glasses, and the like, which represent the figure of an object in no part exactly and truly, but very deformed and mis-shaped: also a Glass that is flaw'd, crack'd, or broke, or cut into the figure of Lozanges, Triangles, Squares, or the like, will present numerous picture of one object. Besides, there are so many alterations made by several lights, their shadows, refractions, reflexions, as also several lines, points, mediums, interposing and intermixing parts, forms and positions, as the truth of an object will hardly be known; for the perception of sight, and so of the rest of the senses, goes no further than the exterior Parts of the object presented; and though the Perception may be true, when the object is truly presented, yet when the presentation is false, the information must be false also . . .

For example; a Lowse by the help of a Magnifying-glass, appears like a Lobster, where the Microscope enlarging and magnifying each part of it, makes them bigger and rounder than naturally they are. The truth is, the more the figure by Art is magnified, the more it appears mis-shapen from the natural, in so much as each joynt will appear as a diseased, swell'd and tumid body, ready and ripe for incision. But mistake me not; I do not say, that no Glass presents the true picture of an object; but onely that Magnifying, Multiplying, and the like optick Glasses, may, and do oftentimes present falsely the picture of an exterior object; I say, the Picture, because it is not the real body of the object which the Glass presents, but the Glass onely figures or patterns out the picture presented in and by the Glass, and there may easily mistakes be committed in taking Copies from Copies. Nay, Artists do confess themselves, that Flies, and the like, will appear of several figures or shapes, according to the several reflections, refractions, mediums and positions of several lights; which if so, how can they tell or judge which is the truest light, position, or medium, that doth present the object naturally as it is? . . .

Wherefore those that invented Microscopes, and such like dioptrical Glasses, at first, did, in my opinion, the world more injury than benefit; for this Art has intoxicated so many mens brains, and wholly imployed their thoughts and bodily actions about phænomena, or the exterior figures of objects, as all better Arts and Studies are laid aside; nay, those that are not as earnest and active in such imployments as they, are, by many of them, accounted unprofitable subjects to the Commonwealth of Learning. But though there be numerous Books written of the wonders of these Glasses, yet I cannot perceive any such, at best, they are but superficial wonders, as I may call them.

Sources: Robert Hooke, *Micrographia; or, some Physiological Descriptions of Minute Bodies made by Magnifying Glasses, with Observations and Inquiries thereupon* (London: John Martyn, 1665), pp. 112–14; Margaret Cavendish, *Observations upon Experimental Philosophy; to which is added, The Description of a New Blazing World* (London: A. Maxwell, 1666), pp. 7–10.

MAYFLIES

Mayflies are among the oldest winged insects on the planet – they emerged during the Late Carboniferous, some 300 million years ago – yet the brevity of their adult existence has turned them into a melancholy symbol of the fleetingness of earthly life. The mayfly metaphor arose in the late seventeenth century, in the work of the Dutch biologist Jan Swammerdam (1637–80), whose scientific practice fused a rigorous Baconian materialism with a profoundly emotional and empathetic response to nature. The mayfly, for Swammerdam, was more than a natural puzzle; it was an object of moral contemplation, and the following extract, from an eighteenth-century translation of his posthumously published *Biblia naturae* (*The Book of Nature*), contains an impressive piece of scientific detective work, revealing that what appeared to be a series of different organisms – egg, larva, pupa and adult winged insect – were really different forms of the same delightful animal, whose fate supplies 'us wretched mortals with a lively image of the shortness of this present life'. (Poignantly, the species that was studied by Swammerdam, *Palingenia* sp., is now extinct in the Netherlands and most of Western Europe, due to its sensitivity to industrial pollution):

This insect has four wings, two little antennæ or horns, six legs, and a very long and hairy tail; it lives at the utmost but five hours. This surprising creature appears every year for three days successively fluttering on the surface of the water, at the mouths of the Rhine, the Meuse, the Wael, the Leck and the Ysel, about the feast of St. Olophius and St. John; but this continued appearance is kept up by a succession of them, for those which begin to live and flutter towards the noon of the first day, are dead before night, and the same happens the second and third days; which being expired, no more of them are to be seen till the returning year again renews this three days wonderful sight.

At this time the female of the Ephemerus, after having thrown off her coat or slough, issues from the water and discharges into it both her ovaries. But this is not done till she has for some time fluttered upon the surface of the water in a very curious and surprising manner, beating it with her wings all the time.

It is at this time that the female like a fish discharges her eggs, which the male, who first quits the water, and afterwards divests himself on the land of a very thin skin, invigorates by ejecting on them his melt or seminal milky substance . . .

The eggs of the female Ephemerus dropt into the water, and there impregnated by an effusion of the male's sperm, in the manner we have described, gradually sink to the bottom. But this is effected in such a manner, that the eggs are scattered over the muddy bottoms of rivers by the motion of the waters. The figure of the eggs themselves does not a little contribute thereto, as they are of a plain convex shape, and are thereby the apter to disperse in their descent. This appears by placing a few of them on the point of a knife, and then letting them fall gently into water, for they immediately separate of themselves in a very curious manner.

It is hard to say, and God only, who gives these insects life and motion, knows how long their eggs lie at the bottom of the waters where they are deposited, and how long it is before the contained insects break through the skin that surrounds them, and perform as it may be called their first moulting. It is not improbable that these things might be experimentally ascertained by dredging for them at different seasons, or keeping their eggs in a vessel full of water with a sufficient quantity of mud at the bottom. For the present we must be satisfied to observe that the eggs of the Ephemerus produce, after an unknown time, a little Worm with six legs. This is the creature fishermen call the bank-bait.

Three years later, these mud-dwelling worms are ready to spread their wings and fly for their few brief hours of airborne adult life:

When the time of the change of the Worm of the Ephemerus is approaching, and the wings, hidden in the cases or husks, have acquired their due strength and form, and that it is no longer in the power of the Worm to delay its change, those which have their parts thus disposed and prepared, march out of their habitations into the water. This usually happens in the evening between six of the clock and half an hour after. This I observed on the thirteenth of the month of June, in the year 1671, pursuing the change of the Worms of the Ephemerus.

The other Worms, which are not as yet come to this state of growth, remain in their cells. Those which have crept into the water move forward, and make all the haste they can from the bottom to the surface; which, when some more swiftly and others more slowly are arrived at, each of them, is changed into a winged insect. But this change or casting of the skin is so suddenly performed, that even the most attentive person cannot otherwise judge, than that the Worm breaks or bursts his way and swiftly flies out of the middle of the water.

Every insect that I have hitherto observed has a certain and determined time appointed for it by the omnipotent God, to expand its wings and dry them, that they may become smooth and polished, before they are able to prepare themselves for flight. But the Ephemerus, on the contrary, is almost at one and the same point of time a reptile and a flying creature. Wherever one sees at this time a little water bubble up, if we cast our eyes on the surface, there is immediately a winged insect observed to issue out of the middle of the water. Argus would want eyes, if he should attempt to trace these miracles of the adorable Creator of the universe . . .

The Ephemerus thus flying about and wandering over the surface of the water, and moving sometimes up and sometimes down through the air, never lives more than four or at most five hours, that is from six of the clock in the evening, or half an hour after, until eleven at night. This I say from experience, because I have carried some of them enclosed in a box into my chamber, and there accurately observed the length of their lives. All die in this very short space of time, nor do any of them, which is a matter very worthy of observation, die a natural death on land. All of them invariably go to the water again, after they have gone through the second change of their skin. God therefore, the supreme artist, has been pleased to assign this insect a short life that surpasses all adoration.

Who has so great a genius, or is so conversant in the art of writing, as to be able to describe, with a due sense, the trouble, and misfortunes this creature is subject to, during the short continuance of its flying life. For my part, I confess I am by no means able to execute this task. Nor do I know whether nature ever produced a more innocent and simple little creature, which is, notwithstanding, destined to undergo so many miseries and horrible dangers.

Besides, that the life of the Ephemerus is short, nay, amazingly and incomprehensibly so, an infinite number of them are always destroyed in the birth, being devoured by fish. Nor does [Theodorus]

Clutius acquit any species of fish of this barbarity except the Perch and Pike. Though the rest of the Ephemeri have escaped this cruel danger, yet on land, when they are engaged in the great work of changing their skin, they are barbarously devoured by Swallows and other birds. Nay, if they escape this danger, when they afterwards approach again to the surface of the water, and carelessly sport and play there with their wings and tails, they a second time become a prey to the fish, which drag them away to the dark bottom of the water and devour them. If they fly higher into the air, another kind of torment attends them, for then they are persecuted with a different barbarity by other kinds of birds, which tear their limbs asunder and devour them. Though these insects are then the most innocent, perhaps, of all others, they are more cruelly treated or used than the most mischievous of wild beasts . . .

One may ask further this question, why, exclusive of all those dangers and misfortunes, the life of the Ephemerus should be so short? In answer to this let it be observed, that the eggs of the Ephemerus, whilst it still swims as a Worm, are arrived to their perfection, so that as soon as the insect is increased and perfected by changing and extending its limbs, those eggs are instantly fit for production or birth: to which may be added, that the Ephemerus has not the nourishing of its offspring; wherefore God has made this creature likewise, more than others, void of reason, as the Ostrich among birds, that He, from whom springs all reason and knowledge, might take upon himself the care of nourishing its progeny.

Since therefore this creature assumes its winged form only to propagate its species, it follows, that when this is done, its death is naturally near at hand, and for this purpose, it seems to remain three years hidden in the water and mud, and to undergo after that time its change, and get wings in that form living, till this business of generation is performed, and then it dies.

Source: Jan Swammerdam, *The Book of Nature: or, the History of Insects*, trans. Thomas Flloyd (London: C. G. Seyffert, 1758), pp. 103–18.

SUNSPOTS

'Heretofore 'twas believ'd, that the Sun was a pure fire; but the error of this opinion was found out in the beginning of this age, by Spots which were discover'd upon the Surface of the Sun.' So says the narrator of Bernard de Fontenelle's Copernican tour of the universe, *Entretiens sur la Pluralité des Mondes* (*A Discovery of New Worlds*, 1686). Unusually for the time, de Fontenelle (1657–1757) wrote the book in idiomatic French rather than in scientific Latin, and approached natural philosophy 'in an altogether unphilosophical manner' as a means of reaching as wide an audience as possible. Like Algarotti's later *Newton for the Ladies* (see p. 110), the book takes the form of a flirtatious dialogue between a male philosopher and an aristocratic female pupil ('the Marquiese'), and Fontenelle's intention, which he outlined in his preface, was to encourage women to engage with scientific ideas rather than 'lose so much time at their toylets in a less charming study'.

It was an idea that caught the imagination of the British author and playwright Aphra Behn (1640–89), whose fine English translation of the work appeared in 1688. 'I thought an English Woman might adventure to translate any thing a French Woman may be supposed to have spoken,' she wrote in her preface to the English edition, and the wit and verve that illuminate her interpretation did much to popularize the book outside France. The reference to the new planets named after princes is a reference to Galileo's 'Medicean stars' (the moons of Jupiter) that had been discovered in 1610, at around the same time that the existence of magnetic sunspots was instrumentally confirmed:

After *Mercury*, you know, the next Planet we find in our journey, is the Sun; and if we judge by the Earth (which is inhabited) that other bodies of the same kind may be so too, we are mistaken, and the *Why not* will fail us here; for the Sun is a body of a quite different nature from the Earth, and other Planets: He is the source and fountain of all that light, which the other Planets do only reflect from one to another, after having receiv'd it from him; and so they can exchange light one with another, but are incapable of producing it: The Sun alone draws from it self this precious substance, which he darts around him with great force and violence, and which is intercepted by every body that is solid; so that there is reflected from one Planet

to another long streams and streaks of light, which crossing and traversing each other in the air, are interwoven a thousand different
ways, and so form a mixture of the richest substance in nature: For
this end the Sun is plac'd in the centre, which is the situation most
proper and commode; from whence he may equally dispence and dis-
tribute his light and heat, for the livening and enlightening all things
round him. The Sun is therefore a body of a particular substance; but
what kind of body, or what kind of substance, is all the difficulty:
Heretofore 'twas believ'd, that the Sun was a pure fire; but the error
of this opinion was found out in the beginning of this age, by Spots
which were discover'd upon the surface of the Sun; as a little after
that time, there were new Planets discover'd, of which I shall speak
hereafter: The learned part of the World were full of nothing else but
these new Planets; and discourses of 'em were so much in fashion,
that they believ'd the spots in the Sun were nothing else but these
new Planets, moving round 'em, which necessarily hid a part of his
body from our sight, when their obscure side was turn'd towards us.
The learned men of the world made their court to most Kings and
Princes, with these new discover'd Planets; giving the name of one
Prince to one, and of another Prince to another; so that they were like
to quarrel, to whom they shou'd belong.

I am not pleas'd with that at all, said the *Marquiese*: You told me
the other night, that the Philosophers and learned men had given
names of Philosophers, Astronomers and Mathematicians to the
several countries of the Moon, and I was very well satisfy'd, and
think it but just, that since the Kings and Princes possess the Earth,
that they ought to suffer Philosophers and Astronomers to rule in
the Moon and the Stars without incroaching upon 'em. What, said
I, Madam, will not you allow Kings and Princes some corner of
the Moon, or some Star, to take their part in time of need? As to the
Spots in the Sun, they can be of no manner of use to 'em; for it has
been found, they are not Planets, but clouds of smoak and vapours,
and, as it were, a scum arising from the surface of the Sun; for sometimes they appear in great quantities, sometimes in less, and at other
times they disappear; sometimes they join in one, and other times
they are dispers'd and dissipated; so that it shou'd seem, the Sun is a
liquid substance; some say, 'tis of melted gold, which boils incessantly, and produces those impurities; and by the force of its motion,
throws upon the Surface its scum and dross; and as those consume,

new ones are produc'd. Pray, Madam, fancie to your self what strange bodies these Spots of the sun must be; there are some of 'em full as big as the globe of the Earth; judge then what a great quantity there must be of this melted gold, and of the extent of this vast ocean of light and fire, which we call the Sun. They say, the Sun appears, through telescopes, to be full of great mountains which vomit flames, and that it is, as it were, a million of Mount Ætna's, join'd together; but at the same time they acknowledge that these mountains may be altogether visionary, and that they are nothing else but the defects of the glasses of the telescopes. To whom shall we trust then, when these very glasses, to which we owe so many new discoveries, deceive us? In fine, let the Sun be what it will, it does not at all seem proper to be inhabited; and yet 'tis pity, for the situation wou'd be extreamly fine; its Inhabitants wou'd be plac'd in the Centre of the Universe, and wou'd see all the other Planets turn regularly round 'em, whereas we observe infinite irregularities in their course; and 'tis only because we are not in a proper situation to consider 'em, as not being in the centre of their motion. Is it not hard, that there is but one place in the Universe, where the Study of the Stars wou'd be easie, and that that place alone shou'd be uninhabitable? You do not think, whilst you speak, said the Marquiese, were any living creature in the Sun, he wou'd see neither Planets, nor fixed Stars; nor, indeed, any thing; the brightness of the Sun wou'd render all things else invisible; and if there were inhabitants in the Sun, they wou'd apt to believe themselves the only people in nature.

I confess, said I, I am mistaken, I consider'd only the situation of the Sun, without the effects of its light. But, pray, Madam, allow me to tell you, that you who have corrected me so justly, may also be mistaken your self: The inhabitants of the Sun would not so much as see itself; for they would be incapable to support the dazzling of his light, or unable to receive it, by being too near; and all things well consider'd, the Sun would be a Country of blind men only. So that, once for all, I conclude, the Sun cannot be inhabited; and if you please, Madam, we will continue our journey to the other Worlds.

Source: Bernard de Fontenelle, *A Discovery of New Worlds, Made English by Mrs. A. Behn* (London: William Ganning, 1688), pp. 103–8.

TEMPERATURE SCALES

Though the first thermometers began to appear at the end of the sixteenth century, the concept of a standard calibration took another century to appear. Among the first to propose a fixed temperature scale was the Danish astronomer Ole Rømer (1644–1710), who in 1701 set the freezing point of brine as zero, and the boiling point of water at just over 60 degrees. A few years later the German instrument maker Daniel Fahrenheit (1686–1736) introduced the mercury-in-glass thermometer, with which he expanded the Rømer scale beyond the freezing point (32 °F) and the boiling point (212 °F) of water, establishing an interval between them of 180 degrees, an easily divisible sexagesimal number with which the Babylonians would have been familiar (see p. 1). Fahrenheit's system was adopted throughout the world following the publication of his temperature scale in the *Philosophical Transactions* of the Royal Society in 1724:

About ten years ago I read in the History of the Sciences issued by the Royal Academy of Paris, that the celebrated Amontons [Guillaume Amontons (1663–1705), who devised the first thermometer by which temperature was measured by air pressure], using a thermometer of his own invention, had discovered that water boils at a fixed degree of heat. I was at once inflamed with a great desire to make for myself a thermometer of the same sort, so that I might with my own eyes perceive this beautiful phenomenon of nature, and be convinced of the truth of the experiment.

I therefore attempted to construct a thermometer, but because of my lack of experience in its construction, my efforts were in vain, though they were often repeated; and since other matters prevented my going on with the development of the thermometer, I postponed any further repetition of my attempts to some more fitting time. Though my powers and my time failed me, yet my zeal did not slacken, and I was always desirous of seeing the outcome of the experiment. It then came into my mind what that most careful observer of natural phenomena had written about the correction of the barometer; for he had observed that the height of the column of mercury in the barometer was a little (though sensibly enough) altered by the varying temperature of the mercury. From this I gathered that a thermometer might perhaps be constructed with mercury, which would not be so hard to construct, and by the use of which it might be possible to carry out the experiment which I so greatly desired to try.

When a thermometer of that sort was made (perhaps imperfect in many ways) the result answered to my prayer; and with great pleasure of mind I observed the truth of the thing.

Three years then passed, in which I was occupied with optical and other work, when I became anxious to try by experiment whether other liquids boiled at fixed degrees of heat.

The results of my experiments are contained in the following table, of which the first column contains the liquids used, the second, their specific gravity, the third, the degree of heat which each liquid attains when boiling.

Liquids	Specific Gravity of Liquids at 48° of heat	Degree Attained by Boiling
Spirits of Wine or Alcohol	8260	176
Rain Water	10000	212
Spirits of Niter	12935	242
Lye prepared from wine lees	15634	240
Oil of Vitriol	18775	546

I thought it best to give the specific gravity of each liquid, so that, if the experiments of others already tried, or which may be tried, give different results, it might be determined whether the difference should be looked for as resulting from differences in the specific gravities or from other causes. The experiments were not made at the same time, and hence the liquids were affected by different degrees of temperature or heat, but since their gravity is altered in a different way and unequally, I reduced it by calculation to the degree 48, which in my thermometers holds the middle place between the limit of the most intense cold obtained artificially in a mixture of water, of ice and of sal-ammoniac or even of sea-salt, and the limit of the heat which is found in the blood of a healthy man.

Two decades later, an alternative centigrade ('hundred steps') temperature scale was proposed by the Swedish astronomer Anders Celsius (1701–44), who set the freezing point of water at 100° and its boiling point at 0°; after Celsius's death in 1744, the centigrade scale was reversed to its present form (with 0° as the freezing point). In 1948 the centigrade scale was formally renamed the Celsius scale (°C) and adopted for international use. At least, that was the intention: the problem today is that there are two competing temperature scales at large in the world, Fahrenheit and Celsius, and they are sometimes used interchangeably as though they were part of the same continuum; so in cold weather we like to say that it is 'minus 5' (Celsius), because that sounds colder than '23°' (Fahrenheit), while in hot weather we prefer to complain that it is '92 in the shade' (Fahrenheit), because that sounds hotter than '33°' (Celsius).

However, the picture is more complicated still, because scientists prefer to use a third scale, the Kelvin scale, which is based on the concept of absolute zero (the theoretical absence of all thermal energy) that was proposed in 1848 by the British physicist William Thomson, Lord Kelvin (1824–1907). Because the Kelvin scale begins at absolute zero (0 kelvin; equivalent to −273.15 °C) there are no negative values to deal with; and because it uses the same degree increment as the Celsius scale, Kelvin is easily converted to Celsius by subtracting 273. (Contrast with the rigmarole of converting Celsius to Fahrenheit: divide by 5, multiply by 9, then add 32.)

Source: D. G. Fahrenheit, 'Experiments on the degree of heat of some boiling liquids', *Philosophical Transactions* 33 (1724), pp. 1–3.

NEWTONIAN APPLES

Newton's falling apple is one of the great enduring images in the history of science. According to the story, Newton was sitting in the garden of his childhood home of Woolsthorpe Manor, Lincolnshire, in 1666 (Cambridge University having closed as a precaution against the spread of plague), when he was startled by the fall of an apple from a tree – the variety has been tentatively identified as Pride of Kent, a now-rare variety of cooking apple – at which point the twenty-two-year-old mathematician wondered to himself: why does it always fall downwards, never sideways or upwards? From the simple answer – that the Earth must somehow attract it – flowed the famous calculations that led to Newton's law of universal gravitation, which states that the gravitational attraction between two bodies (whether apples or planets) is proportional to the product of their masses, and inversely proportional to the square of the distance between them.

Newton's apple has since become a global shorthand for scientific inspiration (some versions of the tale even have the apple landing on his head), though the story only appeared sixty years after the event, at the end of Newton's long life, when he was beginning to think about his posthumous reputation. So was it all just a clever piece of scientific marketing?

William Stukeley

The story appears to have originated in a conversation between the eighty-three-year-old Newton and his friend and biographer William Stukeley (1687–1765) at Newton's London lodgings in April 1726. Having dined indoors, as Stukeley recalled, the pair then headed out to the garden to talk:

After dinner, the weather being warm, we went into the garden, & drank thea under the shade of some appletrees; only he, & myself. Amidst other discourse, he told me, he was just in the same situation, as when formerly, the notion of gravitation came into his mind. "Why should that apple always descend perpendicularly to the ground," thought he to himself; occasion'd by the fall of an apple, as he sat in a contemplative mood. "Why should it not go sideways, or upwards? But constantly to the earth's centre? Assuredly, the reason is, that the earth draws it. There must be a drawing power in matter, & the sum of the drawing power in the matter of the earth must be in the earth's centre, not in any side of the earth. That there is a power like that we here call gravity, which extends its self thro' the universe. Therefore does this apple fall perpendicularly or toward the centre. If matter thus draws matter, it must be in proportion of its quantity. Therefore the apple draws the earth, as well as the earth draws the apple."

& thus by degrees, he began to apply this property of gravitation to the motion of the earth, & of the heavenly bodys: to consider their distances, their magnitudes, their periodical revolutions: to find out, that this property, conjointly with a progressive motion impressed on them in the beginning, perfectly solv'd their circular courses; kept the planets from falling upon one another, or dropping alltogether into one center. & thus he unfolded the Universe. This was the birth of those amazing discoverys, whereby he built philosophy on a solid foundation, to the astonishment of all Europe.

Arthur Eddington

As Newton once said, his gravitational calculations 'seemed to answer, pretty nearly'. It's that 'pretty nearly' that subsequent physics has had problems with, particularly when applied to situations outside the normal range of experiences, such as hypothetical objects travelling at nearly the speed of light. One of those objects was 'the man in the lift', from an ingenious thought experiment proposed in a 1927 lecture by the astrophysicist Sir Arthur Eddington (1882–1944). Eddington had taken part in the famous expedition to the West African island of Principe in 1919 that had successfully upheld Einstein's general theory of relativity during a solar eclipse; Einstein's theory proposed that light from distant stars would be bent by gravity as it travelled past the sun towards earth, an hypothesis that the Principe observations confirmed (well, pretty nearly). Eddington, who became a well-known popular-science writer, spent much of the rest of his career attempting to reconcile Einsteinian relativity with Newtonian mechanics, as in this extract from his celebrated 'man in the lift' lecture, which demonstrates how, under certain circumstances, Newton's apple would have failed to 'fall':

Suppose that this room is a lift. The support breaks and down we go with ever-increasing velocity, falling freely.

Let us pass the time by performing physical experiments. The lift is our laboratory and we shall start at the beginning and try to discover all the laws of Nature – that is to say, Nature as interpreted by the Man in the Lift. To a considerable extent this will be a repetition of the history of scientific discovery already made in the laboratories on *terra firma*. But there is one notable difference.

I perform the experiment of dropping an apple held in the hand. The apple cannot fall any more than it was doing already. You remember that our lift and all things contained in it are falling freely. Consequently the apple remains poised by my hand. There is one incident in the history of science which will not repeat itself to the men in the lift, viz. Newton and the apple tree. The magnificent conception that the agent which guides the stars in their courses is the same as that which in our common experience causes apples to drop, breaks down because it is our common experience *in the lift* that apples do not drop.

I think we have now sufficient evidence to prove that in all other respects the scientific laws determined in the lift will agree with those determined under more orthodox conditions. But for this one omission the men in the lift will derive all the laws of Nature with which we are acquainted, and derive them in the same form that we have derived them. Only the force which causes apples to fall is not present in their scheme.

I am crediting our observers in the lift with the usual egocentric attitude, viz. the aspect of the world to *me* is its natural one. It does not strike them as odd to spend their lives falling in a lift; they think it much more odd to be perched on the earth's surface. Therefore, although they perhaps have calculated that to beings supported in this strange way apples would seem to have a perplexing habit of falling, they do not take our experience of the ways of apples any more seriously than we have hitherto taken theirs . . .

I perform another experiment. This time I take two apples and drop them at opposite ends of the lift. What will happen? Nothing much at first; the apples remain poised where they were let go. But let us step outside the lift for a moment to watch the experiment. The two apples are pulled by gravity towards the centre of the earth. As they approach the centre their paths converge and they will meet at

the centre. Now step back into the lift again. To a first approximation the apples remain poised above the floor of the lift; but presently
we notice that they are drifting towards one another, and they will
meet at the moment when (according to an outside observer) the lift
is passing through the centre of the earth. Even though apples (in the
lift) do not tend to fall to the floor there is still a mystery about their
behaviour; and the Newton of the lift may yet find that the agent
which guides the stars in their courses is to be identified with the
agent which plays those tricks with apples nearer home.

It comes to this. There are both relative and absolute features
about gravitation. The feature that impresses us most is relative –
relative to a frame that has no special importance apart from the
fact that it is the one commonly used by us. This feature disappears
altogether in the frame of the man in the lift and we ought to disre-
gard it in any attempt to form an absolute picture of gravitation. But
there always remains something absolute, of which we must try to
devise an appropriate picture. For reasons which I shall presently
explain we find that it can be pictured as a curvature of space and
time.

Lewis Carroll

Eddington's lecture is a good example of how a deceptively populist explanation can draw the non-specialist reader into an arcane corner of recent
scientific thought. But did Eddington draw inspiration in turn from Lewis
Carroll's celebrated episode from *Sylvie and Bruno* (1889), in which Lady
Muriel, her father the Earl, and the narrator are conversing over cups of tea
with a young doctor (also named Arthur), while wondering whether tea-time
in a gravity-free universe was a realistic proposition:

"How convenient it would be," Lady Muriel laughingly remarked,
à propos of my having insisted on saving her the trouble of carrying a cup of tea across the room to the Earl, "if cups of tea had no
weight at all! Then perhaps ladies would *sometimes* be permitted
to carry them for short distances!"

"One can easily imagine a situation," said Arthur, "where things
would *necessarily* have no weight, relatively to each other, though
each would have its usual weight, looked at by itself."

"Some desperate paradox!" said the Earl. "Tell us how it could be. We shall never guess it."

"Well, suppose this house, just as it is, placed a few billion miles above a planet, and with nothing else near enough to disturb it: of course, it falls *to* the planet?"

The Earl nodded. "Of course – though it might take some centuries to do it."

"And is five-o'clock-tea to be going on all the while?" said Lady Muriel.

"That, and other things," said Arthur. "The inhabitants would live their lives, grow up and die, and still the house would be falling, falling, falling! But now as to the relative weight of things. Nothing can be *heavy*, you know, except by *trying* to fall, and being prevented from doing so. You all grant that?"

We all granted that.

"Well, now, if I take this book, and hold it out at arm's length, of course I feel its *weight*. It is trying to fall, and I prevent it. And, if I let go, it falls to the floor. But, if we were all falling together, it couldn't be *trying* to fall any quicker, you know: for, if I let go, what more could it do than fall? And, as my hand would be falling too – at the same rate – it would never leave it, for that would be to get ahead of it in the race. And it could never overtake the falling floor!"

"I see it clearly," said Lady Muriel. "But it makes me dizzy to think of such things! How *can* you make us do it?"

"There is a more curious idea yet," I ventured to say. "Suppose a cord fastened to the house, from below, and pulled down by someone on the planet. Then of course the *house* goes down faster than its natural rate of falling: but the furniture – with our noble selves – would go on falling at their old pace, and would therefore be left behind."

"Practically, we should rise to the ceiling," said the Earl. "The inevitable result of which would be concussion of brain."

"To avoid that," said Arthur, "let us have the furniture fixed to the floor, and ourselves tied down to the furniture. Then the five-o'clock-tea could go on in peace."

"With one little drawback!" Lady Muriel gaily interrupted. "We should take the *cups* down with us: but what about the *tea*?"

"I had forgotten the *tea*," Arthur confessed. "*That*, no doubt, would rise to the ceiling – unless you chose to drink it on the way!"

"Which, I think, is *quite* enough nonsense for one while!" said the Earl.

Sources: William Stukeley, *Memoirs of Sir Isaac Newton's Life* (unpublished MS, Royal Society of London, 1752), ff. 15–16; Sir Arthur Eddington, *The Nature of the Physical World* (London: Dent, 1935), pp. 115–18; Lewis Carroll, *Sylvie and Bruno* (London: Macmillan, 1889), pp. 101–4.

NEWTON FOR THE LADIES

Newton's writing was notoriously difficult for non-mathematicians to understand – he once claimed to have made the *Principia* deliberately offputting 'to avoid being baited by little Smatterers in Mathematicks' – but such was the public interest in Newtonian science that a profitable new publishing genre sprang up in its wake. One of the most successful of the many eighteenth-century books on the subject was Francesco Algarotti's *Il Newtonianismo per la Dame* (1737), which was translated into English two years later by Elizabeth Carter (1717–1806), 'the most learned female who ever lived', according to Emma Hamilton. Carter's translation successfully reproduced Algarotti's blend of Newtonian ideas and flirtatious banter between the young male teacher and his aristocratic pupil, in which the principles of celestial mechanics are applied to everyday situations in order to bring the reality of universal processes closer to home – a quality shared with Bernard de Fontenelle's similarly instructive dialogue (see p. 98). In the following extract, the decrease of the attractive force on light is demonstrated by the use of candles and a letter in a darkened room:

Suppose one single Candle to be placed in a Room, and recede from it to such a Distance as not to be able to distinguish the Characters of a Book or a Letter, unless perhaps it were a Billet-doux, which may be read at any Distance.

Place yourself afterwards at a Distance twice as far from the Candle as you were at first. In this Situation the Force of the Light must be, according to the established Law, four times less than it was at the first Distance. The Letter then cannot be read with the same Distinctness as it was as first, unless the Light be *quadrupled*: That is, the Law requires that in the Proportion as the Light grows weaker, the Square of the Distance must increase. And this proves the Experiment to be true; for the Letter at the second Distance is then only read with the same Distinctness as at the first, when three more Candles are added to the single one, or, in other Words, when the Light is *quadrupled*.

Considering how very easy People are apt to forget those Objects in their Absence which made the greatest Impressions upon their Mind when present, I cannot help thinking, said the Marchioness, that this Proportion in the Squares of the Distances of Places, or rather of Times, is observed even in Love. Thus after eight Days Absence Love becomes sixty four times less than it was the first Day, and according to this Proportion it must soon be entirely obliterated; I fancy there will be found, especially in the present Age, very few Experiments to the contrary. I believe, said I, that both Sexes are included in this Theorem, which seems rather to follow the Cubes of the Times, which is certainly more convenient, and requires only four Days for an intire Oblivion. But, in general, I believe we may without Scruple establish the Proportion of the Squares, for eight Days are commonly enough to cure the most vehement Passion. You alone have Power to reverse this Theorem, and make the Remembrance of you, and with that a Desire of seeing you, instead of diminishing, increase according to the Squares, or rather the Cubes of the Times. No! no! said the Marchioness, Gallantry must never destroy a Theorem. I am willing to enter into the general Rule, and shall think myself exceedingly happy if I have been able to establish any thing fixed and constant, in an Affair so inconsistent and wavering as Love. If Geometry, answered I, was permitted to get some Footing there, it would in a little Time produce Wonders. The Conclusions would be the most ready and elegant imaginable.

But to be serious, said she: Our Conclusion in Natural Philosophy was, That the attractive Force of the Sun diminishes in Proportion as the Squares of the Distances increase. I suppose the attractive Force of the Planets will follow the same Proportion with Regard to their Satellites. The Satellites, answered I, which turn about any Planet, observe the same Relation between the Distances and the Times of their Revolutions, as the Planets themselves do that turn about the Sun. This is evident in *Jupiter* and *Saturn*, who have more than one Satellite, and consequently the Law of their attractive Force will be the same as that of the Sun.

In the Earth, who has only one single Satellite to her Share, this is not altogether so evident. But what Reason is there why it should not be the same in one as in the other? . . .

If we could raise Bodies from the Earth to very considerable Distances (compared with that at which we stand from the Center which is very great,) we should see the Force of Gravity prodigiously diminished in them. A Man of War of a hundred Guns, for whose Formation a whole Forest was cut down, and a whole Mine exhausted, would be overset by the slightest Breeze of a Zephyr. The famous *Stone-Henge* upon *Salisbury* Plain, the fruitful Source of Fables both to the Learned and Ignorant, those *Colossian* Heaps which are held together by the Force of Gravity, would be no more than Houses built of Card. The Velocity in the Fall of heavy Bodies would be considerably retarded. Bombs, those artificial Thunders, would not be more terrible than so many Flakes of Snow. But these Experiments are impracticable; one of the greatest Distances we can attain is *Pike Tenneriff*, which is only about three Miles perpendicularly high. Besides the Air would be too thin for Respiration, and the Cold, which must be exceedingly sharp at a greater Height, would render any Experiment fatal to the Philosopher who had the Courage to undertake it.

Nature, replied the Marchioness, has in this Case denied us the Means of becoming compleat *Newtonians*. She has here confined us within the Bounds of Probability. If the attractive Force observes a certain Law in the *Sun*, *Jupiter* and *Saturn*, why should not the same Force observe it here on our Earth? In this Point, answered I, we have no Reason to complain; we have no Need of higher Mountains, and a different Constitution of Air, in the present Case. All these Things, and the Defect of another Moon, are, as I before observed, supplied by the Bodies which fall upon the Surface of the Earth. We may compare these Bodies to the Moon herself; and thus instead of Probability we shall have Evidence, and even in this Point be good *Newtonians*.

It has been deduced from Observation, that if the Moon should lose her Motion, and fall towards the Earth, that Force which would set her a falling would be three thousand six hundred times less than the Force which makes Bodies fall upon the Surface of the Earth. You see how well this agrees with Principle. The Moon is distant from the Center of the Earth (where the attractive Force chiefly resides) sixty times as far as these Bodies: The Square of sixty is 3600: The Attraction then of the Earth to the Moon is diminished in the same Proportion as the Square of the Distance is increased; and this is exactly agreeable to the Established Law in *Jupiter*, *Saturn*, and the *Sun*.

If the Moon should happen to fall upon the Earth, replied the Marchioness, it would present a fine and agreeable Sight to the *Newtonians*: They would certainly have neither Curiosity, Eyes, nor Calculations for any thing else.

A scientific flirtation between a young male tutor and his older, aristocratic pupil was later recast by Thomas Hardy in one of his minor Wessex novels, *Two on a Tower* (1882), in which the ambitious young astronomer Swithin St. Cleeve tutors the troubled Viviette Constantine as they take turns to look through his telescope:

'You said you would show me the heavens if I could come on a starlight night. I have come.'

Swithin, as a preliminary, swept round the telescope to Jupiter, and exhibited to her the glory of that orb. Then he directed the instrument to the less bright shape of Saturn.

'Here,' he said, warming up to the subject, 'we see a world which is to my mind by far the most wonderful in the solar system. Think of streams of satellites or meteors racing round and round the planet like a fly-wheel, so close together as to seem solid matter!' He entered further and further into the subject, his ideas gathering momentum as he went on, like his pet heavenly bodies.

When he paused for breath she said, in tones very different from his own, 'I ought now to tell you that, though I am interested in the stars, they were not what I came to see you about . . .'

She spoke in so low a voice that he might not have heard her. At all events, abstracted by his grand theme, he did not heed her. He continued, –

'Well, we will get outside the solar system altogether, – leave the whole group of sun, primary and secondary planets quite behind us in our flight, as a bird might leave its bush and sweep into the whole forest. Now what do you see, Lady Constantine?' He levelled the achromatic at Sirius.

She said that she saw a bright star, though it only seemed a point of light now as before.

'That's because it is so distant that no magnifying will bring its size up to zero. Though called a fixed star, it is, like all fixed stars, moving with inconceivable velocity; but no magnifying will show that velocity as anything but rest.'

And thus they talked on about Sirius, and then about other stars . . . till he asked her how many stars she thought were visible to them at that moment.

She looked around over the magnificent stretch of sky that their high position unfolded. 'O, thousands, – hundreds of thousands,' she said absently.

'No. There are only about three thousand. Now, how many do you think are brought within sight by the help of a powerful telescope?'

'I won't guess.'

'Twenty millions. So that, whatever the stars were made for, they were not made to please our eyes. It is just the same in everything; nothing is made for man.'

'Is it that notion which makes you so sad for your age?' she asked, with almost maternal solicitude. 'I think astronomy is a bad study for you. It makes you feel human insignificance too plainly.'

'Perhaps it does. However,' he added more cheerfully, 'though I feel the study to be one almost tragic in its quality, I hope to be the new Copernicus. What he was to the solar system I aim to be to the systems beyond.'

Then, by means of the instrument at hand, they travelled together from the earth to Uranus and the mysterious outskirts of the solar system; from the solar system to a star in the Swan, the nearest fixed star in the northern sky; from the star in the Swan to remoter stars; thence to the remotest visible; till the ghastly chasm which they had bridged by a fragile line of sight was realised by Lady Constantine.

'We are now traversing distances beside which the immense line stretching from the earth to the sun is but an invisible point,' said the youth. 'When, just now, we had reached a planet whose remoteness is a hundred times the remoteness of the sun from the earth, we were only a two thousandth part of the journey to the spot at which we have optically arrived now.'

'O, pray don't; it overpowers me!' she replied, not without seriousness. 'It makes me feel that it is not worth while to live; it quite annihilates me.'

Sources: Francesco Algarotti, Sir Isaac Newton's Philosophy, Explain'd for the Use of the Ladies; in Six Dialogues on Light and Colours, trans. Elizabeth Carter, 2 vols (London: E. Cave, 1739), II, pp. 169–76; Thomas Hardy, 'The Wessex Edition of Thomas Hardy's Works', 20 vols (London: Osgood, McIlvaine, 1895), V, pp. 31–6.

INOCULATION

Smallpox inoculation, in which a small amount of infectious material is deliberately introduced into a healthy body in order to stimulate the production
of protective antibodies, had been practised in China and the Middle East
since medieval times, although the procedure was not entirely safe, since it
carried the risk of inadvertently passing on the full-blown disease through
the vaccine. Yet it was better than doing nothing, for smallpox was one of the
world's biggest killers, claiming the lives of millions of people in epidemics
that occurred routinely until the disease was finally eradicated in the late
twentieth century.

Rhazes

One of the earliest descriptions of smallpox was given by the Persian physician Rhazes (Muhammad ibn Zakariyyā al-Rāzī, c. 860–925), whose clinical
practice predated the era of widespread inoculation in the East:

The symptoms of the Small-Pox and of the Measles are an acute
fever at the very beginning, with head-ache, and redness of the eyes.
The eruption generally appears on the third day from the beginning
of the fever, but sometimes on the first or second day. One of the
most favourable symptoms is the appearance of the eruption on
the third day, or at the time when the fever is mild; and *vice versa*.
When, however, they appear on the first day, it is from the violence
of the fever or intense mental application.

The fever attending the Small-Pox is a continued fever, and for
the most part attacks children. When you see the eruption appear,
do not give the patient medicines to extinguish the fever, but rather
let him drink a decoction of the seeds of sweet fennel and smallage,
in order to bring it out from the interior.

Before the eruption begins to appear you should bleed the
patient; but not afterwards.

Let the patient keep in his mouth a decoction of lentils and roses,
and make a collyrium for his eyes of antimony and camphor, in order
to prevent the pustules from coming out in his mouth, throat, and
eyes.

When the pustules of the Small-Pox are of a violet or greenish colour, the disease is of a fatal kind. When the pustules are ripened, let the patient sleep upon rice meal, and let fumigations be made with myrtle leaves or olive leaves; for these things tend to dry the pustules. Modern physicians are accustomed to give to drink butter-milk from which the butter has been removed.

Lady Mary Wortley Montagu

The appointment of her husband as British ambassador to Istanbul intro-duced Lady Mary Wortley Montagu (1689–1762) to a new mode of 'Oriental' living that she found highly agreeable. There was much about the Ottoman approach to life that she admired, not least their practice of inoculating against smallpox, which she described in a letter to her friend Sarah Chiswell, written in April 1717:

A propos of distempers, I am going to tell you a thing that I am sure will make you wish yourself here. The small-pox, so fatal, and so general amongst us, is here entirely harmless by the invention of *ingrafting*, which is the term they give it. There is a set of old women who make it their business to perform the operation every autumn, in the month of September, when the great heat is abated. People send to one another to know if any of their family has a mind to have the small-pox: they make parties for this purpose, and when they are met (commonly fifteen or sixteen together), the old woman comes with a nut-shell full of the matter of the best sort of small-pox, and asks what veins you please to have opened. She immediately rips open that you offer to her with a large needle (which gives you no more pain than a common scratch), and puts into the vein as much venom as can lie upon the head of her needle, and after binds up the little wound with a hollow bit of shell; and in this manner opens four or five veins. The Grecians have commonly the superstition of open-ing one in the middle of the forehead, in each arm, and on the breast to mark the sign of the cross; but this has a very ill effect, all these wounds leaving little scars, and is not done by those that are not superstitious, who choose to have them in the legs, or that part of the arm that is concealed. The children or young patients play together all the rest of the day, and are in perfect health till the eighth. Then the fever begins to seize them, and they keep their beds two days, very seldom three. They have very rarely above twenty or thirty in their faces, which never mark; and in eight days' time they are as well as before their illness. Where they are wounded, there remain running sores during the distemper, which I don't doubt is a great relief to it. Every year thousands undergo this operation; and the French ambassador says pleasantly, that they take the small-pox here by way of diversion, as they take the waters in other countries. There is no example of any one that has died in it; and you may believe I am very well satisfied of the safety of the experiment, since I intend to try it on my dear little son.

I am patriot enough to take pains to bring this useful invention into fashion in England; and I should not fail to write to some of our doctors very particularly about it, if I knew any one of them that I thought had virtue enough to destroy such a considerable branch of their revenue for the good of mankind. But that distemper is too beneficial to them not to expose to all their resentment the hardy wight that should undertake to put an end to it. Perhaps, if I live to return, I may, however, have courage to war with them. Upon this occasion admire the heroism in the heart of your friend, &c.

Voltaire

When the Wortley Montagus returned to England in 1718 Lady Mary campaigned energetically on behalf of inoculation, but her attempts met with resistance from surgeons and clergymen, though she did manage to persuade Princess Caroline to have the royal children treated, after which inoculation began to spread among the English upper classes. Wortley Montagu was widely admired for her intelligence and tenacity, and the French writer Voltaire (François-Marie Arouet, 1694–1778), sang her praises in the course of an article in which he despaired of his own countrymen's stubborn opposition to smallpox inoculation:

It is inadvertently affirm'd in the Christian Countries of *Europe*, that the *English* are Fools and Madmen. Fools, because they give their Children the Small-Pox to prevent their catching it; and Madmen, because they wantonly communicate a certain and dreadful Distemper to their Children, merely to prevent an uncertain Evil. The

English, on the other Side, call the rest of the *Europeans* cowardly and unnatural. Cowardly, because they are afraid of putting their Children to a little Pain; unnatural, because they expose them to die one Time or other of the Small-Pox. But that the Reader may be able to judge, whether the *English* or those who differ from them in opinion, are in the right, here follows the History of the fam'd Inoculation, which is mention'd with so much Dread in *France*.

The *Circassian* Women have, from Time immemorial, communicated the Small-Pox to their Children when not above six Months old, by making an Incision in the arm, and by putting into this Incision a Pustle, taken carefully from the Body of another Child. This Pustle produces the same Effect in the arm it is laid in, as Yeast in a Piece of Dough: It ferments, and diffuses through the whole Mass of Blood, the Qualities with which it is impregnated. The Pustles of the Child, in whom the artificial Small-Pox has been thus inoculated, are employ'd to communicate the same Distemper to others. There is an almost perpetual Circulation of it in *Circassia*; and when unhappily the Small-Pox has quite left the Country, the Inhabitants of it are in as great Trouble and Perplexity, as other Nations when their Harvest has fallen short . . .

Some pretend, that the *Circassians* borrow'd this Custom anciently from the *Arabians*; but we shall leave the clearing up of this Point of History to some learned Benedictine, who will not fail to compile a great many Folio's on this Subject, with the several Proofs or Authorities. All I have to say upon it, is that in the beginning of the Reign of King *George* the First, the Lady *Wortley Mountague*, a Woman of as fine a Genius, and endu'd with as great a Strength of Mind, as any of her Sex in the *British* Kingdoms, being with her Husband who was Ambassador at the Port, made no scruple to communicate the Small-Pox to an Infant of which she was deliver'd in *Constantinople*. The Chaplain represented to his Lady, but to no purpose, that this was an unchristian Operation, and therefore that it cou'd succeed with none but Infidels. However, it had the most happy Effect upon the Son of the Lady *Wortley Mountague*, who, at her Return to *England*, communicated the Experiment to the Princess of *Wales*, now Queen of *England*. It must be confess'd that this Princess, abstracted from her Crown and Titles, was born to encourage the whole Circle of Arts, and to do good to Mankind . . . The Moment this Princess heard of Inoculation, she caus'd an Experiment of it to be made on four Criminals sentenc'd to die, and by that means preserv'd their Lives doubly; for she not only sav'd them from the Gallows, but by means of this artificial Small-Pox, prevented their ever having that Distemper in a natural Way, with which they would very probably have been attack'd one Time or other, and might have died of in a more advanc'd Age.

The Princess being assur'd of the Usefulness of this Operation, caus'd her own Children to be inoculated. A great Part of the Kingdom follow'd her Example, and since that Time ten thousand Children, at least, of Persons of Condition owe in this Manner their Lives to her Majesty, and to the Lady *Wortley Mountague*; and as many of the Fair Sex are oblig'd to them for their Beauty.

Upon a general Calculation, threescore Persons in every hundred have the Small-Pox. Of these threescore, twenty die of it in the most favourable Season of Life, and as many more wear the disagreeable Remains of it in their Faces so long as they live. Thus, a fifth Part of Mankind either die, or are disfigur'd by this Distemper. But it does not prove fatal to so much as one, among those who are inoculated in *Turkey* or in *England*, unless the Patient be infirm, or would have died had not the Experiment been made upon him. Besides, no one is disfigur'd, no one has the Small-Pox a second Time, if the Inoculation was perfect. 'Tis therefore certain, that had the Lady of some *French* Ambassador brought this Secret from *Constantinople* to *Paris*, the Nation would have been for ever oblig'd to her . . .

I am inform'd that the *Chinese* have practis'd Inoculation these hundred Years, a Circumstance that argues very much in its Favour, since they are thought to be the wisest and best govern'd People in the World. The *Chinese* indeed don't communicate this Distemper by Inoculation, but at the Nose, in the same Manner as we take Snuff. This is a more agreeable way, but then it produces the like Effects; and proves at the same Time, that had Inoculation been practis'd in *France*, 'twould have sav'd the Lives of Thousands.

Forty years later, the informal procedure of smallpox inoculation was replaced by the safer alternative of cowpox vaccination by the English country doctor Edward Jenner (1749–1823), who noticed that dairymaids who had contracted cowpox – a far milder disease – rarely went on to contract smallpox, even during epidemics. In May 1796 Jenner embarked on his famous clinical trial: he scratched some fluid from a milkmaid's cowpox blister into the arms of the eight-year-old son of his gardener, thereby performing the world's first vaccination. He later attempted to infect the boy with smallpox, who thankfully proved to be immune. Jenner's new form of immunization was safer than the earlier method of inoculation, as it carried no risk of small-pox infection from the vaccine itself. The original source of the dairymaid's vaccine was a cow named Blossom, whose hide now hangs in the library of St George's teaching hospital in London.

Sources: Rhazes, *A Treatise on the Small-Pox and Measles*, trans. William Alexander Greenhill (London: Sydenham Society, 1848), pp. 104–7; *The Letters and Works of Lady Mary Wortley Montagu*, ed. Lord Wharncliffe, 2 vols (London: Swan Sonnenschein, 1893), I, pp. 307–9; Voltaire, 'On Inoculation', *Letters Concerning the English Nation* (London: C. Davis, 1733), pp. 73–82.

WALKING PLANTS

In 1740 a young Swiss tutor named Abraham Trembley (1710–84) went to work for a wealthy family in Holland, where he spent much of his free time examining the plants and animals he found on his employers' estate. There, in one of the ponds, he discovered a small aquatic life form that he named a 'polyp' (it is referred to today as the green hydra, *Chlorohydra viridissima*). At first Trembley assumed that the organism was a plant, but as he began to dissect it, he realized that he was dealing with an animal, related to the jellyfish, albeit one that exhibited plant-like features and behaviour. As can be seen in the following extract from Trembley's account of the discovery, his calm in the face of this apparent blurring of nature's boundaries was impressive, as was his capacity for recognizing something that he could never have expected to see. His memoir allows us to follow his thoughts as they unfold, culminating in his troubling realization that his polyp was in fact an unusual plant/animal hybrid. We join the story as Trembley begins his close observation of the polyps he has collected in a glass box; at this point he still believed they were plants:

One day I moved the glass containing them extremely slightly, to see what effect the resulting movement of the water would have on their arms. I did not in the least anticipate the effect it produced. Instead of seeing, as I expected, the arms and even the body of the *Polypus* simply agitated in the water and carried forward by its movement, I saw them contract suddenly, and so vigorously that the body of each *Polypus* appeared only like a grain of green matter, and the arms disappeared entirely from my sight. I was surprised, and this surprise only served to excite my curiosity and redouble my attention. As I unceasingly ran my eye, aided by a magnifying glass, over several of the *Polyps* that I had caused to contract, I soon saw that some of them were beginning to extend: their arms became visible again, and little by little these *Polyps* took on once more their original form.

This contraction of the *Polyps*, and all the movements that I saw them make when they extended themselves again, awakened sharply in my mind the idea of an animal. I compared them at first to snails and to other animals that contract and expand.

One will perhaps be astonished that I did not decide definitely that the *Polyps* were animals. I confess that their shape and colour still made some impression on me. I thought that it was not impossible that they were sensitive plants; and I found nothing more extraordinary in their contraction and expansion than in the movements that are to be observed when one touches the plant to which this name has been given. This idea held me therefore in some doubt, and I did not wish to decide anything until new observations should persuade me.

A few days later I found several *Polyps* fixed against the sides of the glass at a place where I had not seen them before and where certainly there had not been any of them originally. I soon perceived how they had come there. Several were walking on the sides of the glass while I was observing them. I shall describe elsewhere their manner of walking, and I shall here content myself by saying that they walk nearly in the same way as the caterpillars called geometers, and like various aquatic insects, which fix successively their anterior and posterior ends, the latter after having brought it close to the anterior, and the anterior after having removed it from the posterior.

The sight of this progressive movement of the *Polyps* succeeded in persuading me that they were animals; and when I was convinced, I stopped observing them. I had found out what I was seeking; for up till then I had not proposed anything else than to discover whether they were animals. Almost the whole of the month of September 1740 passed without my giving them the least attention. I was then occupied with other insects that I had been observing for a long time. But towards the end of this month of September the *Polyps* attracted my attention anew, and so strongly that since then I have never ceased observing them . . .

They are voracious animals: their arms extended into the water are so many snares which they set for numbers of small insects that are swimming there. As soon as any of them touches one of the arms, it is caught.

The *Polypus* being seized of a Prey, conveys it to his mouth, by contracting or bending his arm. If the prey be strong enough to make resistance, he makes use of several arms.

A *Polypus* can master a Worm twice or thrice as long as himself. He seizes it, he draws it to his mouth, and what is more, swallows it whole.

6. Trembley's microscope, specially designed to observe the behaviour of aquatic organisms such as freshwater hydra.

If the Worm comes endways to the mouth, he swallows it by that end; if not, he makes it enter double into his stomach, and the skin of the *Polypus* gives way. The size of the stomach extends itself so as to take in a much larger bulk than that of the *Polypus* itself, before it swallowed that Worm. The Worm is forced to make several windings and folds in the stomach, but does not keep there long alive; the *Polypus* sucks it, and after having drawn from it what it serves for his nourishment, he voids the remainder by his mouth, and these are his excrements.

Later that year Trembley decided to dissect some of his pet polyps in order to deepen his understanding of their structure; so, on 25 November 1740, he cut his first specimen, expecting to see it quickly die. The results of the experiment came as a complete surprise:

I now proceed to the Singularities resulting from the Operations I have tried upon them.

If the body of a *Polypus* is cut into two parts transversely, each of those parts becomes a complete *Polypus*. On the very day of the operation, the first part, or anterior end of the *Polypus*, that is, the head, the mouth, and the arms, this part, I say, lengthens itself, it creeps, and eats.

The second part, which has no head, gets one; a mouth forms itself, at the anterior end; and shoots forth arms . . . Each of those parts, thus becomes a perfect *Polypus*, performs absolutely all its functions. It creeps, it eats, it grows, and it multiplies; and all that, as much as a *Polypus* which never had been cut.

In whatever place the body of a *Polypus* is cut, whether in the middle, or more or less near the head, or the posterior part, the experiment has always the same success.

If a *Polypus* is cut transversely, at the same moment, into three or four parts, they all equally become so many complete ones.

The animal is too small to be cut at the same time into a great number of parts; I therefore did it successively. I first cut a Polypus into four parts, and let them grow; next, I cut those quarters again; and at this rate I proceeded, till I had made 50 out of one single one. And here I stopp'd, for there would have been no end of the experiment.

When Trembley published his account of his hybrid discovery, it caused some theological as well as biological disquiet, though it also won him the Royal Society's Copley Medal – the Nobel Prize of its day – while turning him into a minor scientific celebrity. But, like many others in that age of satire, he soon found himself the target of mockery by metropolitan wits such as Henry Fielding, Tobias Smollett and Oliver Goldsmith, who were happy to play their part in the long-standing literary assault on science that characterized the culture wars of the early eighteenth century.

Henry Fielding

Henry Fielding (1707–54) got in first, with this affectionate lampoon on 'the Polypus, which hath this surprising Property, that being cut into several Pieces, each Piece becomes a perfect Animal, or Vegetable, as complete as that of which it was originally only a Part':

I have not, after the minutest Observation, been able to settle with any degree of Certainty, whether this be really an Animal or a Vegetable, or whether it be not strictly neither, or rather both . . .

I have now actually by me several Parts of the same *Polypus*, cut into Pieces above a Year ago; since which time, they have produced a great Number of Young-ones . . .

I have cut a *Polypus*, lengthways, between Seven and Eight in the Morning; and between Two and Three in the Afternoon, *each of the Parts has been able to eat a Worm as long as itself* . . .

In order to comprehend the Experiment I am now going to speak of, one should recollect, that the whole Body of a *Polypus* forms only one Pipe, a sort of Gut, or Pouch.

I have been able to turn that Pouch, that Body of the Polypus, INSIDE-OUTWARDS; AS ONE MAY TURN A STOCKING.

I have several by me, that have remained turned in this manner; THEIR INSIDE IS BECOME THEIR OUTSIDE, AND THEIR OUTSIDE THEIR INSIDE: They eat, they grow, and they multiply, as if they had never been turned.

Oliver Goldsmith

The Anglo-Irish playwright Oliver Goldsmith (c. 1730–74) adopted a less friendly tone when he characterized naturalists such as Trembley as 'the puny pedant, who finds one undiscovered property in the polype, describes an unheeded process in the skeleton of a mole, and whose mind, like his microscope, perceives nature only in detail.' A few years later he had another go at the doings of naturalists, in his epistolary satire, *A Citizen of the World* (1762), in which his fictitious Chinese traveller, Lien Chi Altangi, confides that:

I am amused, my dear Fum, with the labours of some of the learned here. One shall write you a whole folio on the dissection of a caterpillar. Another shall swell his works with a description of the plumage on the wing of a butterfly; a third shall see a little world on a peach leaf, and publish a book to describe what his readers might see more clearly in two minutes, only by being furnished with eyes and a microscope.

I have frequently compared the understandings of such men to their own glasses. Their fields of vision are too contracted to take in the whole of any but minute objects; they view all nature bit by bit; now the proboscis, now the antennæ, now the pinnæ of – a flea. Now the polypus comes to breakfast upon a worm; now it is kept up to see how long it will live without eating; now it is turned inside outward; and now it sickens and dies. Thus they proceed, laborious in trifles, constant in experiment, without one single abstraction, by which alone knowledge may be properly said to increase, till, at last their ideas, ever employed upon minute things, contract the size of the diminutive object, and a single mite shall fill their whole mind's pacity.

Goldsmith, however, was an impecunious hack writer as well as a sardonic wit, and later in his career he was commissioned to write an eight-volume popular survey of natural knowledge, his *History of the Earth, and Animated Nature* (1774–7), in the final volume of which he narrated the natural history of the polyp, noting, rather more respectfully than before, that 'Mr. Trembley was the person to whom we owe the first discovery of the amazing properties and powers of this little vivacious creature; he divided this class of animals into four different kinds; into those inclining to green, those of a brownish cast, those of flesh colour, and those which he calls the polype de panache.'

Sources: Abraham Trembley, 'Observations and Experiments upon the Freshwater Polypus', trans. P.H.Z., *Philosophical Transactions of the Royal Society*, vol. 42 (1742–43), pp. iii–xi; Henry Fielding, *Miscellanies*, 3 vols (London: for the author, 1743), I, pp. 263–73; Oliver Goldsmith, *The Bee: Being essays on the most interesting subjects* (London: Wilkie, 1759), p. 188; Goldsmith, *A Citizen of the World; or, Letters from a Chinese Philosopher, Residing in London, to his Friends in the East*, 2 vols (London: Newbery, 1762), II, pp. 102–6.

SCIENTIFIC STYLE

Science has always been part of culture, reflecting the social and intellectual outlooks that prevailed in any given age. But from time to time science has sought to separate itself from its cultural surroundings, defining itself as an exceptional endeavour with its own paradigms, its own rules and, most importantly, its own specialist languages. One of the first acts of the Royal Society of London (established in 1660), for example, was to prescribe to its members a 'natural' style of thinking, speaking and writing that avoided all use of metaphor and allusion, and thereby approached what Thomas Sprat described as the ideal of 'Mathematical plainness'. Later critics, however, have justly complained that modern scientific papers are more or less incomprehensible to all but a tiny readership, not only because of their abstruse subject matter, but also because of the poor writing skills of their authors.

Thomas Sprat

The Revd Thomas Sprat (1635–1713) was one of the founders of the Royal Society of London, and the author of its first official history, which appeared in 1667. As well as offering a summary of the membership's published research to date, the book explained the principles by which the society was run, including, as can be seen in the following extract, the precise spoken form recommended to anyone thinking of reading a paper at one of the Society's weekly meetings (let alone publishing it in its prestigious monthly journal, the *Philosophical Transactions*):

Their purpose is, in short, to make faithful *Records*, of all the Works of *Nature*, or *Art*, which can come within their reach: that so the present Age, and posterity, may be able to put a mark on the Errors, which have been strengthned by long prescription: to restore the Truths, that have lain neglected: to push on those, which are already known, to more various uses: and to make the way more passable, to what remains unreveal'd. This is the Compass of their Design. And to accomplish this, they have indeavor'd, to separate the knowledge of *Nature*, from the colours of *Rhetorick*, the devices of *Fancy*, or the delightful deceit of *Fables*. They have labor'd to inlarge it, from being confin'd to the Custody of a few, or from Servitude to private Interests. They have striven to preserve it from being over-pres'd by a confus'd Heap of vain and useless Particulars; or from being streightned and bound too much up by general Doctrines. They have tried to put it into a Condition of perpetual Increasing; by settling an inviolable Correspondence between the Hand and the Brain. They have studied, to make it not only an Enterprise of one Season, or of some lucky Opportunity; but a Business of Time; a steady, a lasting, a popular, an uninterrupted Work . . . And lastly, they have begun to establish these Reformations in Philosophy, not so much, by any solemnity of Laws, or Ostentation of Ceremonies, as by solid Practice and Examples; not by a glorious Pomp of Words; but by the silent, effectual, and unanswerable Arguments of real Productions . . .

There is one thing more, about which the *Society* has been most sollicitous; and that is, the manner of their *Discourse*: which, unless they had been very watchful to keep in due temper, the whole spirit and vigour of their *Design*, had been soon eaten out, by the luxury and redundance of *speech*. The ill effects of this superfluity of talking, have already overwhelm'd most other *Arts* and *Professions*; insomuch, that when I consider the means of *happy living*, and the causes of their corruption, I can hardly forbear recanting what I said before; and concluding, that *eloquence* ought to be banish'd out of all *civil Societies*, as a thing fatal to Peace and good Manners . . .

Who can behold, without indignation, how many mists and uncertainties, these specious *Tropes* and *Figures* have brought on our Knowledge? How many rewards, which are due to more profitable, and difficult *Arts*, have been still snatch'd away by the easie vanity of *fine speaking*? For now I am warm'd with this just Anger, I cannot with-hold my self, from betraying the shallowness of all these seeming Mysteries; upon which, *we Writers*, and *Speakers*, look so bigg. And, in few words, I dare say; that of all the Studies of men, nothing may be sooner obtain'd, than this vicious abundance of *Phrase*, this trick of *Metaphors*, this volubility of *Tongue*, which makes so great a noise in the World. But I spend words in vain; for the evil is now so inveterate, that it is hard to know whom to *blame*, or where to begin to *reform* . . . It will suffice my present purpose, to point out, what has been done by the *Royal Society*, towards the correction of its excesses in *Natural Philosophy*; to which it is, of all others, a most profest enemy.

They have therefore been most rigorous in putting in execution, the only Remedy that can be found for this *extravagance*: and that has been, a constant Resolution, to reject all the amplifications, digressions, and swellings of style: to return back to the primitive purity, and shortness, when men deliver'd so many *things*, almost in an equal number of *words*. They have extracted from all their members, a close, naked, natural way of speaking; positive expressions; clear senses; a native easiness; bringing all things as near the Mathematical plainness as they can: and preferring the language of Artizans, Countrymen, and Merchants, before that of Wits, or Scholars.

John R. Baker

Three centuries later, and the subject of scientific style is as hotly debated as ever. The following article, 'English Style in Scientific Papers', by the biologist John R. Baker (1900–84), was published in the journal *Nature* in 1955, but even though it has aged in interesting ways – note the tetchy reference to those 'who think it polite to call a napkin a "serviette"' – it remains just as pertinent today. It is among my favourite pieces of writing about the writing of science, and, as a science writer myself, I try my best to live up to its injunctions:

Many scientific papers published in Great Britain are written in a style quite different from that adopted by good English authors. There would be no disadvantage in that, if an improved version of our language had been invented for scientific papers. This, however, is not so. For clarity and directness one must turn to non-scientific authors. One hesitates to direct attention to this subject lest one be accused of setting oneself up as a stylist. There is no such intention here. It is proposed merely to suggest a few ways in which scientific papers could be made simpler, clearer, and more pleasant to read.

To prevent any misunderstanding, it must be remarked at the outset that it is no part of my purpose to try to oppose the use of necessary scientific words. On the contrary, the precise use of certain technical terms, carefully defined and internationally understood, is an obvious necessity. We shall be concerned here with the words that stand between the technical terms.

If one examines all those passages in scientific papers that are least in accord with good English style, one finds that there are three main kinds of error, which may be described briefly as those of grammar, grandiloquence and German construction.

It would be absurd to treat the subject of grammar at any length here, since good advice on this subject is so readily available. Grammatical errors in scientific papers sometimes make it impossible to be sure what a sentence is intended to mean. In other cases the mistakes are troublesome only because they distract the reader's attention from the subject under discussion. The words ending in -*ing* (especially *using*) are very frequently misused. One repeatedly finds such statements as this: 'Examining a capillary under the microscope, after staining with carmine, it resembles a homogeneous hyaline tube.' Yet anyone who is intelligent enough to carry out scientific research at a university can easily grasp everything that it is essential to know about the use of present participles and gerunds in fifteen minutes.

The menace of grandiloquence seems to derive from the time when doctors used many Latin words; and indeed these errors are much more common in medical and biological than in chemical and physical writings. It is difficult to account for the fact that an Englishman will write '*vide supra*' when he means 'see above'. It is noticeable that those who use Latin tags most freely do not always know enough of that language to realize that the word 'data' is plural.

Long words derived from Greek or Latin roots are often used for expressing very simple ideas. When a man wants to say that something is visible to the naked eye on merely opening the body-cavity, he tries to make it sound grand by saying, 'this phenomenon can be macroscopically observed upon laparotomy'. Sometimes he uses short words, but chooses an elaborate way of expressing himself. Thus, if he wants to refer to a sheet of metal about $^1/_{100}$ in. thick, he says that it is 'of thickness $\sim 10^{-2}$ in.'. If his intention is to tell his readers that the longest of the six pairs of chromosomes in a cell is about three times as long as the shortest, he writes, 'the maximum size range of the six pairs as represented by the ratio length of chromosome 1/length of chromosome 6 is about 3/1'.

The grandiloquent writer brings in abstract words where none is required. Instead of writing 'because the surface of the retina is spherical', he substitutes 'because of the sphericity of the retinal surface'. Instead of saying that something is near the nucleus, he says that it 'occupies a juxta-nuclear position'.

Genteelism is allied to grandiloquence. People who think it polite to call a napkin a 'serviette' have their counterparts in science, who cannot soil their mouths with paste or mash but must say 'Brei' instead. Another example is 'sacrificed' for killed. Allied to genteelism is the deliberate use – not just occasionally, but over and over again – of archaic words instead of their exact equivalents in plain modern English (for example 'save' for except).

Vogue-words have their place in the vocabulary of the grandiloquent; but what may be called negative vogue-words must also be mentioned here – perfectly good words that for an inexplicable reason must not be used. Why must the word 'about' never be used by the grandiloquent? Why must they always say *'circa'*, *'ca.'*, *'c.'*, 'approximately', 'around', 'of the order of', and even '~'? What is the objection to 'separate'? Why must 'discrete' invariably be substituted? (A case can be made for the occasional use of 'discrete', when special emphasis on complete separation is necessary.) 'After' is another negative vogue-word . . .

A strange fact that gives some basis for optimism is this. When an author has finished the scientific part of his paper, he often addresses a note of general information to the reader. At this point he suddenly discloses for the first time that he can write English, for his ideas are clearly expressed. If one wished to translate what he now says into the style adopted in the rest of the paper, one would have to write something like this: 'Some related interest possessing observations by the present writer *et al.* will be the subject of *ca.* $10^{0.3010}$ discrete communications.' The fact that he does not write like that shows that he need not have done so in the body of the paper.

One last, necessary word. The best English writers occasionally use some of the strange constructions mentioned in this article, often to produce a special effect for a particular occasion. They do so, however, only at long intervals. The greater part of their writing is so smooth and fluent that the reader forgets that he is reading and knows only that he is absorbing ideas.

Note. With the exception of the passage in inverted commas in the last paragraph but one, all the examples given in this article are genuine: none was invented by myself or modified to suit my purpose.

Sources: Thomas Sprat, *The History of the Royal-Society of London, for the Improving of Natural Knowledge* (London: J. Martyn, 1667), pp. 61–2; 111–13; John R. Baker, 'English Style in Scientific Papers', *Nature* 176 (5 November 1955), pp. 851–2.

INSIDERS AND OUTSIDERS

As was seen in the previous entry, the attempt to fashion a new scientific language for use among the initiates of the Royal Society was part of a wider attempt to uncouple the scientific endeavour from mainstream intellectual culture. And as the sciences became increasingly professionalized, institutional membership became an ever more important way for an individual to define himself or herself as a scientific insider or outsider.

Addison and Steele

Literary satirists were not slow to react to what they perceived as the pretensions of the new scientific elite. Joseph Addison and Richard Steele, writing in the *Tatler* in October 1710, made the following observations about the pedantry and dullness that, they claimed, characterized the officially sanctioned man of science:

There is no Study more becoming a rational Creature, than that of Natural Philosophy; but as several of our Modern Virtuoso's manage it, their Speculations do not so much tend to open and enlarge the Mind, as to contract and fix it upon Trifles.

This in *England* is in a great Measure owing to the worthy Elections that are so frequently made in our Royal Society. They seem to be in a Confederacy against Men of polite Genius, noble Thought, and diffusive Learning; and chuse into their Assemblies such as have no Pretence to Wisdom, but Want of Wit; or to natural Knowledge, but Ignorance of every Thing else. I have made Observations in this Matter so long, that when I meet with a young Fellow that is an humble Admirer of the Sciences, but more dull than the rest of the Company, I conclude him to be a Fellow of the Royal Society.

'Sir' John Hill

The English botanist 'Sir' John Hill (1716–75) spent much of his career raging at the Royal Society for repeatedly blackballing his applications to join. The rejection was the central fact of his life, and he wrote many books in which he reminded his readers that he 'had the honour *not* be a Fellow of the Royal Society'. His 'knighthood', of which he was immensely proud, was in fact the Swedish Order of Vasa, which he received from the King of Sweden in recognition of his support for the Linnean classification system (see p. 163). The following short extract is from the final page of his *Dissertation on Royal Societies* (1750), in which he fantasizes, with some bitterness, that a senior societarian's dog has just been elected as a Fellow:

It was with some Amazement that, as I sat at the Coffee-House I have already mentioned to you, I read this Name, with *F. R. S.* at the End of it, on the Collar of a fine *Newfoundland* Spaniel, that had good-naturedly laid his shaggy Nose upon my Knee. I was not suffered long to continue in the easy Mistake of supposing this the Name of the Master of the Creature; that stood in Capitals under it, and was one that I was very well acquainted with: On expressing my Astonishment at the *F. R. S.* at the End of the other Denomination on the Collar, I was informed that it belonged to the Creature who wore it, that it was one of two Names by which he was indifferently called by his Master, that it never had belonged to any living Creature except the Dog, and that he had been fairly balloted for, and elected a Fellow, upon the Recommendation of five of the Members of the Body, who knew about as much of his Qualifications as the People who sign Certificates of that Kind among them usually do of the Persons they belong to.

It seems it is a sort of Law in the Regulations of this Body, to have but one of a Family a Member of it at a Time, and that the least qualified; in Consequence of this, they have elected this Gentleman's Dog.

Charles Babbage

The mathematician Charles Babbage (1791–1871), who we will meet again in connection with his Analytical Engine (p. 222), was an outspoken critic of what had become the British scientific establishment. He was particularly critical of the Royal Society (of which he was a Fellow), blaming it for the declining status of British science and engineering, and in this extract from his *Reflections on the Decline of Science in England* (1830), he considered the absurdity and unjustness of the Society's voting procedures:

I have no intention of stating what *ought* to be the qualifications of a Fellow of the Royal Society; but, for years, the practical mode of arriving at that honour, has been as follows: – A. B. gets any three Fellows to sign a certificate, stating that he (A. B.) is desirous of becoming a member, and likely to be a useful and valuable one. This is handed in to the Secretary, and suspended in the meeting-room. At the end of ten weeks, if A. B. has the good fortune to be perfectly unknown by any literary or scientific achievement, however small, he is quite sure of being elected as a matter of course. If, on the other hand, he has unfortunately written on any subject connected with science, or is supposed to be acquainted with any branch of it, the members begin to inquire what he has done to deserve the honour; and, unless he has powerful friends, he has a fair chance of being black-balled.

In fourteen years' experience, the few whom I have seen rejected, have all been known persons; but even in such cases a hope remains; – perseverance will do much, and a gentleman who values so highly the distinction of admission to the Royal Society, may try again; and even after being twice black-balled, if he will a third time condescend to express his desire to become a member, he may perhaps succeed, by the aid of a hard canvass. In such circumstances, the odds are much in favour of the candidate possessing great scientific claims; and the only objection that could then reasonably be suggested, would arise from his estimating rather too highly a distinction which had become insignificant from its unlimited extension.

It should be observed, that all members contribute equally, and that the sum now required is fifty pounds. It used, until lately, to be ten pounds on entrance, and four pounds annually. The amount of this subscription is so large, that it is calculated to prevent many men of real science from entering the Society, and is a very severe tax on those who do so; for very few indeed of the cultivators of science rank amongst the wealthy classes. Several times, whilst I have been consulting books or papers at Somerset House, persons have called to ask the Assistant-secretary the mode of becoming a member of the Royal Society. I should conjecture, from some of these applications, that it is not very unusual for gentlemen in the country to order their agents in London to take measures for putting them up at the Royal Society.

Women in Science

Women scientists have fared particularly badly when it comes to institutional acceptance. Even Marie Curie, with her two Nobel prizes, was never admitted to the French Academy of Sciences, while the Royal Society of London continued to reject female applicants until 1945. As will often be seen in the course of this collection (and I am acutely conscious that only twenty-four of its one hundred and one pieces were written or co-written by women),* repeated attempts were made by women to gain professional recognition. Caroline Herschel, in the 1780s, was the first woman scientist to receive state funding, but, unlike her brother William, she was disbarred from presenting her work in person at Royal Society meetings. Mary Somerville, in the 1820s, had to have her husband read her papers aloud on her behalf; later, when she was at the height of her scientific fame, the Royal Society fudged the issue by placing a marble bust of her in the lobby. She was, however, elected to the Royal Irish Academy in 1834, and to the Royal Astronomical Society the following year. Then there was Hertha Ayrton in the 1900s – she succeeded in becoming the first woman to read a scientific paper in person at a Royal Society meeting, but she was still turned down when it came to the matter of a fellowship. The reason given: she was married. In the United States, meanwhile, the astronomer Henrietta Swan Leavitt (1868–1921) gained menial employment as a human 'computer' at the Harvard College Observatory, working solely from photographs because women were forbidden to use the telescopes. Nevertheless, her work on the luminosity of variable stars earned her a nomination for a Nobel Prize in 1924 – a nomination that came three years too late, as the Swedish Academy of Sciences discovered, for she had died of cancer in 1921, and Nobels are not awarded posthumously.

Two decades later, Kathleen Lonsdale (see p. 362) and Marjory Stephenson became the first women to be elected as Royal Society fellows, on 22 March 1945, a date that ought to be better known, not least for its revealing lateness: the American National Academy of Sciences had elected its first woman Fellow (the histologist Florence Sabin) in 1925, though the French Academy of Sciences did not admit women until 1967.

---

* Though this figure does not include the numerous translations by women authors that feature in the book, such as those by Lucy Hutchinson (p. 19), Aphra Behn (p. 98), Elizabeth Carter (p. 110) or Elizabeth Sabine (p. 232); translation being one of the many unacknowledged ways in which women have contributed to science over the centuries. — R.H.

There was more than simple kudos at stake in all this. Joining a learned society puts you in the intellectual swim by granting direct access to the latest ideas and research. Being denied such access remained the biggest frustration of being a woman scientist in the early twentieth century, as the chemical crystallographer Dorothy Hodgkin (1910–94) recalled in a radio interview in the late 1980s: 'In Oxford, the worst thing was that the Alembic club, which was the chemical club of the university, was not open to the membership of women when I was an undergraduate. We were allowed to go to the general meetings but not to the small meetings discussing research week by week. This I minded a great deal.' Indeed her fellowship of the Royal Society from 1947 meant more to her than the Nobel Prize that was awarded to her in 1964 for her work on the structure of vitamin B12. 'You see, the thing that one missed out on as a woman was a certain amount of the absorption into scientific societies . . . societ[ies] of people who like talking to one another about scientific affairs, which was what I wished to do. Nobel Prizes were a little bit out of my knowledge.' As will be seen in later entries, the issue of Nobel Prize decisions is a whole other can of worms, and there have been many women scientists, such as the nuclear physicist Lise Meitner (p. 354) or the astrophysicist Jocelyn Bell Burnell (p. 425), who were denied the prizes they clearly deserved, though as Bell Burnell has always maintained, 'I am not myself upset about it – after all, I am in good company, am I not!'

Sources: *Tatler* 236 (12 October 1710), p. 2; John Hill, *A Dissertation on Royal Societies*; in *Three Letters from a Nobleman on his Travels, to a Person of Distinction in Sclavonia* (London: John Doughty, 1750), pp. 46–7; Charles Babbage, *Reflections on the Decline of Science in England, and on some of its causes* (London: B. Fellowes, 1830), pp. 50–52; 'Finding What's There: Dorothy Hodgkin, Chemist', in Lewis Wolpert and Alison Richards (eds), *A Passion for Science* (Oxford: Oxford University Press, 1988), pp. 76–8.

GLACIERS

Glaciers have been objects of fascination for centuries, although it was not until the early nineteenth century that the scale of their impact on mountainous landscapes began to be appreciated. Several generations of Alpine scientists contributed to the growing understanding of the slow processes of glacial advance and retreat, which culminated in the idea of cyclical ice ages that was first proposed by the palaeontologist Louis Agassiz (1807–73) in the 1830s.

Pierre Martel

One of those early Alpine scientists was Pierre Martel (c. 1701–61), a Genevan engineer and instrument-maker who took a particular interest in the nearby glaciers of Savoy. He and four companions, including 'a Goldsmith very well skilled in Minerals' and 'an Apothecary who was a good Chemist and Botanist', made the first barometrical ascent of the Mer de Glace in the summer of 1742, which Martel wrote up in the following account, which is filled with a confusing array of measurements taken with an arsenal of non-standard instruments. But in spite of his enthusiasm for trying out all the various bits of kit that he had insisted on hauling up the mountain, Martel made some important observations, not least that the glacier appeared to have moved up and down the valley over time:

I took with me a good Barometer, included in a wooden Tube, which I filled at every Station, according to the Method of *Torricelli*, with all possible Precaution; and for this Purpose, carried a good deal of Mercury to be always provided in Case of Accidents: I had with me my Semicircle of ten *English* Inches Radius, with some sea Compasses, a Camera-Obscura, and all Implements for Drawing. I took also a Thermometer of my own make, filled with Mercury, divided into a hundred equal Parts, from the freezing Point, to boiling Water, answering to 180 Parts of *Farenheit*'s Thermometer, beginning at 32, and ending at 212. I divided my Barometer into Inches and Lines, *French* Measure, to have at once the Height of the Mercury. I left at *Geneva* with Baron *Rotberg* a Barometer and Thermometer, similar to those I took along with me; that I might compare our Experiments, with the Variations of the Barometer, in case the Weather had varied; but the Weather being all the Time fine, the changes were not sensible . . .

We left *Geneva Aug.* 20, 1742. Before we set out I tried my Barometer, which upon the Side of the *Rhone* stood at 27 Inches 2/12; and the Thermometer at 18 D. above the freezing Point which answers to 60. of *Farenheit*. We dined at *Bonneville*; where the Barometer was at 26 8/12; which makes 6 Lines less than at the Side of the *Rhone* at *Geneva*; thus it appears that the *Arve* is at *Bonneville*, above the Level of the *Rhone* 403 F. 10 I. 5 L. *French* Measure; which is not a very great Rise, considering the Distance is 5 Leagues, or 15 Miles *English* . . .

The Night between the 22d and 23d I hung out my Thermometer in the open Air, and found it in the Morning two Degrees above the freezing Point, which answers to 35 1/2 Degrees of *Farenheit*. This made us cloath ourselves warmer, in order to go up the Mountain; for which Place we set out about six in the Morning, having with us seven Men both to assist us in climbing, and to carry Provisions; we took in other Respects the same Precautions as you did, and I carried my Instruments with me. At a Halt which we made after three Hours climbing, I tried my Barometer, which gave me 1 I. 4/12 lower than at *Chamouny*; and by the Table, I found that we had mounted 1179 F. 0. I. from the *Arve* at *Chamouny*; after two Hours and half more very difficult climbing, we got on the Top of the Mountain called *Montanver*; from whence we saw the Ice Valley, and were struck with Astonishment at so extraordinary a Sight. After having taken a View of it while we rested, I tried the Barometer, which stood at 22 8/12 which was 22 8/12 less than at *Chamouny*, which gives for the Height of the Mountain 2427 F. 8. 10. and above the Level of the Lake F. 3947. 2. 3.

In order to find a Place to Dine in we descended towards the Ice, and got behind a kind of Mound, of great Stones which the Ice had raised, as I will explain hereafter. The Barometer rose two Lines, which shewed us we had gone down F. 159. 7. 8.; in this Place we dressed our Victuals, and dined under the Shade of a great Rock. The Thermometer was got down to only one Degree above the freezing Point, which answers to about 33 1/2 of *Farenheit*. We were not able to stay here long by reason of the Cold, which obliged us to get into the Sunshine, altho' we were dressed as in the Middle of Winter; and after Dinner every one went according to his Inclinations, some upon the Ice, others to look for Crystal; for my Part I took two Men with me, and returned to *Montanvert*, where I remained near three Hours, which time I employed in making a Plan of the *Glacieres*, which I have put at the End of this Account.

Glacial Erratics

Like many Cumbrians before and since, the poet William Wordsworth was intrigued by the isolated boulders that lie scattered over the hills and valleys of the Lake District. In his poem *Revolution and Independence* (1802), he had described how

> A huge stone is sometimes seen to lie
> Couched on the bald top of an eminence;
> Wonder to all who do the same espy,
> By what means it could thither come,
> And whence ...

Once the idea had arisen that vast rivers of ice once flowed down those English valleys, the answer to Wordsworth's question was not long in coming: the stranded boulders had been left behind by the retreat of long-vanished glaciers. As shown in the following extracts from a series of letters sent to Wordsworth by the geologist Adam Sedgwick (1785–1873), the mystery of these erratic boulders – 'sometimes many tons in weight, and in positions most strange and difficult to be accounted for' – were amenable not only to the language of poetry, but to the new and equally vivid language of geology. His first letter, however, began with a defence of the new science against what he saw as unjust literary attacks on 'my brethren of the hammer':

My Dear Sir, – In writing these letters, I am only endeavouring to perform a promise made many years since when I had the happiness of rambling with you through some of the hills and valleys of your native country. One of your greatest works seems to contain a poetic ban against my brethren of the hammer, and some of them may well have deserved your censures: for every science has its minute philosophers, who neither have the will to soar above the material things around them, nor the power of rising to the contemplation of those laws by which Nature binds into union the different portions of her kingdom. But geology has now a different form and stature from what she had in earlier days: she is the handmaid of labourers who are toiling, as they believe, for the good of their fellow men: she claims kindred with all the offspring of exact knowledge: and she lends no vulgar help to the loftiest investigations of human thought. To reject her altogether, can only be done consistently by one who shuts his eyes to the light of material science; and this, I know, is no part of your philosophy; for no one has put forth nobler views of the universality of nature's kingdom than yourself.

Sedgwick then turned to what he called the 'great difficulty: By what power were these "erratic blocks" scattered over the north of England, and lodged in positions that seem so utterly strange and anomalous':

Late observations on the marine shells derived from the Crag of Norfolk and Suffolk, and other very recent marine deposits on the eastern coast of England, make it probable, that during a period not long before the great diluvial drift our climate was much *colder* than it is at the present day. The appearances on the coast of North America have given rise to the same conclusion; and the labours of M. M. Agassiz, Charpentier, and other Swiss naturalists, have, I think, clearly proved, that, just before the historic time, the glaciers of the Alps were far more extended than they are now. If this be true, may we not suppose that, at the same period, the higher valleys of England and Scotland were filled with glaciers, and that numberless blocks of stone which had rolled down the mountain sides, or been torn off from the neighbouring precipices, were then packed up in thick-ribbed ice?

No one will, I trust, be so bold as to affirm that an uninterrupted glacier could ever have extended from Shap Fells to the coast of Holderness, and borne along the blocks of granite through the whole distance, without any help from the floating power of water. The supposition involves difficulties tenfold greater than are implied in the phenomenon it pretends to account for. The glaciers descending through the valleys of the higher Alps have an enormous transport-ing power: but there is no such power in a great sheet of ice expanded over a country without mountains, and at a nearly dead level.

The period of refrigeration (if such indeed there were) had at length an end; and we can hardly conceive any general change of climate without some great oscillation in the water level. Let us then suppose the earth to sink, or the ocean to rise up, so that the coast line may reach our higher valleys, and then currents of the sea may float away the ancient glaciers with their imbedded fragments of rock. In this way we can conceive it possible that blocks of Shap granite may have been stranded on the side of Cross Fell, or floated over the top of Stainmoor and the crest of the Hambleton hills; and dropped, by the gradual melting of the ice-bergs, on the spots where we now find them. Soon afterwards, our island may have gained a condition of equilibrium, and the land may have risen, or the sea descended, to its present level; in which there appears to have been very little change during the period of modern authentic history.

The previous hypothesis is not new. It was first started, forty or fifty years since, to explain the transporting power which had brought away millions of bowlders and fragments of rock from the Scandinavian chain, and scattered them over the plains in the north of Germany, and in Poland and a part of Russia. But it seemed to be entangled in the greatest difficulty, for how were we to find *the ice*, which was the most important part of the machinery? Geological phenomena appeared to indicate a gradual lowering of temperature, from the oldest epoch down to the present period: and hence it was inferred, that in the epoch just before the historic time, the earth must have been warmer than in our days. But no analogy can stand against the direct evidence of facts; and if there has been a period of refrigeration, accompanied by a great oscillation in the level of land and water, the glacial theory will then lend itself readily to the transport of the "erratic blocks," and it involves no supposition which is in antagonism with the known workings of nature. For sea and land have changed their relative levels many times; and ice-bergs, year by year, do bear away great blocks of stone from the arctic regions, and drop them in the sea many hundred miles from the shores they first started from. But whether the glacial theory truly accounts for all the strange movements of the Shap granite, above described, is a question on which I wish not to offer any decided opinion.

One thing at least is certain, that, by whatever cause the "erratic blocks" were floated across our valleys and over our mountains, their dispersion took place at a comparatively recent time. For many of them, though lying bare on the surface, and exposed to all the action of our climate, still clink under the hammer, and hardly shew more signs of decay than the granite of an Egyptian obelisk. I see no reason for supposing that the movement of the great bowlders necessarily took place before the existence of the human race. On this question there seems no direct or conclusive evidence leading to one side or the other. We know, indeed, that bowlders, like those above described, are often associated with ancient marine drift, containing bones of mammals of *extinct species* (such as Mammoth, Mastodon, Rhinoceros, Hippopotamus, &c. &c.) – and we believe that no human bones have been found in the old gravel of Europe, except in situations which seem to shew that they were introduced at a more recent date. But allowing the negative conclusion, that no human bones were entombed, along with the extinct mammals, in the old gravel of Europe, it does not thence follow, that the human race was in no other part of the world ever coeval with the Mastodon and the Mammoth. Whatever may become of such a question, the direct evidence remains untouched; and the condition of the travelled bowlders of Shap granite proves that they were not floated away from the hills of Westmorland during any ancient and indefinite period of time long before the creation of our species.

Sources: Peter Martel, *An Account of the Glacieres or Ice Alps in Savoy, in Two Letters* (London: Peter Martel, 1744), pp. 13–28; Adam Sedgwick, 'Three Letters on the Geology of the Lake District', in John Hudson (ed.), *A Complete Guide to the Lakes, comprising Minute Directions for the Tourist, with Mr. Wordsworth's Description of the Scenery of the Country, &c.* (Kendal: Hudson and Nicholson, 1842), pp. 3–13.

ELECTRICITY

Electricity was the Enlightenment's first great scientific fashion, propelling an esoteric research topic into the mainstream of eighteenth-century culture. Electrical machines proved to be popular drawing-room novelties on both sides of the Atlantic, offering highly visual effects tinged with mystery and magic – a perfect evening's entertainment for all the family. Yet electricity was also hailed as the force of the future, the means by which we might light our homes, cook our food, heal the sick, and even bring the dead back to life. Two of the principal champions of the new electrical age were the American polymath Benjamin Franklin (1706–90) and his English associate, the chemist Joseph Priestley (1733–1804).

Benjamin Franklin

Until the early 1740s electricity had been thought of as two distinct fluids – 'vitreous' and 'resinous' – but Franklin's pioneering experiments conducted at his home in Philadelphia revealed that electricity was a single flowing phenomenon made up of two inseparable charges that he termed 'positive' (+) and 'negative' (–), with the flow tending to travel from a positively charged to a negatively charged body. Other terms introduced by Franklin included 'battery', 'conductor', and 'charged', all of which are still in use today.

The three short extracts below are taken from the letters on electricity that Franklin sent to Peter Collinson, a Fellow of the Royal Society in London who had sent Franklin his first piece of electrical equipment. The first introduces the idea from which Franklin went on to develop the lightning rod; the second shows him playing with republican sentiments in the context of a scientific parlour game; while the third describes the famous kite experiment that, along with Newton's falling apple, went on to become one of the enduring images of scientific discovery:

The Doctrine of Points

Sir,   July 11, 1747
    In my last I informed you that, in pursuing our electrical enquiries, we had observed some particular phænomena, which we

looked upon to be new, and of which I promised to give you some
account, though I apprehended they might not possibly be new to
you, as so many hands are daily employed in electrical experiments
on your side of the water, some or other of which would probably hit
on the same observations.

The first is the wonderful effect of pointed bodies, both in *drawing off* and *throwing off* the electrical fire. For example,

Place an iron shot of three or four inches diameter on the mouth
of a clean dry glass bottle. By a fine silken thread from the ceiling,
right over the mouth of the bottle, suspend a small cork-ball, about
the bigness of a marble; the thread of such a length, as that the
cork-ball may rest against the side of the shot. Electrify the shot, and
the ball will be repelled to the distance of four or five inches, more or
less, according to the quantity of Electricity. – When in this state, if
you present to the shot the point of a long, slender, sharp bodkin, at
six or eight inches distance, the repellancy is instantly destroyed, and
the cork flies to the shot. A blunt body must be brought within an
inch, and draw a spark to produce the same effect. To prove that the
electrical fire is *drawn off* by the point, if you take the blade of the
bodkin out of the wooden handle, and fix it in a stick of sealing-wax,
and then present it at the distance aforesaid, or if you bring it very
near, no such effect follows; but sliding one finger along the wax till
you touch the blade, and the ball flies to the shot immediately. – If
you present the point in the dark, you will see, sometimes at a foot
distance and more, a light gather upon it, like that of a fire-fly, or
glow-worm; the less sharp the point, the nearer you must bring it to
observe the light; and at whatever distance you see the light, you may
draw off the electrical fire, and destroy the repellancy. – If a cork-ball
so suspended be repelled by the tube, and a point be presented quick
to it, though at a considerable distance, 'tis surprizing to see how
suddenly it flies back to the tube. Points of wood will do near as well
as those of iron, provided the wood is not dry; for perfectly dry wood
will no more conduct electricity than sealing-wax.

'The Conspirators'

The magical picture is made thus. Having a large mezzotinto with a
frame and glass, suppose of the King, (God preserve him) take out
the print, and cut a pannel out of it, near two inches distant from the

frame all round. If the cut is through the picture 'tis not the worse. With thin paste or gum-water, fix the border that is cut off on the inside of the glass, pressing it smooth and close; then fill up the vacancy by gilding the glass well with leaf gold or brass. Gild likewise the inner edge of the back of the frame all round except the top part, and form a communication between that gilding and the gilding behind the glass: then put in the board, and that side is finished. Turn up the glass, and gild the fore side exactly over the back gilding, and when it is dry, cover it by pasting on the pannel of the picture that had been cut out, observing to bring the corresponding parts of the border and picture together, by which the picture will appear of a piece as at first, only part is behind the glass, and part before.

Hold the picture horizontally by the top, and place a little moveable gilt crown on the king's-head. If now the picture be moderately electrified, and another person take hold of the frame with one hand, so that his fingers touch its inside gilding, and with the other hand endeavour to take off the crown, he will receive a terrible blow, and fail in the attempt. If the picture were highly charged, the consequence might perhaps be as fatal as that of high-treason; for when the spark is taken through a quire of paper laid on the picture, by means of a wire communication, it makes a fair hole through every sheet, that is through 48 leaves, (though a quire of paper is thought good armour against the push of a sword or even against a pistol bullet) and the crack is exceeding loud. The operator, who holds the picture by the upper-end, where the inside of the frame is not gilt, to prevent its falling, feels nothing of the shock, and may touch the face of the picture without danger, which he pretends is a test of his loyalty. – if a ring of persons take the shock among them, the experiment is called *The Conspirators*.

The Kite Experiment

As frequent mention is made in public papers from Europe of the success of the *Philadelphia* experiment for drawing the electric fire from clouds by means of pointed rods of iron erected on high buildings, &c. it may be agreeable to the curious to be informed that the same experiment has succeeded in *Philadelphia*, though made in a different and more easy manner, which is as follows:

Make a small cross of two light strips of cedar, the arms so long as to reach to the four corners of a larger thin silk handkerchief when extended; tie the corners of the handkerchief to the extremities of the cross, so you have the body of a kite; which being properly accommodated with a tail, loop, and string, will rise in the air, like those made of paper; but this being of silk is fitter to bear the wet and wind of a thunder gust without tearing. To the top of the upright stick of the cross is to be fixed a very sharp pointed wire, rising a foot or more above the wood. To the end of the twine, next the hand, is to be tied a silk ribbon, and where the silk and twine join, a key must be fastened. This kite is to be raised when a thunder-gust appears to be coming on, and the person who holds the string must stand within a door or window, or under some cover, so that the silk ribbon may not be wet; and care must be taken that the twine does not touch the frame of the door or window. As soon as any of the thunder clouds come over the kite, the pointed wire will draw the electric fire from them, and the kite, with all the twine, will be electrified, and the loose filaments of the twine will stand out every way, and be attracted by an approaching finger. And when the rain has wetted the kite and twine, so that it can conduct the electric fire freely, you will find it stream out plentifully from the key on the approach of your knuckle. At this key the phial may be charged; and from electric fire thus obtained, spirits may be kindled, and all the other electric experiments be performed, which are usually done by the help of a rubbed glass globe or tube, and thereby the sameness of the electric matter with that of lightning completely demonstrated.

Joseph Priestley

In the summer of 1766, the chemist and natural philosopher Joseph Priestley began a series of electrical experiments on live animals. The experiments formed part of the research for his book, *The History and Present State of Electricity* (1767), which introduced a new kind of popular science writing, filled with anecdotes and observations designed to engage a non-specialist audience. It was Priestley's book that made Franklin's kite experiment famous throughout the world. But at the same time, an appreciation of animal intelligence and suffering was beginning to emerge, and Priestley's electrical torments caused some disquiet among his readers, who felt that even if such experiments were necessary, they should be kept to the barest minimum.

As Priestley himself concluded, at the end of the following passage, 'it is paying dear for philosophical discoveries, to purchase them at the expense of humanity':

As I have constructed an electrical battery of considerably greater force than any other that I have yet heard of, and as I have sometimes exposed animals to the shock of it, and have particularly attended to several circumstances, which have been overlooked, of misapprehended by others; it may not be improper to relate a few of the cases, in which the facts were, in any respect, new or worth notice.

June the 4th. I killed a rat with the discharge of two jars, each containing three square feet of coated glass. The animal died immediately, after being universally convulsed, at the instant of the stroke. After some time, it was carefully dissected; but there was no internal injury perceived, particularly no extravasation, either in the abdomen, thorax, or brain.

June the 19th. I killed a pretty large kitten with the discharge of a battery of thirty-three square feet; but no other effect was observed, except that a red spot was found on the pericranium, where the fire entered. I endeavoured to bring it to life, by distending the lungs, blowing with a quill into the trachea, but to no purpose. The heart beat a short time after the stroke, but respiration ceased immediately.

June the 21st. I killed a small field-mouse with the discharge of a battery of thirty-six square feet, but no other effect was perceived, except that the hair of the forehead was singed, and in part torn off. There was no extravasation any where, though the animal was so small, and the force with which it was killed so great. This fact, and many others of a similar nature, make me suspect some mistake, in cases where larger animals are said to have had all their blood vessels burst by a much inferior force.

In all the accounts that I have met with of animals killed by the electric shock, the victims were either small quadrupeds, or fowls; and they are all represented as killed so suddenly, that it could not be seen how they were affected previous to their expiration. In some of my experiments, the great force of my battery has afforded me a pretty fair opportunity of observing in what manner the animal system is affected by the electric shock, the animals which I have exposed it to being pretty large; so that a better judgement may be formed of their sensations, and consequently of the immediate cause

of their death, by external signs. I do not pretend to draw any conclusion myself from the following facts. I have only noted them as
carefully as I could for the use of physicians and anatomists.

June the 26th. I discharged a battery of thirty-eight square feet of
coated glass, through the head, and out at the tail of *a full grown cat*,
three or four years old. At that instant she was violently convulsed
all over. After a short respite, there came on smaller convulsions, in
various muscles, particularly on the sides; which terminated in a violent convulsive respiration, attended with a rattling in the throat.
This continued five minutes, without any motion that could be called
breathing, but was succeeded by an exceedingly quick respiration,
which continued near half an hour. Towards the end of this time, she
was able to move her head, and fore feet, so as to push herself backwards on the floor; but she was not able to move her hind feet in
the least, notwithstanding the shock had not passed through them.
While she continued in this condition, I gave her a second stroke,
which was attended, as before, with the violent convulsion, the short
respite, and the convulsive respiration; in which, after continuing
about a minute, she died.

Being willing to try, for once, the effect of a much greater shock
than that which killed the cat upon a large animal, I gave an explosion of sixty-two square feet of coated glass to a dog the size of a
common cur. The moment he was struck, which was on the head
(but, not having a very good light, I could not tell exactly where) all
his limbs were extended, he fell backwards, and lay without any
motion, or sign of life for about a minute. Then followed convulsions, but not very violent, in all his limbs; and after that a convulsive
respiration, attended with a small rattling in the throat. In about four
minutes from the time that he was struck, he was able to move,
though he did not offer to walk till about half an hour after; in all
which time, he kept discharging a great quantity of saliva; and there
was also a great flux of rheum from his eyes, on which he kept putting his feet; though in other respects he lay perfectly listless. He
never opened his eyes all the evening in which he was struck, and the
next morning he appeared to be quite blind, though seemingly well
in every other respect.

Having dispatched the dog, by shooting him through the hinder
part of his head, I examined one of his eyes (both of which had an
uniform blueish cast, like a film over the pupil) and found all the

three humours perfectly transparent, and, as far as could be judged, in their right state; but the *cornea* was throughout white and opaque, like a bit of gristle, and remarkably thick.

Before this experiment, I had imagined that animals struck blind by lightning had probably a *gutta serena*, on account of the concussion which is seemingly given to the nervous system by the electric shock; whereas this case was evidently an inflammation, occasioned by the explosion being made so near the eyes, terminating in a species of the *albugo*; but which I suppose would have been incurable. One of the eyes of this dog was affected a little more than the other; owing, probably, to the stroke being made a little nearer to one eye than the other. I intended to give the stroke about an inch above the eyes . . .

Being willing to observe, if possible, the immediate effect of the electric shock on the heart and lungs of animals, I gave, June the 5th, a shock from six square feet to a frog, in which the thorax had been previously laid open, so that the pulsation of the heart might be seen. Upon receiving the stroke, the lungs were instantly inflated; and, together with the other contents of the thorax, thrown quite out of the body. The heart, however, continued to beat, though very languidly, and there was no other sign of life for about ten minutes. After that, a motion was first perceived under its jaws; which was propagated, by degrees, to the muscles of the sides; and at last the creature seemed as if it would have come to life, if it had not been so much mangled. The stroke entered the head, and went out at the hind feet.

June the 6th. I discharged a battery of thirty-three square feet through the head and whole extended body of another frog. Immediately upon receiving the stroke, there was, as it were, a momentary distention of all the muscles of the body, and it remained shrivelled up in a most surprising manner. For about five minutes there appeared no sign of life, and the pulsation of the heart could not be felt with the finger. But afterwards, there first appeared a motion under the jaws, then all along the sides, attended with convulsive motions of the other parts, and in about an hour it became, to all appearance, as well as ever.

The same day, I gave the same stroke to two other frogs. They were affected in the same manner, and perfectly recovered in less than three hours.

These facts surprised me very much. I attribute the recovery of the frogs partly to the moisture, which always seems to cover their body, and which might transmit a good part of the shock; and partly to that provision in their constitution, whereby they can subsist a long time without breathing. To ascertain this, I would have given the shock to toads, serpents, fishes, &c. and various other exanguinous animals, but I had not an opportunity. Besides, it is paying dear for philosophical discoveries, to purchase them at the expense of humanity.

Sources: Benjamin Franklin, *Experiments and Observations on Electricity, made at Philadelphia in America*, 5th edn (London: F. Newbery, 1774), pp. 3–4; 29–30; 117–18; Joseph Priestley, from *The History and Present State of Electricity, with Original Experiments*, 4th edn (London: C. Bathurst & T. Lowndes, 1775), pp. 597–601.

BLUNT *VERSUS* POINTED

As was seen in the previous entry, one immediate practical outcome of Benjamin Franklin's electrical researches was the invention of the lightning rod, designed to protect vulnerable buildings from the 'sudden and terrible mischief' of an electrical strike. They worked by conducting the charge safely down to earth, usually into a shallow well dug for the purpose. Pointed iron rods were soon being fitted to public buildings throughout the United States, but sceptical Europeans took much longer to be convinced of the benefits. It was not until 1769 that the dome of St Paul's Cathedral in London was fitted with a lightning conductor, but three years later, in 1772, controversy broke out when Franklin was asked by the Board of Ordnance for his opinion on the best means of protecting the royal gunpowder stores at Purfleet, a British naval depot near the mouth of the Thames. Franklin recommended the installation of a pointed lightning rod: 'Its upper extremity should be carried ten feet above the summit of the roof, and taper off gradually till it ends in a sharp point', he wrote, but a rival British electrical researcher named Benjamin Wilson (1721–88) disagreed with Franklin's prescriptions, and urged the Board of Ordnance to install round-capped rods instead, arguing that pointed rods actually attracted lightning from the air, 'frequently occasioning a discharge where it might not otherwise have happened'. Wilson was overruled, however, and Franklin's pointed lightning rods were duly installed.

Five years later, on 15 May 1777, the Purfleet stores were struck by lightning, and the roof of the board-house slightly damaged. A Royal Society committee was hastily convened to investigate the incident; the first witness statement they heard was from Edward Nickson, the store-keeper at Purfleet:

Purfleet, May 16, 1777

Yesterday afternoon we had much rain and distant thunder; but at six a very heavy cloud, in passing over the house, presented us with part of its contents, which struck the North-east corner of the house on one of the cramps that held the coping stones together, forced off about a square foot of that stone and one brick, and has displaced about a cube foot of brick-work underneath. It has not been yet discovered that any of the conductors have acted during the passage of that cloud, although the flash and report were both very great . . . My son says that there is a dent in the cramp, on which the lightning fell, and I intend to preserve it as a curiosity. If the conductor on the house has acted, it is imperceptible as I am informed. I thought this account would be acceptable to you from,

Honourable sir, &c.

Edward Nickson

The next statement came from the aggrieved electrician Benjamin Wilson, who delivered a lengthy I-told-you-so rebuke to those who had commissioned the Purfleet rods five years before, as well as to the absent Franklin, who was now the untrustworthy representative of an enemy power (the American War of Independence having broken out the previous year). Wilson took full advantage of a welcome opportunity to make his feelings known:

When this important subject was first debated in the Purfleet committee of 1772, a passage was quoted from Dr. Franklin's philosophical publications, respecting the nature of such buildings as were secure from attacks by lightning.

The passage alluded to is this: 'Buildings that have their roofs covered with lead or other metal, and spouts of metal continued from the roof into the ground to carry off the water, are never hurt by lightning; as, whenever it falls on such a building, it passes in the metals, and not in the walls.'

With this idea the building at Purfleet, called the Board-house, was considered by that committee to be in a similar situation, and consequently secure from such attacks, without having any other conductors than the leaden gutters, pipes, &c. . . .

I proposed that the magazines themselves should be put into the same circumstances; otherwise there would appear to be an inconsistency in the different methods of securing those buildings.

My argument had no other effect than to occasion, at the next meeting of the committee, a resolution for fixing pointed conductors to all the buildings.

From this resolution I dissented, and gave in writing my reasons at large for differing in opinion, which are printed in your Transactions.

What has been the consequence since the conductors were put up?

Behold! this very Board-house, which was never attacked before by lightning, hath very lately been struck, and that within a few inches of the conductor; contrary to Dr. Franklin's assertion, which positively says, that in such circumstances the lightning passes in the metals, and not in the walls.

We may refine in our reasoning upon the philosophy of this event as much as we please; but let me tell you, gentlemen, there is no getting rid of the fact: which, according to my judgement, appears to be truly alarming. And as, I apprehend, the reputation of this learned Society is greatly concerned therein, we ought immediately to avail ourselves of this providential warning, and reject an apparatus which threatens us every hour with some unhappy consequences.

In considering the propriety of pointed conductors, I think it necessary to observe, that increasing the number of them in any given space does not by any means, in my opinion, lessen the risk of accidents by lightning; but on the contrary (at least in many cases) a greater number of such conductors will necessarily invite a larger quantity of lightning. At Purfleet there are several of those conductors; and by the storekeeper's letter sent to the Board of Ordnance, which was lately read before us, it appears that he himself observed a very heavy cloud hanging over the house for some time before the stroke happened.

According to Dr. Franklin's idea, this event ought never to have happened; because he says that pointed conductors will draw all the lightning out of the clouds, and carry it away into the earth silently.

This philosophy I never had any faith in, unless the quantity of lightning contained in the clouds happens to be very little, and incapable of producing any fatal consequences.

I have now only to add, that I did not propose to have troubled this Society any more, had I not thought, upon this great occasion, it was my duty to stand forth, and give my opinion against the present report; as I know of no possible advantage to be derived from such conductors; at least none that are consistent with true philosophy, and a sincere regard to the welfare of society.

A few months later, Wilson submitted a detailed account of some fifty experiments that he had undertaken in order to disprove the safety of Franklin's pointed rods. The results of these experiments backed up his belief that pointed rods actually 'solicited' lightning from the air, and again he urged that, for safety's sake, rounded copper caps be fitted to the lightning conductors at Purfleet. The scientific committee – under the chairmanship of Sir John Pringle, President of the Royal Society – met once more to consider Wilson's findings, and in March 1778 they delivered their verdict:

We, having attentively examined the experiments and observations of Mr. Wilson, contained in a paper referred to the Society by the Board of Ordnance; and having maturely considered the subject at large, submit it as our opinion:

1. That it is very improbable, that the powder magazines, guarded in the manner in which they are at present, should receive any damage from lightning.

2. That they would be still less liable to be injured if three other elevated pointed rods, similar to those already erected, were to be fixed upon the roof of each of the five magazines, between the extreme rods, at equal distances from each other, with three strips of lead, about one foot in breadth, strongly connected with them, and carried down the roof, from the ridge to the eaves, on each side of the building; thence two of them to be continued into the earth, and to terminate at the bottom of wells; one of which should be dug for that purpose nearly in the middle of each of the intervals between the magazines, deep enough to contain at least four feet of water . . . We also advise that other high pointed rods be erected; one at each of the four corners, and one over each of the metal doors in the middle of the sides; which latter should be bent, so as to avoid the doors, in the same manner as those which are already placed upon the outward side of the outermost magazines. All which rods should be continued into the earth, and be made to communicate with the bottom of the water of the nearest wells, by means of leaden pipes, closely connected with these iron rods . . . We do also advise in general: That the iron rods be painted, except in those places where they are to be in contact with the lead; that they be all ten feet high; and that they be terminated with pieces of copper eighteen inches long, like those already erected: and, that these copper terminations be very finely tapered, and *as acutely pointed as possible.*

We give these directions, being persuaded, that elevated rods are preferable to low conductors terminated in rounded ends, knobs, or balls of metal; and conceiving, that the experiments and reasons, made and alledged to the contrary by Mr. Wilson, are inconclusive.

    J. Pringle, P.R.S. *et al*   12 March 1778

But if Pringle and the rest of the committee thought that that was the end of the matter, they were sorely mistaken. Wilson, a fashionable society portrait painter as well as a noted electrician, was very well connected at court; in fact he had communicated his concerns about the Purfleet lightning rods directly to the King, George III, who had attended some of Wilson's electrical experiments and apparently convinced himself that Franklin's pointed rods were an American Revolutionary conspiracy designed to sabotage British gunpowder stores. The king ordered pointed rods to be removed from all royal buildings and replaced with patriotic blunt ones; he also asked Pringle to reverse the decision of the Purfleet committee in favour of Wilson's design. Pringle's celebrated reply, that 'the laws of nature are not changeable at royal command', cost him the presidency of the Royal Society – he was replaced by Sir Joseph Banks in November 1778, victim of an overheated political dispute that saw pointed and round-headed lightning rods transformed into revolutionary symbols. An epigram that circulated in the wake of the king's decision summed up the episode well:

> While you, great George, for knowledge hunt
> And sharp conductors change for blunt
>   The Empire's out of joint.
> Franklin another course pursues
> And all your thunder heedless views
>   By keeping to the point.

The rival claims of blunt and pointed rods continue to divide meteorological opinion, although recent trials on a range of models at a lightning-prone test site in the mountains of New Mexico found that conductors with a flat, 'moderately blunt' 19mm diameter end outperformed both the tapered and the mushroom-capped designs. It turns out that Franklin and Wilson may well have been right and wrong in equal measures.

Source: 'Sundry Papers relative to an Accident from Lightning at Purfleet, May 15, 1777', *Philosophical Transactions* 68:1 (1778), pp. 232–317.

SEISMIC WAVES

Until the advent of plate tectonic theory in the mid-twentieth century (see p. 380), the cause of earthquakes was a source of disagreement among natural philosophers, who proposed an array of competing mechanisms as their chief seismic culprits, including subterranean fire, the build-up of explosive gases in caves, and even underground electrical storms. As will be seen in the first extract below, one of the most compelling explanations involved trapped subterranean steam that sent powerful vibrations hurtling through the earth. That was in the 1760s, but even by the time Charles Richter devised the famous earthquake scale that bears his name, the origin of seismic waves was still not properly understood. We now know that they are caused by the interactions of tectonic plates as they bump and grind past one another during their slow planetary migrations, but, as the pair of extracts in this section show, you do not always need to understand the science in order to be a brilliant scientist.

John Michell

The destruction of Lisbon in the earthquake of 1 November 1755 caused moral and intellectual consternation all over Europe; it was no longer enough to call such events 'acts of God', and a search for the origins of earthquakes was soon underway. John Michell (1724–93), of Queens' College, Cambridge, studied centuries' worth of earthquake data, and noted that certain regions of the earth seemed more seismically active than others. This, he thought, was due to subterranean steam – generated from sea-water vaporized by molten rock – making its way between geological strata until it reached discontinuities near the surface. The steam would then rush towards those areas, its movements sending powerful waves hurtling through the earth's crust, which would then reach the surface in the form of earthquakes.

Michell's work was significant, not for the red-herring steam hypothesis, but for its emphasis on the elastic propagation of waves, the idea that would establish the science of seismology. Most impressively, Michell demonstrated that it was possible to determine the precise source of an earthquake (its underground 'focus' or 'hypocentre') by measuring the seismic waves' arrival time from a series of different locations. By analysing a set of eyewitness accounts from Lisbon, Michell estimated the source of the 1755 earthquake to have been somewhere in the region of one to three miles (1.6 to 4.8 kilometres) below the Atlantic Ocean, just off the Portuguese coast. He was right:

It has been the general opinion of philosophers, that earthquakes owe their origin to some sudden explosion in the internal parts of the earth. This opinion is very agreeable to the phænomena, which seem plainly to point out something of that kind. The conjectures, however, concerning the cause of such an explosion, have not been yet, I think, sufficiently supported by facts; nor have the more particular effects, which will arise from it, been traced out; and the connexion of them with the phænomena explained. To do this, is the intent of the following pages; and this we are now the better enabled to do, as the late dreadful earthquake of the 1st of November 1755 supplies us with more facts, and those better related, than any other earthquake of which we have an account . . .

As a small quantity of vapour almost instantly generated at some considerable depth below the surface of the earth, will produce a vibratory motion, so a very large quantity (whether it be generated almost instantly, or in any small portion of time) will produce a wave-like motion. The manner in which this wave-like motion will be propagated, may, in some measure, be represented by the following experiment. Suppose a large cloth, or carpet, (spread upon a floor) to be raised at one edge, and then suddenly brought down again on to the floor, the air under it, being by this means propelled, will pass along, till it escapes at the opposite side, raising the cloth in a wave all the way as it goes. In like manner, a large quantity of vapour may be conceived to raise the earth in a wave, as it passes along between the strata, which it may easily separate in an horizontal direction, there being, as I have said before, little or no cohesion between one stratum and another. The part of the earth that is first raised, being bent from its natural form, will endeavour to restore itself by its elasticity, and the parts next to it beginning to have their weight supported by the vapour, which will insinuate itself under them, will be raised in their turn, till it either finds some vent, or is again condensed by the cold into water, and by that means prevented from proceeding any farther.

If a large quantity of vapour should continue to be generated for some time, several waves might be produced by it; and this would be, in some measure, the case, if the quantity at first generated was exceedingly great, though the whole of it was generated in less time, than whilst the motion was propagated through the distance between two waves.

These waves must rise the higher, the nearer they are to the place from whence they have their source; but, at great distances from thence, they may rise so little, and so slowly, as not to be perceived, but by the motions of waters, hanging branches in churches, &c.

The vibratory motion occasioned by the first impulse of the vapour, will be propagated through the solid parts of the earth, and therefore, it will much sooner become too weak to be perceived, than the wave-like motion; for this latter, being occasioned by the vapour insinuating itself between the strata, may be propagated to very great distances; and even after it has ceased to be perceived by the senses, it may still discover itself by the appearances before-mentioned.

Michell's emphasis on tracking the 'vibratory motions' of earthquakes set the seismological agenda for the next two centuries, and earthquake scientists now devote most of their attentions to monitoring seismic-wave activity in an effort to predict major tremors. Grading those tremors became a science in itself, and an array of earthquake intensity scales have been used around the world, but the best known one was devised in 1935 by Charles Richter (1900–85), a seismologist based at the California Institute of Technology. The Richter scale is not in fact an *intensity* scale at all, but a *magnitude* scale: it does not grade damage or other visible effects of earthquakes on the ground, but instead provides an estimate of the amount of energy released during the event.

Richter's system worked by correlating two sets of figures: 1) the amplitude of the maximum earthquake motion, which was established by reading the needle traces recorded by seismographs in a number of different locations (an idea first championed by John Michell in the 1760s); and 2) the square root of the ratio of the actual distances between those seismographs and the earthquake's epicentre. The relationship between the two sets of figures was then expressed logarithmically on a rising scale that soon became known as the Richter magnitude, or, more popularly, as the Richter scale.

The key difference between Richter's magnitude scale and all earlier intensity scales was that its unit changes were no longer linear, like those on a thermometer, but logarithmic, using common base-10 logarithms. Due

to the complex relationship between energy release and magnitude, however, each unit change on the scale represented not a ten-fold increase but a
thirty-fold increase of total earthquake energy; so an earthquake measured
at magnitude 8.0 releases something like thirty times more energy than
one of magnitude 7.0, which in turn releases thirty times more energy
than one of magnitude 6.0. Thus, an earthquake measuring 8.0 on the
Richter scale is not *two times* more powerful as one measuring 6.0, but
around *900 times* more powerful, due to the mathematical ratios involved.
Although Richter suggested that earthquakes of magnitude 10 were possible,
so far the most powerful earthquake ever recorded was the 9.5 magnitude
event that shook the coast of Chile on 22 May 1960, sending a powerful
tsunami racing across the Pacific.

An Ancient Chinese Seismoscope

Nearly two thousand years before the rise of Western seismology, what
Joseph Needham called 'the ancestor of all seismographs' was built by the
Chinese astronomer Zhang Hêng (c. 78–139). It was a brilliantly simple
device that showed the direction of a near or distant earthquake by means
of a series of suspended metal balls that dropped in the event of a tremor.
The following extract is a translation of a description of the celebrated
instrument from the *Hou Han Shu*, a fifth-century historical chronicle of the
later Han dynasty:

In the first Yang-chia year (AD 132) he made an instrument for
judging the direction of earth movements formed of fine cast copper,
eight feet in diameter, fitted with a domed cover, and in shape like
a wine jar. For ornament he used the lines of antique characters and
the shapes of mountains, turtles, birds, and beasts. Inside there was
a control post, which would move sideways in any one of the eight
directions, for the purpose of holding or releasing the mechanism.
Outside the vessel were eight dragons' heads each holding a copper
ball in its mouth, and below (each) was a frog with mouth wide open
to catch the ball. The cogged machinery and cunning constructions
were hidden inside the jar, and the cover fitted down closely all round
without a crevice. If there was an earth movement, then the jar
shook, the mechanism of a dragon was released and vomited the ball,
and the frog caught the ball in its mouth. The sound of shaking was

7. A modern reconstruction of Zhang Hêng's celebrated seismograph. Earth tremors displaced a pendulum linked to a mechanism which opened the jaws of whichever dragon faced the earthquake's source. A ball then fell from the dragon's teeth into the mouth of a frog below, creating a visible record of the quake's direction.

high and loud, and the observer was thus aroused to know of the earthquake. Even though it was the mechanism of one dragon that was released while the seven heads did not move, yet, according to the point of the compass of the dragon which did move, they knew the position of the shock; and they corroborated it by the supernatural way in which the facts tallied.

Judging by the records in the histories such a thing had never before existed. On one occasion the mechanism of a dragon went off, but the earth did not move perceptibly; and all the learned at the capital mocked at the failure. But several days later a messenger arrived, and there really had been an earthquake in Lung-Hsi (about 400 miles to the west); and upon this every one admitted the mysterious power of the instrument. From this time onwards the official

historians were ordered to record the direction from which movements of the earth began.

Sources: John Michell, *Conjectures concerning the Cause, and Observations upon the Phænomena, of Earthquakes; particularly of that great Earthquake of the first of November 1755, which proved so fatal to the City of Lisbon, and whose Effects were felt as far as Africa, and more or less throughout almost all Europe* (London, 1760), pp. 3–39; A. C. Moule (trans.), 'An Ancient Seismometer', *T'oung Pao* 23:1 (1924), pp. 37–9.

CLASSIFICATION

The first step to wisdom, as an old Chinese saying goes, is getting things by their right names. Man has always been a classifying animal, and his appetite for arranging the world into patterns was a contributing factor in the development of the natural sciences. Though many classification schemes have appeared over the centuries, the most lasting and influential was the one devised by the eighteenth-century Swedish naturalist Carl Linnaeus (1707–78), who proposed an all-encompassing system based on shared physiological characteristics. An extract from his work appears below, followed by two other examples of Enlightenment classification that took their cue from the Linnean system.

Linnaeus and the *Systema Naturae*

Linnaeus outlined his classificatory system in a short pamphlet published in 1735; by the tenth edition (1758) the *Systema Naturae* was a sprawling 800-page volume that classified 4,400 species of animals and 7,700 plants in a variety of ways (including, as will be seen, animals from 'speaking' to 'silent'). Linnaeus's new methods met with some resistance at first, and he relied upon a network of influential supporters to convince the scientific world to adopt his new classification. The Swedish playwright August Strindberg later described Linnaeus as 'a poet who happened to become a naturalist', and as this excerpt from a late-eighteenth-century translation of his zoological classification shows, he had a poet's ear for arresting and insightful imagery:

Methodical arrangement, which is the soul of science, indicates every natural body at first sight, so that it may be known by its own name; and this name points out whatever the industry of the age has discovered concerning the body to which it belongs: Thus, amidst the greatest apparent confusion of things, the order of Nature is seen to retain the highest degree of exactness. This systematic arrangement is most conveniently divided into branches, subordinate to each other, which have received various appellations; thus,

Class,	Order,	Genus,	Species,	Variety.
Highest genus,	Intermediate genus,	Proximate genus,	Species,	Individual.
Province,	District,	Parish,	Ward,	Hamlet.
Legion,	Battalion,	Company,	Mess,	Soldier.

"For, unless natural bodies be reduced under regular order, and distributed as in the division of a well regulated camp, every thing that is known concerning them must remain in confusion and uncertainty." *Caesalpinus*.

The names and characters employed in a system must apply accurately to the order of arrangement, and are therefore to be divided, as above, into Classes, Orders, Genera, Species, and Varieties. The differential characters, which distinguish these divisions and subdivisions from each other must likewise occupy a principal part in systematic arrangement; for it is indispensably necessary to the knowledge of any individual, that its name may be readily known and discovered from among the rest; "for if the names of things be confused, the whole science must fall into inextricable perplexity." *Ibid*. Hence one great employment of man, at the beginning of the world, must have been to examine created objects, and to impose on all the species names according to their kinds.

The science of Nature is founded on an exact knowledge of the nomenclature of natural bodies, and of their systematic arrangement; this, like the clew of Ariadne, enables a philosopher to travel alone, and in safety, through the devious meanderings of Nature's labyrinth. In this methodical arrangement, the Classes and Orders are the creatures of human invention, while the division of these into Genera and Species is the work of Nature. All true knowledge refers finally to the species of things, while, at the same time, what regards the generic divisions is substantial in its nature.

One order of things originates from the Creator, while the other is the work of man, and is the subject of our present labour. God, beginning from the most simple terrestrial elements, advances through Minerals, Vegetables, and Animals, and finishes with Man. Man, on the contrary, reversing this order, begins from himself, and proceeds downwards to the materials of the earth. The framer of a systematic arrangement begins his study by the investigation of particulars, from which he ascends to more universal propositions; while the teacher of this method, taking a contrary course, first explains the general propositions, and then gradually descends to particulars. Springs unite together into rivulets, and these conjoin to form rivers; through these the skillful navigator ascends so far as his art allows, but is never able to reach the original fountains. A distinct knowledge of things being given us, we must endeavour to penetrate farther into their particular properties, and to investigate, as far as we are able, their phenomena, their mysterious operations, their natures, their virtues, and their uses.

In the science of Natural History, through its several departments of the three kingdoms of Nature, lies the only sure foundation of Regimen, Medicine, and Economy, both that which regards the arts of life, and that which is followed in the operations of Nature. "Happy are those who cultivate this science, if they know and employ justly the blessings which they enjoy!"

All created things are proofs of the Divine power and wisdom, and fertile sources of human happiness; in their proper use the goodness of God is manifested to man; from their beauty and fitness the wisdom of the Creator shines forth; and, from the admirable economy which appears in their preservation, their just proportions to each other, and in the means employed for their perpetual renovation, the power of the Divine Majesty is most clearly shewn: Therefore the discovery of these things has in all ages been highly esteemed and earnestly prosecuted by the wise and truly learned; while this study hath only been despised by the ignorant and foolish . . .

The natural Division of Animals into Classes, may be formed from a knowledge of the Internal Structure:

1.	{ A heart with two auricles and two ventricles;	{ Viviparous.	Cl. 1. *Mammalia.*
	{ Warm and red blood:	{ Oviparous.	Cl. 2. *Birds.*
2.	{ A heart with one auricle and one ventricle;	{ Voluntary lungs.	Cl. 3. *Amphibia.*
	{ Cold red blood:	{ External gills.	Cl. 4. *Fishes.*

3.
{ A heart with one auricle   { Having antennae. Cl. 5. *Insects.*
  and one ventricle;   {
{ Cold colourless blood: { Having tentacula. Cl. 6. *Worms.*

Thus Nature, in her Menagerie, preserves Animals in six different forms:

MAMMALIA,   covered with hair,   walk on the earth,   speaking.

BIRDS,   covered with feathers, fly in the air,   singing.

AMPHIBIA,   covered with skin,   creep in warm places,   hissing.

FISHES,   covered with scales,   swim in the water,   smacking.

INSECTS,   covered with armour,   skip on dry ground,   buzzing.

WORMS,   without skin,   crawl in moist places, silent.

Under the influence of Linnaeus, the late eighteenth century became the great age of scientific classification, with everything from mammals to microbes being named and arranged for study or stuffing. In fact the taxonomic zeal extended further, to phenomena as fleeting as weather and sounds, as the following pair of pieces show:

The Taxonomy of Birdsong

Ornithology is an inherently taxonomic pursuit, which over the centuries has sought to classify birds according to ever more specialist criteria such as plumage, talon shape and beak curvature. But in the following essay, first published in the *Philosophical Transactions* of the Royal Society in 1773, the naturalist Daines Barrington (1727–1800) attempted to go further still, and in what the historian David Clifford has called a 'joyfully misapplied piece of empiricism', he proposed a comprehensive classification of birds by song. The essay was later reprinted in *The History of Singing Birds* (1791), an eighteenth-century owners' manual for caged-bird enthusiasts, from which this extract has been taken. As Charles Darwin pointed out a hundred years later, there is something rather sobering about the sound of birdsong, which, however beautiful it seems to us, is really the sound of the pitiless struggle for existence:

As the experiments and observations I mean to lay before the Royal Society relate to the singing of birds, which is a subject that hath never before been scientifically treated of, it may not be improper to prefix an explanation of some uncommon terms, which I shall be obliged to use, as well as others which I have been under a necessity of coining.

To chirp, is the first sound which a young bird utters as a cry for food, and is different in all nestlings, if accurately attended to; so that the hearer may distinguish of what species the birds are, though the nest may hang out of his sight and reach.

This cry is, as might be expected, very weak and querulous; it is dropped entirely as the bird grows stronger, nor is afterwards inter-mixed with its song, the chirp of a Nightingale (for example) being hoarse and disagreeable.

To this definition of the chirp, I must add, that it consists of a single sound, repeated at very short intervals, and that it is common to nestlings of both sexes.

The call of a bird, is that sound which it is able to make when about a month old: it is, in most instances (which I happen to recol-lect), a repetition of one and the same note; is retained by the bird as long as it lives; and is common generally to both the cock and hen.

The next stage in the notes of a bird is termed by the bird-catch-ers *recording*; which word is probably derived from a musical instrument formerly used in England, called a *recorder*.

This attempt in the nestling to sing, may be compared to the imperfect endeavour in a child to babble. I have known instances of birds beginning to record when they were not a month old.

This first essay does not seem to have the least rudiments of the future song; but as the bird grows older and stronger, one may begin to perceive what the nestling is aiming at.

Whilst the scholar is thus endeavouring to form his song, when he is once sure of a passage, he commonly raises his tone, which he drops again when he is not equal to what he is attempting; just as a singer raises his voice, when he not only recollects certain parts of a tune with precision, but knows that he can execute them.

What the nestling is not thus thoroughly master of, he hurries over, lowering his tone, as if he did not wish to be heard, and could not yet satisfy himself . . .

A young bird commonly continues to record for ten or eleven

months, when he is able to execute every part of his song, which afterwards continues fixed, and is scarcely ever altered.

When the bird is thus become perfect in his lesson, he is said to sing his song round, or in all its varieties of passages, which he connects together, and executes without a pause.

I would therefore define a bird's song to be a succession of three or more different notes, which are continued without interruption during the same interval with a musical bar of four crochets in an adagio movement, or whilst a pendulum swings four seconds.

By the first requisite in this definition, I mean to exclude the call of a Cuckow or clucking of a hen, as they consist of only two notes; while the short bursts of singing birds, contending with each other (called *jerks* by the bird-catchers), are equally distinguished from what I term *song*, by their not continuing for four seconds.

As the notes of a Cuckow and Hen, therefore, though they exceed what I have defined the call of a bird to be, do not amount to its song, I will, for this reason, take the liberty of terming such a succession of two notes as we hear in these birds, *the varied call*.

Having thus settled the meaning of certain words, which I shall be obliged to make use of, I shall now proceed to state some general principles with regard to the singing of birds, which seem to result from the experiments I have been making for several years, and under a great variety of circumstances.

Notes in birds are no more innate than language is in man, and depend entirely upon the master under which they are bred, as far as their organs will enable them to imitate the sounds which they have frequent opportunities of hearing.

Most of the experiments I have made on this subject have been made with cock Linnets, which were fledged and nearly able to leave their nest, on account not only of this bird's docility and great powers of imitation, but because the cock is easily distinguished from the hen at that early period, by the superior whiteness in the wing.

In many other sorts of singing birds, the male is not at the age of three weeks so certainly known from the female, and if the pupil turns out to be a hen,

- - - *ibi omnis*
Effusus labor.
'then all their work was wasted'

The Naming of Clouds

In a lecture to an amateur science club in the winter of 1802, the Quaker pharmacist Luke Howard (1772–1864) proposed the three-part cloud classification that remains in international use today. Howard named the three main cloud families *cirrus* (Latin for 'fibre' or 'hair'), *cumulus* ('heap' or 'pile'), and *stratus* ('layer' or 'sheet'), but in recognition of their essential instability, he also named a sequence of intermediate and compound forms, such as *cirrostratus* and *stratocumulus*, as a means of accommodating the regular transitions that occur between the cloud types. Thus, by applying Linnean principles to a phenomenon as changeable as the weather, Howard solved the perennial problem of naming transitional forms in nature:

There are three simple and distinct modifications, in any one of which the aggregate of minute drops called a cloud may be formed, increase to its greatest extent, and finally decrease and disappear.

But the same aggregate which has been formed in one modification, upon a change in the attendant circumstances, may pass into another:

Or it may continue for a considerable time in an intermediate state, partaking of the characters of two modifications; and it may also disappear in this stage, or return to the first modification.

Lastly, aggregates separately formed in different modifications may unite and pass into one, exhibiting different characters in different parts, or a portion of a simple aggregate may pass into another modification without separating from the remainder of the mass.

Hence, together with the simple, it becomes necessary to admit intermediate and compound modifications, and to impose names on such of them as are worthy of notice.

The simple Modifications are thus named, and defined:

1. CIRRUS. *Def. Nubes cirrata, tenuissima, quæ undique crescat.*
   Parallel, flexuous, or diverging fibres, extensible in any or in all directions.

2. CUMULUS. *Def. Nubes cumulata, densa, sursum crescens.*
   Convex or conical heaps, increasing upward from a horizontal base.

3. STRATUS. *Def. Nubes strata, aquæ modo expansa, deorsum crescens.*

8. The three main families of cloud: cirrus, cumulus, and stratus, engraved after the original watercolours by London pharmacist Luke Howard, who came up with the terms in a lecture written for an amateur science club in December 1802.

A widely extended, continuous, horizontal sheet, increasing from below . . .

The intermediate Modifications which require to be noticed are:
4. CIRRO-CUMULUS. *Def. Nubeculæ densiores subrotundæ et quasi in agmine appositæ.*
Small, well defined roundish masses, in close horizontal arrangement or contact.
5. CIRRO-STRATUS. *Def. Nubes extenuata sub-concava vel undulata. Nubeculæ hujus modi appositæ.*
Horizontal or slightly inclined masses, attenuated towards a part or the whole of their circumference, bent downward, or undulated, separate, or in groups consisting of small clouds having these characters.

The compound Modifications are:
6. CUMULO-STRATUS. *Def. Nubes densa, basim planam undique supercrescens, vel cujus moles longinqua videtur partim plana partim cumulata.*
The Cirro-stratus blended with the Cumulus, and either appearing intermixed with the heaps of the latter, or super-adding a wide-spread structure to its base.
7. CUMULO-CIRRO-STRATUS *vel* NIMBUS. *Def. Nubes vel nubium congeries pluviam effundens.*
The rain cloud. A cloud or system of clouds from which rain is falling. It is a horizontal sheet, above which the cirrus spreads, while the cumulus enters it laterally and from beneath.

Of the Cirrus.

Clouds in this modification appear to have the least density, the greatest elevation, and the greatest variety of extent and direction. They are the earliest appearance after serene weather. They are first indicated by a few threads pencilled, as it were, on the sky. These increase in length, and new ones are in the mean time added laterally. Often the first-formed threads serve as stems to support numerous branches, which in their turn give rise to others.

The increase is sometimes perfectly indeterminate, at others it has a very decided direction. Thus the first few threads being once formed, the remainder shall be propagated either in one, two, or more directions laterally, or obliquely upward or downward, the direction being often the same in a great number of clouds visible at the same time: for the oblique descending tufts shall appear to converge towards a point in the horizon, and the long straight streaks to meet in opposite points therein; which is the optical effect of parallel extension.

Their duration is uncertain, varying from a few minutes after the first appearance to an extent of many hours. It is long when they appear alone and at great heights, and shorter when they are formed lower and in the vicinity of other clouds.

This modification, although in appearance almost motionless, is intimately connected with the variable motions of the atmosphere. Considering that clouds of this kind have long been deemed a prognostic of wind, it is extraordinary that the nature of this connection should not have been more studied, as the knowledge of it might have been productive of useful results.

In fair weather, with light variable breezes, the sky is seldom quite clear of small groups of the oblique cirrus, which frequently come on from the leeward, and the direction of their increase is to windward. Continued wet weather is attended with horizontal sheets of this cloud, which subside quickly and pass to the cirro-stratus.

Before storms they appear lower and denser, and usually in the quarter opposite to that from which the storm arises. Steady high winds are also preceded and attended by streaks running quite across the sky in the direction they blow in.

The relation of this modification with the state of the barometer, thermometer, hygrometer, and electrometer, have not yet been attended to.

Of the Cumulus.

Clouds in this modification are commonly of the most dense structure: they are formed in the lower atmosphere, and move along with the current which is next the earth.

A small irregular spot first appears, and is, as it were, the nucleus on which they increase. The lower surface continues irregularly plane, while the upper rises into conical or hemispherical heaps; which may afterwards continue long nearly of the same bulk, or rapidly rise to mountains.

In the former case they are usually numerous and near together, in the latter few and distant; but whether there are few or many, their bases always lie nearly in one horizontal plane, and their increase upward is somewhat proportionate to the extent of the base, and nearly alike in many that appear at once.

Their appearance, increase, and disappearance, in fair weather, are often periodical, and keep pace with the temperature of the day. Thus they will begin to form some hours after sun-rise, arrive at their maximum in the hottest part of the afternoon, then go on diminishing, and totally disperse about sun-set.

But in changeable weather they partake of the vicissitudes of the atmosphere; sometimes evaporating almost as soon as formed, at others suddenly forming and as quickly passing to the compound modifications.

The cumulus of fair weather has a moderate elevation and extent, and a well defined rounded surface. Previous to rain it increases more rapidly, appears lower in the atmosphere, and with its surface full of loose fleeces or protuberances.

The formation of large cumuli to leeward in a strong wind, indicates the approach of a calm with rain. When they do not disappear or subside about sun-set, but continue to rise, thunder is to be expected in the night.

Independently of the beauty and magnificence of it adds to the face of nature, the cumulus serves to skreen the earth from the direct rays of the sun, by its multiplied reflections to diffuse, and, as it were, economize the light, and also to convey the product of evaporation to a distance from the place of its origin. The relations of the cumulus with the state of the barometer, &c. have not yet been enough attended to.

Of the Stratus.

This modification has a mean degree of density.

It is the lowest of clouds, since its inferior surface commonly rests on the earth or water.

Contrary to the last, which may be considered as belonging to the day, this is properly the cloud of night; the time of its first appearance being about sun-set. It comprehends all those creeping mists which in calm evening ascend in spreading sheets (like an inundation of water) from the bottom of valleys and the surface of lakes, rivers, &c.

Its duration is frequently through the night.

On the return of the sun the level surface of this cloud begins to put on the appearance of cumulus, the whole at the same time separating from the ground. The continuity is next destroyed, and the cloud ascends and evaporates, or passes off with the appearance of the nascent cumulus.

This has long been experienced as a prognostic of fair weather, and indeed there is none more serene than that which is ushered in by it. The relation of the stratus to the state of the atmosphere as indicated by the barometer, &c. appears notwithstanding to have passed hitherto without due attention.

The influence of Linnean classification continues to this day, though not always entirely helpfully, as the American biologist Edward O. Wilson recalled in the opening of his now-classic book *Consilience: The Unity of Knowledge* (1998). As a nature-loving teenager in 1947, Wilson had entered the University of Alabama to study zoology, his intellectual outlook shaped by the Linnean system, a 'deceptively easy' conceptual world that in its hierarchical, quasi-militaristic pomp was 'made for the mind of an eighteen-year-old'. It was then that he had the luck to discover evolution. 'This epiphany I owed to my mentor Ralph Chermock, an intense, chain-smoking young assistant professor newly arrived in the provinces with a PhD in entomology from Cornell University. After listening to me natter for a while about my lofty goal of classifying all the ants of Alabama, he handed me a copy of Ernst Mayr's 1942 *Systematics and the Origin of Species*. Read it, he said, if you want to become a real biologist.' Young Wilson did as he was advised, and, as he described it, 'a tumbler fell somewhere in my mind, and a door opened to a new world. I was enthralled, couldn't stop thinking about the implications evolution has for classification and for the rest of biology.' He had just been saved from a life spent classifying all the ants of Alabama.

Sources: *The Animal Kingdom, or Zoological System, of the Celebrated Sir Charles Linnaeus*, ed. Robert Kerr (London: J. Murray, 1792), pp. 22–32; Daines Barrington, 'Experiments and Observations on the Singing of Birds', from *The History of Singing Birds, Containing an Exact Description of their Habits and Customs* (Edinburgh: Silvester Doig, 1791), pp. 5–10; Luke Howard, *On the Modifications of Clouds, &c.* (London: J. Taylor, 1804), pp. 4–8.

THE CURE FOR SCURVY

Of all the daily hazards facing the eighteenth-century sailor, among the most feared was scurvy, a condition now known to be caused by a dietary deficiency of vitamin C. Its symptoms were bleeding gums, stiff joints and lassitude, and if left untreated it was usually fatal: Commodore George Anson's four-year circumnavigation of the globe, from 1740 to 1744, had seen a crew of more than 1,800 men reduced to fewer than 200 by the ravages of the disease. Though Admiral Sir Richard Hawkins had advocated the eating of sour oranges and lemons as a preventative as far back as 1593, his advice had been ignored (and subsequently forgotten), and it took another century and a half for the first clinical experiment to be carried out in search of a cure.

The research was conducted by an Edinburgh-born naval surgeon named James Lind (1716–94), while serving on board HMS *Salisbury* in 1746 and 1747. Although Lind had no concept of vitamins (that would come later, in the 1930s), he knew that the condition was related to the sailors' poor diet, especially the lack of fresh meat and vegetables, and his carefully conducted trials, which transformed the vessel into a tightly controlled experimental space, demonstrated that a daily ration of citrus fruit was the most effective remedy of the six he tried. He published his findings in his *Treatise of the Scurvy* (1753), although the emphasis of the book, and its later editions, was on improving living conditions in general. It was only after later physicians had turned their attentions to the prevention of scurvy as a condition that daily rations of lemon juice were issued throughout the Royal Navy, from 1800:

On the 20th of *May* 1747, I took twelve patients in the Scurvy, on board the *Salisbury* at sea. Their cases were as similar as I could have them. They all in general had putrid gums, the spots and lassitude, with weakness of their knees. They lay together in one place, being a proper apartment for the sick in the fore-hold; and had one diet common to all, *viz.* water-gruel sweetened with sugar in the morning; fresh mutton-broth often times for dinner; at other times puddings, boiled biscuit with sugar, *&c.*; and for supper, barley and raisins, rice and currants, sago and wine, or the like. Two of these were ordered each a quart of cyder a-day. Two others took twenty-five gutta of *elixir vitriol* three times a-day, upon an empty stomach; using a gargle strongly acidulated with it for their mouths. Two others took two spoonfuls of vinegar three times a-day, upon an empty stomach; having their gruels and their other food well acidulated with it, as also the gargle with their mouth. Two of the worst patients, with the tendons in the ham rigid, (a symptom none of the rest had), were put under a course of sea-water. Of this they drank half a pint every day, and sometimes more or less as it operated, by way of gentle physic. Two others had each two oranges and one lemon given them every day. These they eat with greediness, at different times, upon an empty stomach. They continued but six days under this course, having consumed the quantity that could be spared. The two remaining patients, took the bigness of a nutmeg three times a-day, of an electuary recommended by an hospital-surgeon, made of garlic, mustard-seed, *rad. raphan,* balsam of *Peru,* and gum myrrh; using for common drink, barley-water well acidulated with tamarinds; by a decoction of which, with the addition of *cremor tartar*, they were gently purged three or four times during the course.

The consequence was, that the most sudden and visible good effects were perceived from the use of the oranges and lemons; one of those who had taken them, being at the end of six days fit for duty. The spots were not indeed at that time quite off his body, nor his gums sound; but without any other medicine, than a gargarism of *elixir vitriol*, he became quite healthy before we came into *Plymouth,* which was on the 16th of *June*. The other was the best recovered of any in his condition; and now being deemed pretty well, was appointed nurse to the rest of the sick.

Next to the oranges, I thought the cyder had the best effects. It was indeed not very sound, being inclinable to be aigre [sour] or pricked. However, those who had taken it, were in a fairer way of recovery than the others at the end of the fortnight, which was the length of time all these different courses were continued, except the oranges. The putrefaction of their gums, but especially their lassitude and weakness, were somewhat abated, and their appetite increased by it.

As to the *elixir* of *vitriol*, I observed that the mouths of those who had used it by way of gargarism, were in a much cleaner and better condition than many of the rest, especially those who used the vinegar; but perceived otherwise no good effects from its internal use upon the other symptoms. I indeed never had a great opinion of the efficacy of this medicine in the scurvy, since our longest cruise in the *Salisbury*, from the 10th of *August* to the 28th *October* 1746; when we had but one scurvy in the ship. The patient was a marine, (one *Walsh*); who after recovering from a quotidian ague in the latter end of *September*, had taken the *elixir vitriol* by way of restorative for three weeks; and yet at length contracted the disease, while under a course of a medicine recommended for its prevention.

There was no remarkable alteration upon those who took the electuary and tamarind decoction, the sea-water, or vinegar, upon comparing their condition, at the end of the fortnight, with others who had taken nothing but a little lenitive electuary and *cremor tartar*, at times, in order to keep their belly open; or a gentle pectoral in the evening, for relief of their breast. Only one of them, while taking the vinegar, fell into a gentle flux at the end of ten days. This I attributed to the genius and course of the disease, rather than to the use of the medicine. As I shall have occasion elsewhere to take notice of the effects of other medicines in this disease, I shall here only observe, that the result of all my experiments was, that oranges and lemons were the most effectual remedies for this distemper at sea. I am apt to think oranges preferable to lemons, though perhaps both given together will be found most serviceable.

Source: James Lind, *A Treatise of the Scurvy* (Edinburgh: Kincaid & Donaldson, 1753), pp. 191–6.

ANIMAL MAGNETISM

Franz Anton Mesmer (1734–1815) was a German physician who claimed to have discovered a hitherto unknown physical property that he called 'animal magnetism', defining it as a vital magnetic force that flowed through all animated matter, and was distinct from other forms of magnetism such as mineral or planetary magnetism. In the late 1770s Mesmer moved to Paris, where he developed a reputation for treating an array of ailments through the channelling of this magnetic force, a technique that became known as 'mesmerism'. Mesmer was not a charlatan; he believed that he had made a valuable scientific discovery, and he sought the backing and approval of the European medical establishment. But he was able to find only one Parisian doctor, Charles d'Eslon, who was prepared to take up mesmerism professionally, and in 1784 it was d'Eslon rather than Mesmer who was investigated by a state-commissioned panel of scientists that included Benjamin Franklin, the chemist Antoine Lavoisier (1743–94) and the astronomer Jean-Sylvain Bailly (1736–93). The panel concluded that mesmerism had no basis in science, and that any positive results from the treatment could be attributed to the power of suggestion:

At the beginning of the year 1778, a German doctor established himself at Paris. This physician could not fail of succeeding in what was then styled high society. He was a foreigner. His government had expelled him; acts of the greatest effrontery and unexampled charlatanism were imputed to him . . .

Mesmer, since we must call him by his name, pretended to have discovered an agent till then totally unknown both in the arts and in physics; an universally distributed fluid, and serving thus as a means of communication and of influence among the celestial globes; — a fluid capable of flux and reflux, which introduced itself more or less abundantly into the substance of the nerves, and acted on them in a useful manner, — hence the name of animal magnetism given to this fluid.

Mesmer said: "Animal magnetism may be accumulated, concentrated, transported, without the aid of any intermediate body. It is reflected like light; musical sounds propagate and augment it."

Properties so distinct, so precise, seemed as if they must be capable of experimental verification. It was requisite, then, to be prepared for some instance of want of success, and Mesmer took good care not to neglect it. The following was his declaration: "Although the fluid be universal, all animated bodies do not equally assimilate it into themselves; there are some even, though very few in number, that by their very presence destroy the effects of this fluid in the surrounding bodies."

So soon as this was admitted, as soon it was allowed to explain instances of non-success by the presence of neutralising bodies, Mesmer no longer ran any risk of being embarrassed. Nothing prevented his announcing, in full security, "that animal magnetism could immediately cure diseases of the nerves, and mediately other diseases; that it afforded to doctors the means of judging with certainty of the origin, the nature, and the progress of the most complicated maladies; that nature, in short, offered in magnetism a universal means of curing and preserving mankind."

Before quitting Vienna, Mesmer had communicated his systematic notions to the principal learned societies of Europe. The Academy of Sciences at Paris, and the Royal Society of London, did not think proper to answer. The Academy of Berlin examined the work, and wrote to Mesmer that he was in error.

Some time after his arrival in Paris, Mesmer tried again to get into communication with the Academy of Sciences. This society even acceded to a rendezvous. But, instead of the empty words that were offered them, the academicians required experiments. Mesmer stated—I quote his words—that *it was child's play*; and the conference had no other result.

Mesmer left Paris in search of a more sympathetic reception, leaving Charles d'Eslon to continue his treatments alone. D'Eslon's technique involved passing iron rods or wands over affected areas of the body, at which his 'mesmerized' patients would usually fall into dramatic seizures and convulsions. The scientific panel that investigated mesmerism in 1784 was particularly interested in these hysterical responses, as described in the following passages from a biography of Jean-Sylvain Bailly, the lead author of the commission's report:

The commissioners, magnetised by d'Eslon, felt no effect. After the healthy people, some ailing ones followed, taken of all ages, and from various classes of society. Among these sick people, who amounted to fourteen, five felt some effects. On the remaining nine, magnetism had no effect whatever.

Notwithstanding the pompous announcements, magnetism already could no longer be considered as a certain indicator of diseases.

Here the reporter made a capital remark: magnetism appeared to have no effect on incredulous persons who had submitted to the trials, nor on children. Was it not allowable to think, that the effects obtained in the others proceeded from a previous persuasion as to the efficacy of the means, and that they might be attributed to the influence of imagination? Thence arose another system of experiments. It was desirable to confirm or to destroy this suspicion; "it became therefore requisite to ascertain to what degree imagination influences our sensations, and to establish whether it could have been in part or entirely the cause of the effects attributed to magnetism."

There could be nothing neater or more demonstrative than this portion of the work of the commissioners. They go first to Dr. Jumelin, who, let it be observed, obtains the same effects, the same crises as d'Eslon and Mesmer, by magnetising according to an entirely different method, and not restricting himself to any distinction of poles; they select persons who seem to feel the magnetic action most forcibly, and put their imagination at fault by now and then bandaging their eyes.

What happens then?

When the patients see, the seat of the sensations is exactly the part that is magnetised; when their eyes are bandaged, they locate these same sensations by chance, sometimes in parts very far away from those to which the magnetiser is directing his attention. The patient, whose eyes are covered, often feels marked effects at a time when they are not magnetising him, and remains, on the contrary, quite passive while they are magnetising him, without his being aware of it.

Persons of all classes offer similar anomalies. An instructed physician, subjected to these experiments, "feels effects whilst nothing is being done, and often does not feel effects while he is being acted upon. On one occasion, thinking that they had been magnetising him for ten minutes, this same doctor fancied that he felt a heat in his lumbi, which he compared to that of a stove."

Sensations thus felt, when no magnetising was exerted, must evidently have been the effect of imagination.

The commissioners were too strict logicians to confine themselves with these experiments. They had established that imagination, in some individuals, can occasion pain, and heat — even a considerable degree of heat — in all parts of the body; but practical female Mesmerisers did more; they agitated certain people to that pitch, that they fell into convulsions. Could the effect of imagination go so far?

Some new experiments entirely did away with these doubts.

A young man was taken to Franklin's garden at Passy, and when it was announced to him that d'Eslon, who had taken him there, had magnetised a tree, this young man ran about the garden, and fell down in convulsions, but it was not under the magnetised tree: the crisis seized him while he was embracing another tree, very far from the former.

D'Eslon selected, in the treatment of poor people, two women who had rendered themselves remarkable by their sensitiveness around the famous rod, and took them to Passy. These women fell into convulsions whenever they thought themselves mesmerised, although they were not. At Lavoisier's, the celebrated experiment of the cup gave analogous results. Some plain water engendered convulsions occasionally, when magnetised water did not.

We must really renounce the use of our reason, not to perceive a proof in this collection of experiments, so well arranged that imagination alone can produce all the phenomena observed around the mesmeric rod, and that mesmeric proceedings, cleared from the delusions of imagination, are absolutely without effect. The commissioners, however, recommence the examination on these last grounds, multiply the trials, adopt all possible precautions, and give to their conclusions the evidence of mathematical demonstrations. They establish, finally and experimentally, that the action of the imagination can both occasion the crises to cease, and can engender their occurrence.

Source: François Arago, *Biographies of Distinguished Men*, trans. W. H. Smyth *et al.* (London: Longman, Brown, 1857), pp. 83–9.

ON HIS COLOUR-BLINDNESS

The English chemist John Dalton (1766–1844) and his older brother Jonathan were both affected by red-green colour-blindness, although neither was aware of it until the younger brother began to take a close interest in botany. As Dalton recalled in this landmark paper from 1794, it was the task of comparing the colours of flowers that exposed the anomaly in his perception, though whenever he asked for a second opinion on whether a given flower was blue or pink, 'I was generally considered to be in jest.' Dalton began a series of experiments upon his own vision, using a prism to divide white light into the universally agreed spectrum of colours; comparing his own (and his brother's) perceptions with those of others led him to identify a particular defect in his colour vision, which he believed was caused by a discoloration of the aqueous humour of the eye. Although Dalton was mistaken about the cause – red-green colour-blindness is passed on through a faulty gene on an X-chromosome, which is why it is far more prevalent in men than in women – his remarkable act of self-examination successfully revealed the existence of a hitherto unsuspected condition:

I was always of opinion, though I might not often mention it, that several colours were injudiciously named. The term *pink*, in reference to the flower of that name, seemed proper enough; but when the term *red* was substituted for pink, I thought it highly improper; it should have been *blue*, in my apprehension, as pink and blue appear to me very nearly allied; whilst pink and red have scarcely any relation.

In the course of my application to the sciences, that of optics necessarily claimed attention; and I became pretty well acquainted with the theory of light and colours before I was apprized of any peculiarity in my vision. I had not, however, attended much to the practical discrimination of colours, owing, in some degree, to what I conceived to be a perplexity in their nomenclature. Since the year 1790, the occasional study of botany obliged me to attend more to colours than before. With respect to colours that were *white*, *yellow*, or *green*, I readily assented to the appropriate term. *Blue*, *purple*, *pink*, and *crimson* appeared rather less distinguishable; being, according to my idea, all referable to *blue*. I have often seriously asked a person whether a flower was blue or pink, but was generally considered to be in jest. Notwithstanding this, I was never convinced of a peculiarity in my vision, till I accidentally observed the colour of the flower of the *Geranium zonale* by candle-light, in the Autumn of 1792. The flower was pink, but it appeared to me almost an exact sky-blue by day; in candle-light, however, it was astonishingly changed, not having then any blue it, but being what I called red, a colour which forms a striking contrast to blue. Not then doubting but that the change of colour would be equal to all, I requested some of my friends to observe the phænomenon; when I was surprised to find they all agreed, that the colour was not materially different from what it was by day light, except my brother who saw it in the same light as myself. This observation clearly proved, that my vision was not like that of other persons; — and, at the same time, that the difference between day-light and candle-light, on some colours, was indefinitely more perceptible to me than to others.

Two years later, Dalton engaged upon some practical experiments with the assistance of an acquaintance with a good knowledge of optics, making use of a Newtonian glass prism:

My observations began with the solar *spectrum*, or coloured image of the sun, exhibited in a dark room by means of a glass prism. I found that persons in general distinguish six kinds of colour in the solar image; namely, *red, orange, yellow, green, blue,* and *purple*. Newton, indeed, divides the purple into *indigo* and *violet*; but the difference between him and others is merely nominal. To me it is quite otherwise:— I see only *two* or at most *three* distinctions. These I should call *yellow* and *blue*; or *yellow, blue,* and *purple*. My yellow comprehends the *red, orange, yellow,* and *green* of others; and my *blue* and *purple* coincide with theirs. That part of the image which others call red, appears to me little more than a shade, or defect of light; after that the orange, yellow, and green seem *one* colour, which descends pretty uniformly from an intense to a rare yellow, making what I should call different shades of yellow. The difference between the green part and the blue part is very striking to my eye: they seem to be strongly contrasted. That between the blue and the purple is much less so. The purple appears to be blue much darkened and condensed. In viewing the flame of a candle by night through the prism, the appearances are pretty much the same, except that the red extremity of the image appears more vivid than that of the solar image . . .

Reflecting upon those facts, I was led to conjecture that one of the humours of my eye must be a transparent, but *coloured*, medium, so constituted as to absorb *red* and *green* rays principally, because I obtain no proper ideas of these in the solar spectrum; and to transmit blue and other colours more perfectly. What seemed to make against this opinion however was, that I thought red bodies, such as vermilion, should appear black to me, which was contrary to fact. How this difficulty was obviated will be understood from what follows.

Newton has sufficiently ascertained, that opake bodies are of a particular colour from their reflecting the rays of light of that colour more copiously than those of the other colours; the unreflected rays being absorbed by the bodies. Adopting this fact, we are insensibly led to conclude, that the more rays of any one colour a body reflects, and the fewer of every other colour, the more perfect will be the colour. This conclusion, however, is certainly erroneous. Splendid coloured bodies reflect light of every colour copiously; but that of their own most so. Accordingly we find, that bodies of all colours, when placed in homogeneal light of any colour, appear of that particular colour. Hence a body that is red may appear of any other colour to an eye that does not transmit red, according as those other colours are more copiously reflected from the body, or transmitted through the humours of the eye.

It appears therefore almost beyond a doubt, that one of the humours of my eye, and of the eyes of my fellows, is a coloured medium, probably some modification of blue. I suppose it must be the vitreous humour; otherwise I apprehend it might be discovered by inspection, which has not been done. It is the province of physiologists to explain in what manner the humours of the eye may be coloured, and to them I shall leave it.

Source: John Dalton, 'Extraordinary Facts relating to the Vision of Colours', *Memoirs of the Manchester Literary and Philosophical Society* 5:1 (1798), pp. 28–45.

COMET SWEEPER

The astronomer William Herschel (1738–1822) became internationally famous in the wake of his discovery of the planet Uranus in March 1781, but five years later his sister and co-worker Caroline Herschel (1750–1848) won fame in her own right by discovering the first of a number of comets that now bear her name. In 1787 she was awarded £50 a year state funding for her work, thereby becoming Britain's first salaried woman scientist. From then on she and her brother, who originally came to England as itinerant musicians, worked together on a range of astronomical projects, presenting their work to learned societies all over Europe. Although Sir William, as he became in 1816, was elected to the fellowship of the Royal Society while Caroline was not, she was, at the age of eighty-five, elected to the Royal Astronomical Society in a joint ceremony with Mary Somerville (see p. 213).

The following selection from her letters and memoirs narrates some of the key events in her scientific career, while offering insights into her frequent tussles with uncooperative equipment, as well as the sense of common purpose she shared with astronomical friends and colleagues.

The Hazards of Astronomy

According to Caroline Herschel's autobiography, she and her brother spent the last night of 1783 sweeping the sky with their newly constructed twenty-foot telescope, when Caroline tripped over part of the scaffolding, spearing her leg on a protruding metal hook:

That my fears of danger and accidents were not wholly imaginary I had an unlucky sample of on the night of the 31st of Decr. The evening had been cloudy but about 10 o'Clock a few stars became visible, and in the greatest hurry all was got ready for observing. My Brother at the front of the Telescope directing me to make some alteration in the lateral motion which was done by a machinery in which the point of support of the tube and mirror rested; at each end of the machine or trough was an iron hook such as butchers use for hanging their joints upon, and having to run in the dark on ground covered foot deep with melting snow, I fell on one of these hooks which entered my right leg about 6 inches above the knee, my brothers call make haste I could only answer by a pittiful crey I am hooked. He and the workman were instantly with me, but they could not lift me without leaving near 2 oz. of my flesh behind. The workmans Wife was called but was affraid to do anything, and I was obliged to be my own surgeon by applying aquabaseda and tying a kerchief about it for some days; till Dr Lind hearing of my accident brought me ointment and lint and told me how to use it. But at the end of six weeks I began to have fears about my poor Limb and had Dr Lind's opinion, who on seeing the wound found it going on well; but said, if a soldier had met with such a hurt he would have been entituled to 6 weeks nursing in a hospital.

I had the comfort to know that my Brother was no loser through this accident for the remainder of the night was cloudy and several nights afterwards afforded only a few short intervals favourable for sweeping until the 16 of Jany before there was any necessity for exposing myself for a whole night to the severity of the season.

I could give a pretty long list of accidents of which my Brother as well as myself narrowly escaped of proving fatal for observing with such large machineries, where all around is in darkness is not unattended with danger; especially when personal safety is the last thing with which the mind is occupied at such times, even poor Piazi* did not go home without getting broken shins by falling over the rackbarr which projects in high altitudes in front of the telescope when in the hurry the cap had been forgotten to be put over it.

Her First Comet

The following letter, dated 2 August 1786, was sent by Caroline Herschel to Charles Blagden, Secretary of the Royal Society of London, announcing the discovery of the comet now known as Comet C/1786 P1 (Herschel). Her brother was away at the time, as he often was when she made her most important discoveries, though as his footnote to her letter shows, he always checked over her findings upon his return. Comet Herschel remained visible

---

* Giuseppe Piazzi (1746–1826), Director of the Palermo Observatory from 1790. — R.H.

for nearly two months, brightening towards the end of August before fading and then disappearing from view on 26 October:

Sir,

In consequence of the friendship which I know to exist between you and my Brother, I venture to trouble you in his absence with the following imperfect account of a comet.

The employment of writing down the observations, when my Brother uses the 20-feet reflector, does not often allow me time to look at the heavens; but as he is now on a visit to Germany, I have taken the opportunity of his absence to *sweep* in the neighbourhood of the sun, in search of comets; and last night, the 1st of August, about 10 o'clock, I found an object very much resembling in colour and brightness the 27th nebula of the *Connoissance des Temps*, with the difference however of being round. I suspected it to be a comet; but a haziness coming on, it was not possible intirely to satisfy myself as to its motion till this evening. I made several drawings of the stars in the field of view with it, and have inclosed a copy of them, with my observations annexed, that you may compare them together.

August 1, 1786, 9 h. 50', the object in the center is like a star out of focus, while the rest are perfectly distinct, and I suspect it to be a comet. Tab. I. fig. 1.

10 h. 33', fig. 2. the suspected comet makes now a perfect isosceles triangle with the two stars *a* and *b*.

11 h. 8', I think the situation of the comet is now as in fig. 3.; but it is so hazy that I cannot sufficiently see the small star *b* to be assured of the motion.

By the naked eye the comet is between the 54th and 53rd Ursæ majoris, and the 14th, 15th, and 16th Comæ Berenices, and makes an obtuse triangle with them, the vertex of which is turned towards the south.

August 2. 10 h. 9', the comet is now, with respect to the stars *a* and *b**, situated as in fig. 4. therefore the motion since last night is evident.

---

* A doubt having arisen about the identity of the stars marked a and b in the figures, I have examined that part of the heavens in which the comet was the 1st of August, in order to settle this point, but find so many small stars in that neighbourhood that I have not been able to fix on any of them that will exactly answer these figures; and as they

9. William and Caroline Herschel at work with their home-made reflecting telescope in the early 1780s.

10 h. 30', another considerable star c may be taken into the field with it, by placing a in the center; when the comet and the other star will both appear in the circumference, as in fig. 5.

These observations were made with a Newtonian sweeper of 27 inches focal length, and a power of about 20, the field of view is 2° 12'. I cannot find the stars a and c in any catalogue; but suppose they may be easily traced in the heavens; whence the situation of the comet, as it was last night at 10 h. 33', may be pretty nearly ascertained.

You will do me the favour of communicating these observations to my brother's astronomical friends.

I have the honour to be, &c.

Caroline Herschel

Slough, near Windsor,

Aug. 2, 1786.

Her Fourth Comet

Herschel's first comet won her fame and funding, after which she went on to discover seven more over the course of the following decade. Her fourth, Comet C/1790 H1 (Herschel), was spotted on 17 April 1790 – again when her brother was away – and remained observable until the end of June that year. The following letters, to and from the astronomer Alexander Aubert (1730–1805) and Sir Joseph Banks (1743–1820), President of the Royal Society, shed interesting light on the scientific hierarchies of the time; Banks, who by then had been President of the Royal Society for thirty years, expected to be kept personally informed of any worthwhile discovery, hence Herschel's somewhat hesitant tone in her letter to him; Aubert, on the other hand, was an old friend and colleague, and her candour concerning

---

were drawn from observations made by moonlight, twilight, hazy weather, and very near the horizon, it would not be at all surprising if a mistake had been made: however, as these figures were only given with a view to show the motion of the comet, the conclusion of the change of place, which was drawn from them, was equally good whether these stars were the same or different.

Dec. 14, 1786.   William Herschel.

her defective telescope, and his offer to lend her his, speaks of her membership of a supportive, informal network of like-minded observers:

— *Caroline Herschel to Alexander Aubert* —

Slough, *April* 18, 1790

Dear Sir,—

I am almost ashamed to write to you, because I never think of
doing so but when I am in distress. I found last night, at 16^h 24',
sidereal time, a comet, and do not know what to do with it, for my
new sweeper is not half finished; and besides, I broke the handle of
the perpendicular motion in my brother's absence (who is on a little
tour into Yorkshire). He has furnished me to that instrument a
Rumboides, but the wires are too thin, and I have no contrivance for
illuminating them. All my hopes were that I should not find anything
which would make me feel the want of these things in his absence;
but, as it happens, here is an object in a place where there is no nebula, or anything which could look like a comet, and I would be much
obliged to you, sir, if you would look at the place where the annexed
eye-draft will direct you to. My brother has swept that part of the
heavens, and has many nebulæ there, but none which I must expect
to see with my instrument. I will not write to Sir J. Banks or Dr.
Maskelyne, or anybody, till you, sir, have seen it; but if you could,
without much trouble, give my best respects and that part of this
letter which points out the place of the comet to Mr. Wollaston, you
would make me very happy.

I am, dear sir, &c., &c.,

C. H.

— *Caroline Herschel to Sir Joseph Banks* —

April 19*th*, 1790

Sir,—

I am very unwilling to trouble you with incomplete observations,
and for that reason did not acquaint you yesterday with the discovery of a comet. I wrote an account of it to Dr. Maskelyne and
Mr. Aubert, in hopes that either of those gentlemen, or my brother,

whom I expect every day to return, would have furnished me with the means of pointing it out in a proper manner.

But as perhaps several days might pass before I could have any answer to my letters, or my brother return, I would not wish to be thought neglectful, and therefore if you think, sir, the following description is sufficient, and that more of my brother's astronomical friends should be made acquainted with it, I should be very happy if you would be so kind as to do it for the sake of astronomy.

The comet is a little more than 3 1/2° following α Andromedæ, and about 1 1/2° above the parallel of that star. I saw it first on April 17th, 16^h 24' sidereal time, and the first view I could have of it last night was 16^h 5'. As far as I am able to judge, it has decreased in P. D. nearly 1°, and increased in A. R. something above 1'.

These are only estimations from the field of view, and I only mention it to show that its motion is not so very rapid.

I am, &c.,

C. H.

— *Alexander Aubert to Caroline Herschel* —

London, the 21*st April*, 1790

Dear Miss Herschel, —

I am much obliged to you for your kind letter. The night before last was cloudy. Last night, or rather this morning, about half-past two, I got up to look for the phenomenon; it was somewhat hazy. I observed with a common night-glass of Dollond's *a faint something* in a line between α and π Andromedæ, much like a faint star; it had no coma nor fuzzy appearance. By looking at Flamsteed's Atlas I find no small star there. I was preparing to attack it with a good magnifying power, and to get its place with my Smeaton's equatorial micrometer, but when I was ready a haze came on, and soon after too much daylight, so I can say no more to it as yet. If I saw what you judged a comet, it must have moved but little since you saw it; it was as large as a star of 7th magnitude, but rather faint. I sent this morning to Dr. Maskelyne: he says he could see nothing *with a good night glass*, but will try again the next fair morning, and after trying he will answer you; in the meanwhile he begs his best compliments. I will also try again. Pray let me know if you think it was the comet I saw. I have mentioned it to no one but to Mr. Wollaston, who thanks you sincerely, but did not find himself well enough to observe . . .

You cannot, my dear Miss Herschel, judge of the pleasure I feel when your reputation and fame increase; everyone must admire you and your brother's knowledge, industry, and behaviour. God grant you many years health and happiness. I will soon pay you a visit, as soon as your brother returns. If I have any instrument you wish to use, it is at your service.

Believe me, &c., &c.,
Alexander Aubert

— *Sir Joseph Banks to Caroline Herschel* —

Soho Square, *April* 20, 1790

Madam,—

I return you many thanks for the communication you were so good as to make me this day of your discovery of a comet. I shall take care to make our astronomical friends acquainted with the obligations they are under to your diligence.

I am always happy to hear from you, but never more so than when you give me an opportunity of expressing my obligations to you for advancing the science you cultivate with so much success.

Dear Madam,
Your faithful servant,
J. Banks.

Sources: *Caroline Herschel's Autobiographies*, ed. Michael Hoskin (Cambridge: Science History Publications, 2003), pp. 76–7; Caroline Herschel, *An Account of a New Comet*, read at the Royal Society, Nov. 9, 1786 (London: J. Nichols, 1787), pp. 1–3; *Memoir and Correspondence of Caroline Herschel* (London: John Murray, 1876), pp. 85–8.

BOTANICAL POEMS

By the late eighteenth century, botany had evolved into a suitable science for a genteel woman to pursue. In spite of Linnaeus's sexual classification scheme – the language of which added a dangerous frisson to the job of cataloguing garden plants – women had staked a successful claim on botanical culture. The poet Sarah Hoare (1777–1856) first turned to the subject in 1818 in what is still her best-known work, 'The Pleasures of Botanical Pursuits', which was appended to later editions of Priscilla Wakefield's *An Introduction to Botany* (1796). Her only solo volume was the *Poems on Conchology and Botany* (1831), from which the following stanzas have been chosen. The detailed notes that accompanied each verse were compiled by Hoare from a variety of sources, including William Withering's *Botanical Arrangement* (1776), Samuel Rootsey's *Botanical Lectures* (1818), and the near-ubiquitous *Encyclopædia Britannica*:

I.

Science, illuminating ray!
Fair mental beam, extend thy sway,
   And shine from pole to pole!
From thy accumulated store,
O'er every mind thy riches pour,
Excite from low desires to soar,
   And dignify the soul.

II.

Science! thy charms will ne'er deceive,
But still increasing pleasure give,
   And varied joys combine;
Nor ever leave on memory's page
A pang repentance would assuage;
But purest, happiest thoughts engage
   To sweeten life's decline.

III.

To thee, oh Botany! I owe
Of pure delight, the ardent glow,
        Since childhood's playful day;
E'en then I sought the sweet perfume,
Exhal'd along the banks of Froome;
Admir'd the rose's opening bloom,
        And nature's rich array.

VI.

But not alone for pleasure's sake,
We search the thicket, copse, and brake,
        Or rove from clime to clime;
Nor yet for the abundant store
Of plants, that fragrant balsams pour,
Whether they deck the valley's o'er,
        Or mountain's brow sublime.

VII.

'Tis that with scientific eye
We explore the vast variety,
        To find the hidden charm:
'Tis to allay the fever's rage,
The pang arthritic to assuage,
To aid the visual nerve of age,
        And fell disease disarm.

VIII.

Linné, by thy experience taught,
And ample page so richly fraught
        With scientific lore:
I scann'd thy curious systems clear,
Of plants that court the mountain air,
That bloom o'er hills, o'er meadow fair,
        The forest and the moor.

XII.

Verbena,[1] once of high renown,
Of charms medicinal the crown,
    So ancient tales attest;
And *Salvia*[2] too, in days of yore,
As efficacious to restore
Lost health, and heal the rankling sore,
    All potent was confest.

XXXII.

Papaver![3] thou "pale misery's friend"!
The soothing lymph thy fibres send
    Through devious veins to creep,
With care we seek, for sorrow knows
Thy power to tranquillize her woes;
To give the wearied soft repose,
    And sweetly lull to sleep.

[1] Verbena Officinalis, or Vervain. Class Diandr. Monog. In former times the Verbena seems to have been held sacred, and was employed in celebrating the sacrificial rites. It was worn suspended about the neck as an amulet: this practice, thus founded in superstition, was, however, in process of time, adopted in medicine, and therefore, in order to obtain its virtues more effectually, the Vervain was directed to be bruised before it was suspended to the neck. Mr. Morley, who has written an Essay on Scrofula, directs that the Vervain be tied with a yard of satin ribbon round the neck, where it is to remain till the patient shall have recovered! !

[2] Class Diandr. Monog. Salvia Officinalis, or Sage. It appears that the Salvia Officinalis was, anciently, as highly esteemed as the Vervain, hence the old saying,

    "Cur moriatur homo, cui Salvia crescit in horto."
    "Why dies the man, whose garden, sage affords."

[3] Papaver Rhœas. Corn, or Red Poppy; gives out a fine color when infused; and a syrup prepared from the infusion is kept in the shops, under the name of Diacodion; it partakes in a small degree of the property of opium. *Withering's Bot.*

XXXIII.

And *Digitalis*[4] wisely given,
Another boon of favoring Heaven
   Will happily display;
The rapid pulse it can abate,
The hectic flush can moderate,
And blest by him, whose will is fate,
   May give a happier day.

XXXV.

Curiously form'd thy spiral seed,
Eurodium![5] — and the wise who read,—
   Artificer Divine!
Thy skill, amidst these sweets immense,
These gems of nature's elegance,
Admire the vast intelligence,
   Th' unfathomable mine!

---

[4] Class Digyn. Angiosp. Digitalis Purpurea, or Common Foxglove. In this class is found Nepeta Cataria, or Cat Mint. Cats are so delighted with this plant, that they can hardly be out of the garden where it grows. Millar says, that cats will not meddle with it if raised from seeds; and in support of this opinion, quotes an old saying,

    "If you set it, cats will eat it;
      If you sow it, cats won't know it,"

Mentha Arvensis, or Corn Mint, is another plant belonging to this class. It prevents the coagulation of milk; and when cows have eaten of it, as they will do largely at the end of the summer when the pastures are bare, and hunger distresses them, their milk can be hardly made to yield cheese, a circumstance that sometimes puzzles the dairy maid. *Withering's Bot.*

[5] Class Monadelphia Pentandr. Erodium Cicutarium. In the Class Monadelphia, there is not a more interesting, or more beautiful plant, than the Erodium Cicutarium. The seeds, which have been noticed by Dr. Arnold, to twist and untwist when wetted and dried, are considered by Samuel Rootsey, as the most curious and accurate of all natural hygrometers.

XXXVI.

Lathyrus![6] from Sicilian plain,
Here in thy radiant purple reign,
    Secure from Etna's storm;
Shrink not from our inclement sky,
No fiery cloud will close thine eye;
No burning lava passes by,
    Thy beauty to deform.

XLVI.

Thy fine furcated leaves to find,
Acrosticum![7] I'd leave behind
    The fields of richest flowers;
E'en climb the rocks, — wind Arthur's seat,
To which Duneden's sons retreat
And converse hold, refin'd and sweet,
    To improve the leisure hours.

XLVII.

You, who the curious search pursue,
Proclaim, does not a closer view,
    The patient toil repay,
And prove the gracious end design'd, —
The chief, tho' latent, good to find,
And for the love of human kind,
    The wond'rous work display.

[6] Class Diadelph. Decandr. Lathyrus Odoratus, or Sweet Pea; said to be indigenous to Sicily.

[7] Class cryptogramia. Acrosticum Septentrionale, or Forked Maidenhair. It grows out of the rocks, about Arthur's seat, near Edinburgh.

XLVIII.

For not alone to please the eye,
Or deck our fields, this rich supply
    Of ornaments profuse;
Medicinal their juices flow,
Nor idly do their colors glow,
For He, who dress'd the beauteous show
    Assign'd to each its use.

XLIX.

And not to casual glance display'd
Alone; — by microscopic aid,
    We view a wond'rous store;
The cups nectareous now appear,
The fringe, the down, the gland'lar hair,
The germ enclos'd with curious care,
    And petals spangled o'er.

Source: Sarah Hoare, *Poems on Conchology and Botany, with plates and notes* (London: Simpkin & Marshall, 1831), pp. 61–85.

PHOSPHORUS

Jane Marcet (1769–1858) was a professional science writer who produced a series of popular textbooks marketed at women, 'whose education is seldom calculated to prepare their minds for abstract ideas, or scientific language', as she observed. Having attended Humphry Davy's incendiary chemistry lectures at the Royal Institution, she hit upon the idea of replicating, in dialogue form, the contents of an entire course of chemical demonstrations. Thus was her fictitious instructor, 'Mrs B.', born, and what a wonderful addition she was to the canon of literary educators. Patient, hands-on, and remarkably up-to-date with the latest chemical theories (Marcet took care to update the scientific content of every new edition of her *Conversations on Chemistry*), she was responsible for introducing chemical ideas to generations of women and men throughout the English-speaking world. The book had a particular influence on the young Michael Faraday (1799–1867) when a copy was delivered to the bookbinder's shop where he worked as an apprentice. Reading it, he said, opened his eyes to the promise of a life of laboratory research, and in later years Faraday often referred to Jane Marcet as his first science teacher:

MRS. B.

Phosphorus is a simple substance that was formerly unknown. It was first discovered by Brandt, a chemist of Hamburgh, whilst employed in researches after the philosopher's stone; but the method of obtaining it remained a secret till it was a second time discovered both by Kunckel and Boyle, in the year 1680. You see a specimen of phosphorus in this phial; it is generally moulded into small sticks of a yellowish colour, as you find it here . . . It is found in all animal substances, and is now chiefly extracted from bones, by a chemical process. It exists also in some plants, that bear a strong analogy to animal matter in their chemical composition.

EMILY.

But it is never found in its simple state?

MRS. B.

Never, and this is the reason of its having remained so long undiscovered.

EMILY.

It is possible, then, that in course of time we may discover other new simple bodies?

MRS. B.

Undoubtedly; and we may also learn that some of those, which we now class among the simple bodies, may, in fact, be compounded; indeed, you will soon find that discoveries of this kind are by no means infrequent.

Phosphorus is eminently combustible; it melts and takes fire at the temperature of 100°, and absorbs in its combustion nearly once and a half its own weight of oxygen.

CAROLINE.

What! will a pound of phosphorus consume a pound and a half of oxygen?

MRS. B.

So it appears from accurate experiments. I can show you with what violence it combines with oxygen, by burning some of it in that gas. We must manage the experiment in the same manner as we did the combustion of sulphur. – You see I am obliged to cut this little bit of phosphorus under water, otherwise there would be danger of its taking fire by the heat of my fingers. – I now put it into the receiver, and kindle it by means of a hot wire.

EMILY.

What a blaze! I can hardly look at it. I never saw any thing so brilliant. Does it not hurt your eyes, Caroline?

CAROLINE.

Yes; but still I cannot help looking at it. A prodigious quantity of oxygen must indeed be absorbed, when so much light and caloric [heat] are disengaged!

MRS. B.

In the combustion of a pound of phosphorus, a sufficient quantity of caloric is set free to melt upwards of a hundred pounds of ice; this has been computed by direct experiments with the calorimeter.

EMILY.

And is the result of this combustion, like that of sulphur, an acid?

MRS. B.

Yes; phosphoric acid. And had we duly proportioned the phosphorus and the oxygen, they would have been completely converted into phosphoric acid, weighing together, in this new state, exactly the sum of their weights separately. The water would have ascended into the receiver, on account of the vacuum formed, and would have filled it entirely . . .

CAROLINE.

Is it not very singular that phosphorus should burn at so low a temperature in atmospherical air, whilst it does not burn in pure oxygen without the application of heat?

MRS. B.

So it at first appears. But this circumstance seems to be owing to the nitrogen gas of the atmosphere. This gas dissolves small particles of phosphorus, which being thus minutely divided and diffused in the atmospherical air, combines with the oxygen, and undergoes this slow combustion. But the same effect does not take place in oxygen gas, because it is not capable of dissolving phosphorus; it is therefore necessary, in this case, that heat should be applied to effect that division of particles, which, in the former instance, is produced by the nitrogen.

EMILY.

I have seen letters written with phosphorus, which are invisible by day-light, but may be read in the dark by their own light. They look as if they were written with fire; yet they do not seem to burn.

MRS. B.

But they do really burn; for it is by their slow combustion that the light is emitted; and phosphoric acid is the result of this combustion.

Phosphorus is sometimes used as a test to estimate the purity of atmospherical air. For this purpose, it is burnt in a graduated tube called an *eudiometer*, and from the quantity of air which the phosphorus absorbs, the proportion of oxygen in the air examined, is deduced; for the phosphorus will absorb all the oxygen, and the nitrogen alone will remain.

EMILY.

And the more oxygen is contained in the atmosphere, the purer, I suppose, it is esteemed?

MRS. B.

Certainly. Phosphorus, when melted, combines with a great variety of substances. With sulphur it forms a compound so extremely combustible, that it immediately takes fire on coming in contact with the air. It is with this composition that the phosphoric matches are prepared, which kindle as soon as they are taken out of their case and are exposed to the air.

EMILY.

I have a box of these curious matches; but I have observed, that in very cold weather, they will not take fire without being previously rubbed.

MRS. B.

By rubbing them you raise their temperature; for, you know, friction is one of the means of extricating heat.

EMILY.

Will phosphorus combine with hydrogen gas, as sulphur does?

MRS. B.

Yes; and the compound gas which results from this combination has a smell still more fetid than the sulphurated hydrogen; it resembles that of garlic.

The *phosphorated hydrogen gas* has this remarkable peculiarity, that it takes fire spontaneously in the atmosphere, at any temperature. It is thus that are produced those transient flames, or flashes of light, called by the vulgar *Will-of-the-Wisp*, or more properly *Ignes-fatui*, which are often seen in church-yards, and places where the putrefaction of animal matter exhales phosphorus and hydrogen gas.

CAROLINE.

Country people, who are so much frightened by those appearances, would soon be reconciled to them, if they knew from what a simple cause they proceed!

Source: Jane Marcet, *Conversations on Chemistry; in which the Elements of that Science are Familiarly Explained and Illustrated by Experiments*, 2 vols (London: Longman, Hurst, Rees & Orme, 1806), I, pp. 191–9.

THE MINER'S SAFETY LAMP

Sir Humphry Davy (1778–1829) was a Cornish-born chemist who rose to fame during the first great era of public science in Britain. Celebrated as a lecturer and a leading scientific personality, his many achievements included the isolation of an array of elements including sodium and barium, as well as the invention for which he remains best known, the miner's safety lamp (still widely referred to as the Davy lamp). The lamp, which was first trialled in January 1816, was intended to allow miners to work safely in the presence of methane and other flammable gases, which are referred to as 'fire-damp' in the following essay. A mesh screen surrounding the wick allowed air to flow in, keeping the flame inside the lamp burning, while preventing the combustion of any ambient methane present in the mine. Davy's refusal to patent the safety lamp ensured that it would be widely used in mines throughout the world:

Since the earliest period of the application of mineral coal to the purposes of fuel, the explosions in coal-mines from inflammable air, or fire-damp, have been regarded as the greatest evil occurring in the working of the mines. The strata of coal lie usually parallel, or nearly parallel to the surface, at certain depths beneath it; and this coal, when the pressure of the superincumbent material is removed, affords inflammable air; which is disengaged, not only in the common operations of mining, when the coal is broken and removed, but is likewise permanently evolved, often in enormous quantities, from fissures in the strata.

When it has accumulated in any part of the gallery or chamber of a mine, so as to be mixed in certain proportions with common air, the presence of a lighted candle or lamp causes it to explode, and to destroy, injure, or burn whatever is exposed to its violence.

To give detailed accounts of the tremendous accidents, owing to this cause, would be merely to multiply pictures of death, and of human misery. The phenomena are always of the same kind. The miners are either immediately destroyed by the explosion, and thrown with the horses and machinery through the shaft into the air, the mine becoming as it were an enormous piece of artillery, from which they are projected; or they are gradually suffocated, and undergo a more painful death, from the carbonic acid and azote remaining in the mine, after the inflammation of the fire-damp; or what, though it appears the mildest, is perhaps the most severe fate, they are burnt or maimed, and often rendered incapable of labour, and of healthy enjoyment for life.

The fire-damp is found in the greatest quantity, and is the most dangerous in the deepest mines; but it likewise often occurs in superficial excavations; and I have now a letter, of the date of June 8, 1816, in my possession, in which it is stated that in the very commencement of working a coal-mine in Shropshire, several miners were killed, and others severely burnt . . .

The number of dreadful accidents, indeed, which had happened within the last three or four years in the last mentioned districts, particularly that by which ninety-six persons were destroyed in the Felling colliery, had so strongly impressed the minds of a number of benevolent persons belonging to or connected with the coal districts, that it was said to be in their contemplation to bring the subject before Parliament, that by making it a national question, it might obtain that consideration which its importance demanded.

After describing a number of earlier attempts to devise non-flammable lamps, including one designed by Alexander von Humboldt in 1796, Davy described the course of his own first efforts 'to find a light, which, at the same time that it enabled the miner to work with security in explosive atmospheres, should likewise consume the fire-damp':

I first began with a minute chemical examination of the substance with which I had to contend. The analysis of various specimens of fire-damp shewed me that the pure inflammable part of it was light carburetted hydrogen, as Dr. Henry had before stated, hydrogen or pure inflammable air combined with charcoal or carbon.

I made numerous experiments on the circumstances under which it explodes, and the degree of its inflammability. I found that it required to be mixed with very large quantities of atmospheric air to produce explosion; even when mixed with three or nearly four times its bulk of air, it burnt quietly in the atmosphere, and extinguished a taper. When mixed with between five and six times its volume of air it exploded feebly: it exploded with most energy when mixed with seven or eight times its volume of air, and mixtures of fire-damp and air retained their explosive power when the proportions were one of gas to fourteen of air. When the air was in larger quantity the flame of a taper was merely enlarged in the mixture, an effect which was still perceived in thirty parts of air to one of gas.

I found the fire-damp much less combustible than other inflammable gases. It was not exploded or fired by red-hot charcoal or red-hot iron; it required iron to be white-hot, and itself in brilliant combustion, for its inflammation. The heat produced by it in combustion was likewise much less than that of most other inflammable gases, and hence, in its explosion, there was much less comparative expansion . . .

In reasoning upon these various phenomena it occurred to me, as a *considerable* heat was required for the inflammation of the fire-damp, and as it produced in burning comparatively a small degree of heat, that the effect of carbonic acid and azote, and of the surfaces of small tubes in preventing its explosion, depended upon their cooling powers; upon their lowering the temperature of the exploding mixture so much that it was no longer sufficient for its continuous inflammation.

This idea, which was confirmed by various obvious considerations, led to an immediate result – the possibility of constructing a lamp, in which the cooling powers of the azote or carbonic acid, formed by combustion or the cooling powers of the apertures, through which the air entered or made its exit, should prevent the communication of explosion.

I first tried the effects of lamps in which there was a very limited circulation of air; and I found that when a taper in a close lantern was supplied with air so as to burn feebly from very small apertures below the flame, and at a considerable distance from it, it became extinguished in explosive mixtures; but I ascertained that precautions which it would be dangerous to trust to workmen were required to make this form of a lamp safe, and that at best it could give only a feeble light; and I immediately adopted systems of tubes above and below, of that diameter in which I had ascertained that explosions would not take place.

In trying my first tube lamp in an explosive mixture I found that it was safe; but unless the tubes were very short and numerous, the flame could not be well supported; and in trying tubes of the diameter of one-seventh or one-eighth of an inch I determined that they were safe only to small quantities of explosive mixture, and when of a given length; and that tubes even of a much smaller diameter communicated explosion from a close vessel. Hence I took a new method of ascertaining the safety of my apertures, and of trying different forms of apertures.

I had a vessel furnished with wires by which the electrical spark could be taken in an explosive mixture, and which was larger in capacity than a safe lamp or lantern was required to be. I placed my flame sieves, i.e. my systems of apertures, between this jar and a bladder containing likewise an explosive mixture, and I judged the aperture to be safe only when they stopped explosion acting upon them in this concentrated way.

In this mode of experimenting I soon discovered that a *few apertures* even of very small diameter were not safe unless their sides were very deep; that a single tube of one-twenty-eighth of an inch in diameter and two inches long suffered the explosion to pass through it; and that a *great number* of small tubes or of apertures, stopped explosion even when the depths of their sides was only equal to their diameters; and at last I arrived at the conclusion that a *metallic tissue*, however thin and fine, of which the apertures filled more space than the cooling surface, so as to be permeable to air and light, offered a perfect barrier to explosion, from the force being divided between, and the heat communicated to an immense number of surfaces.

My first safety lamps, constructed on these principles, gave light in explosive mixtures containing a great excess of air, but became extinguished in explosive mixtures in which the fire-damp was in sufficient quantity to absorb the whole of the oxygen of the air, so that such mixtures never burnt continuously at the air feeders, which in lamps of this construction was important, as the increase of heat, where there was only a small cooling surface, would have altered the conditions of security.

I made several attempts to construct safety lamps which should give light in all explosive mixtures of fire-damp, and after complicated combinations I at length arrived at one evidently the most simple, that of surrounding the light entirely by wire gauze, and making the same tissue feed the flame with air and emit light.

In plunging a light surrounded by a cylinder of fine wire-gauze into an explosive mixture I saw the whole cylinder become quietly

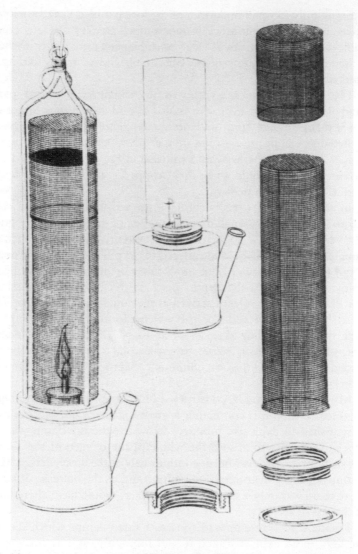

10. After a mine explosion in 1815 that claimed more than fifty lives, Humphry Davy was asked for his help in devising a miner's safety lamp. Davy's design consisted of a cylinder of wire gauze attached to an iron frame containing a wick and an oil reservoir. The gauze served to cool the temperature of the flame, lowering the risk of explosions when mining gaseous seams.

and gradually filled with flame; the upper part of it soon appeared red hot; yet *no* explosion was produced.

It was easy at once to see that by increasing the cooling surface in the top, or in any other part of the lamp, the heat acquired by it might be diminished to any extent; and I immediately made a number of experiments to perfect this *invention*, which was evidently the one to be adopted, as it excluded the necessity of using glass or any fusible or brittle substance in the lamp, and not only deprived the fire-damp of its explosive powers, but rendered it an useful light.

Source: Humphry Davy, *On the Safety Lamp for Coal Miners; with some Researches on Flame* (London: R. Hunter, 1818), pp. 1–15.

BROWNIAN MOTION

Brownian motion, the constant random collisions between particles suspended in a fluid medium, crops up elsewhere in this collection, in association with early atomism (see p. 20), and with the actions of magnetotactic bacteria (on p. 391). It was named after the Scottish botanist Robert Brown, who first described it in a paper published in 1828. Later generations of physicists returned to the phenomenon, and in 1905 Albert Einstein published a theoretical paper on the mathematics of Brownian motion as part of his ongoing attempt to prove that atoms and molecules were not merely hypothetical entities. A series of experiments conducted by the French physicist Jean Perrin (1870–1942) confirmed that Einstein was right, and that Brownian motion is in fact a fundamental property of matter.

Robert Brown

Robert Brown (1773–1858) served as a botanist on Matthew Flinders's scientific expedition to Australia from 1801 to 1803; he returned to Britain with nearly four thousand plant specimens, many of which were new to science, and he went on to become President of the Linnean Society. Yet Brown is best known today for his contribution to physics, in the form of Brownian motion, which he discovered in the summer of 1827 during microscopic observations of pollen grains suspended in water. He noticed that tiny particles within the grains seemed to move around independently, even though the surrounding water was still; later observations of dust grains in water discovered the same mysterious behaviour:

This plant was *Clarckia pulchella*, of which the grains of pollen, taken from antherae full grown, but before bursting, were filled with particles or granules of unusually large size, varying from nearly 1/4000th to about 1/5000th of an inch in length, and of a figure between cylindrical and oblong, perhaps slightly flattened, and having rounded and equal extremities. While examining the form of these particles immersed in water, I observed many of them very evidently in motion; their motion consisting not only of a change of place in the fluid, manifested by alterations in their relative positions, but also not unfrequently of a change of form in the particle itself; a contraction or curvature taking place repeatedly about the middle of one side, accompanied by a corresponding swelling or convexity on the opposite side of the particle. In a few instances the particle was seen to turn on its longer axis. These motions were such as to satisfy me, after frequently repeated observation, that they arose neither from currents in the fluid, nor from its gradual evaporation, but belonged to the particle itself.

Grains of pollen of the same plant taken from antherae immediately after bursting, contained similar subcylindrical particles, in reduced numbers, however, and mixed with other particles, at least as numerous, of much smaller size, apparently spherical, and in rapid oscillatory motion.

These smaller particles, or Molecules as I shall term them, when first seen, I considered to be some of the cylindrical particles swimming vertically in the fluid. But frequent and careful examination lessened my confidence in this supposition; and on continuing to observe them until the water had entirely evaporated, both the cylindrical particles and spherical molecules were found on the stage of the microscope . . .

Having found motion in the particles of the pollen of all the living plants which I had examined, I was led next to inquire whether this property continued after the death of the plant, and for what length of time it was retained.

In plants, either dried or immersed in spirit for a few days only, the particles of pollen of both kinds were found in motion equally evident with that observed in the living plant; specimens of several plants, some of which had been dried and preserved in an herbarium for upwards of twenty years, and others not less than a century, still exhibited the molecules or smaller spherical particles in considerable numbers, and in evident motion, along with a few of the larger particles, whose motions were much less manifest, and in some cases not observable.

Jean Perrin

The classic explanation of Brownian motion was given in 1909 by the Nobel Prize-winning physicist Jean Perrin (1870–1942) whose attention had been drawn to the subject by the paper published by Einstein in 1905. Perrin demonstrated that the movements observed by Brown and others were caused by collisions among the rapidly moving water molecules, which upheld the controversial view that molecules were real physical entities, hence the pointed title of his influential paper, *Brownian Movement and Molecular Reality* (1910), in which he observed that particles suspended in a liquid medium do not assume regular movements according to their density, but on the contrary, appear animated by irregular motions. 'They go and come, stop, start again, *mount*, descend, *remount again*, without in the least tending toward immobility. This is the *Brownian movement*, so named in memory of the naturalist Robert Brown,' he wrote. But what exactly was this movement? Was it really analogous to the quiet movement of dust particles in currents of air, as described by Lucretius, or was there something else going on, something to do with activity at the molecular level?

Perrin's experiments led him to the following conclusions: that Brownian motion has nothing to do with vibrations travelling through the liquid; neither is it to do with convection currents, since the same motions are observed even in mediums where thermal equilibrium has been achieved. On the contrary, 'the Brownian movement is permanently at constant temperature: that is an experimental fact. The motion of the molecules which it leads us to imagine is thus itself also permanent. If these molecules come into collision like billiard balls, it is necessary to add that they are perfectly elastic, and this expression can, indeed, be used to indicate that in the molecular collisions of a thermally isolated system the sum of the energies of motion remains definitely constant. In brief the examination of Brownian movement alone suffices to suggest that every fluid is formed of elastic molecules, animated by a perpetual motion.'

Sources: 'A Brief Account of Microscopical Observations made in the Months of June, July, and August, 1827, on the Particles Contained in the Pollens of Plants; and on the General Existence of Active Molecules in Organic and Inorganic Bodies', *Philosophical Magazine* 4 (1828), pp. 161–73; Jean Perrin, *Brownian Movement and Molecular Reality* (London: Taylor and Francis, 1910), p. 9.

THE VELOCITY OF METEORITES

Like Jane Marcet before her (see p. 199), Mary Somerville (1780–1872) was a leading popular-science writer, whose ability to communicate complex ideas brought her fame and a respectable income at a time when most women faced insurmountable barriers to any kind of scientific career. Largely self-taught from her family library, in 1812 she married her cousin, William Somerville, who greatly encouraged her scientific researches. Her first paper, on sunlight, was accepted by the Royal Society in 1826, but her husband had to read it out on her behalf as women were still barred from presenting their work in person. Somerville's insightful translation of Pierre-Simon Laplace's *Mécanique Céleste* (*The Mechanism of the Heavens*, 1831) made her reputation within scientific circles – and was a university set text for many years – while her next book, *On the Connexion of the Physical Sciences* (1834), from which the following extract comes, was a critical and commercial success, selling more than fifteen thousand copies in her lifetime. Somerville was elected to the Royal Irish Academy in 1834, and to the Royal Astronomical Society in 1835, though never to the Royal Society, which compromised by placing a marble bust of her in the lobby. Today, both an Oxford college and (appropriately, in the context of the following pages) a lunar crater are named after her.

Somerville wrote that her intention was 'to make the laws by which the material world is governed, more familiar to my countrywomen', although as William Whewell noted in his review of the book (see the following entry), there were plenty of men who were also in need of a basic scientific education:

Astronomy affords the most extensive example of the connection of the physical sciences. In it are combined the sciences of number and quantity, of rest and motion. In it we perceive the operation of a force which is mixed up with everything that exists in the heavens or on earth; which pervades every atom, rules the motions of animate and inanimate beings, and is as sensible in the descent of a rain-drop as in the falls of Niagara; in the weight of the air, as in the periods of the moon. Gravitation not only binds satellites to their planet, and planets to the sun, but it connects sun with sun throughout the wide extent of creation, and is the cause of the disturbances, as well as of the order of nature; since every tremor it excites in any one planet is immediately transmitted to the farthest limits of the system, in oscillations which correspond in their periods with the cause producing them, like sympathetic notes in music, or vibrations from the deep tones of an organ.

The heavens afford the most sublime subject of study which can be derived from science. The magnitude and splendour of the objects, the inconceivable rapidity with which they move, and the enormous distances between them, impress the mind with some notion of the energy that maintains them in their motions, with a durability to which we can see no limit. Equally conspicuous is the goodness of the great First Cause, in having endowed man with faculties, by which he can not only appreciate the magnificence of His works, but trace, with precision, the operation of His laws, use the globe he inhabits as a base wherewith to measure the magnitude and distance of the sun and planets, and make the diameter of the earth's orbit the first step of a scale by which he may ascend to the starry firmament. Such pursuits, while they ennoble the mind, at the same time inculcate humility, by showing that there is a barrier which no energy, mental or physical, can ever enable us to pass: that, however profoundly we may penetrate the depths of space, there still remain innumerable systems, compared with which, those apparently so vast must dwindle into insignificance, or even become invisible; and that not only man, but the globe he inhabits — nay, the whole system of which it forms so small a part — might be annihilated, and its extinction be unperceived in the immensity of creation . . .

So numerous are the objects which meet our view in the heavens, that we cannot imagine a part of space where some light would not strike the eye;— innumerable stars, thousands of double and multiple systems, clusters in one blaze with their tens of thousands of stars, and the nebulæ amazing us by the strangeness of their forms and the incomprehensibility of their nature, till at last, from the imperfection of our senses, even these thin and airy phantoms vanish in the distance. If such remote bodies shine by reflected light, we should be unconscious of their existence; each star must then be a sun, and may be presumed to have its system of planets, satellites, and comets, like our own; and for aught we know, myriads of bodies may be wandering in space unseen by us, of whose nature we can form no idea, and still less of the part they perform in the economy of the universe; nor is this an unwarranted presumption; many such do come within the sphere of the earth's attraction, are ignited by the velocity with which they pass through the atmosphere, and are precipitated with great violence on the earth. The fall of meteoric stones is much more frequent than is generally believed; hardly a year passes without some instances occurring, and if it be considered that only a small part of the earth is inhabited, it may be presumed that numbers fall in the ocean or on the uninhabited part of the land, unseen by man. They are sometimes of great magnitude; the volume of several has exceeded that of the planet Ceres,* which is about 70 miles in diameter. One which passed within 25 miles of us was estimated to weigh about 600000 tons, and to move with a velocity of about 20 miles in a second, — a fragment of it alone reached the earth. The obliquity of the descent of meteorites, the peculiar substances they are composed of, and the explosion accompanying their fall, show that they are foreign to our system.

Luminous spots, altogether independent of the phases, have occasionally appeared on the dark part of the moon; these have been ascribed to the light arising from the eruption of volcanos; whence it has been supposed that meteorites have been projected from the moon by the impetus of volcanic eruption. It has even been computed that, if a stone were projected from the moon in a vertical line, with an initial velocity of 10992 feet in a second, — more than four times the velocity of a ball when first discharged from a cannon, — instead of falling back to the moon by the attraction of gravity, it would come within the sphere of the earth's attraction, and revolve about it like a satellite. These bodies, impelled either by the direction of the primitive impulse, or by the disturbing action of the sun, might ultimately penetrate the earth's atmosphere, and arrive at its surface. But from whatever source meteoric stones may come, it seems highly probable that they have a common origin, from the uniformity — we may almost say identity — of their chemical composition.

Source: Mary Somerville, *On the Connexion of the Physical Sciences* (London: John Murray, 1834), pp. 1–2; 361–2; 404–6.

* A dwarf planet located in the asteroid belt between Mars and Jupiter, discovered in 1801. — R.H.

FIRST USE OF THE WORD 'SCIENTIST'

There is a secondary point of interest to Mary Somerville's *On the Connexion of the Physical Sciences* (see previous entry), which is that the polymath William Whewell (1794–1866), in a long and favourable review of the work in the *Quarterly Review* (1834), first employed the word 'scientist' in print. The new word, he explained, had been proposed by 'some ingenious gentleman' (naturally, it was Whewell himself) at the 1833 meeting of the British Association for the Advancement of Science, but, to his dismay, the term was unanimously rejected by the membership. Two decades later, Whewell became President of the British Association, and his campaign on behalf of his pet word 'scientist' was renewed – this time with more success.

The need for such a term illustrated the growing professionalization of science in the period, as well as the recognition that ever-increasing specialization was 'dividing the soil of science into infinitely small allotments'; a single unifying term, hoped Whewell, might prevent the various disciplines from drifting further apart:

Mrs. Somerville's work is, and is obviously intended to be, a popular view of the present state of science, of the kind we have thus attempted to describe. In her simple and brief dedication to the Queen, she says, 'If I have succeeded in my endeavour to make the laws by which the material world is governed, more familiar to my countrywomen, I shall have the gratification of thinking, that the gracious permission to dedicate my book to your Majesty has not been misplaced.' And if her 'countrywomen' have already become tolerably familiar with the technical terms which the history of the progress of human speculations necessarily contains; if they have learned, as we trust a large portion of them have, to look with dry eyes upon oxygen and hydrogen, to hear with tranquil minds of perturbations and eccentricities, to think with toleration that the light of their eyes may be sometimes polarized, and the crimson of their cheeks capable of being resolved into *complementary colours*; – if they have advanced so far in philosophy, they will certainly receive with gratitude Mrs. Somerville's able and *masterly* (if she will excuse this word) exposition of the present state of the leading branches of the physical sciences. For our own parts, however, we beg leave to enter a protest, in the name of that sex to which all critics (so far as we have ever heard) belong, against the appropriation of this volume to the sole use of the author's country*women*. We believe that there are few individuals of that gender which plumes itself upon the exclusive possession of exact science, who may not learn much that is both novel and curious in the recent progress of physics from this little volume. Even those who have most sedulously followed the track of modern discoveries cannot but be struck with admiration at the way in which the survey is brought up to the present day. The writer 'has read up to Saturday night,' as was said of the late Sir Samuel Romilly; and the latest experiments and speculations in every part of Europe are referred to, rapidly indeed, but appropriately and distinctly . . .

We must recollect that her professed object is to illustrate 'The *Connexion* of the Physical Sciences.' This is a noble object; and to succeed in it would be to render a most important service to science. The tendency of the sciences has long been an increasing proclivity to separation and dismemberment. Formerly, the 'learned' embraced in their wide grasp all the branches of the tree of knowledge; the Scaligers and Vossiuses of former days were mathematicians as well as philologers, physical as well as antiquarian speculators. But these days are past; the students of books and of things are estranged from each other in habit and feeling. If a moralist, like Hobbes, ventures into the domain of mathematics, or a poet, like Goethe, wanders into the fields of experimental science, he is received with contradiction and contempt; and, in truth, he generally makes his incursions with small advantage, for the separation of sympathies and intellectual habits has ended in a destruction, on each side, of that mental discipline which leads to success in the other province. But the disintegration goes on, like that of a great empire falling to pieces; physical science itself is endlessly subdivided, and the subdivisions insulated. We adopt the maxim 'one science only can one genius fit.' The mathematician turns away from the chemist; the chemist from the naturalist; the mathematician, left to himself, divides himself into a pure mathematician and a mixed mathematician, who soon part company; the chemist is perhaps a chemist of electro-chemistry; if so, he leaves common chemical analysis to others; between the mathematician and the chemist is to be interpolated a '*physicien*' (we have no English name for *him*), who studies heat, moisture, and the like. And thus science, even mere physical science, loses all trace of unity.

A curious illustration of this result may be observed in the want of any name by which we can designate the students of the knowledge of the material world collectively. We are informed that this difficulty was felt very oppressively by the members of the British Association for the Advancement of Science, at their meetings in York, Oxford, and Cambridge, in the last three summers. There was no general term by which these gentlemen could describe themselves with reference to their pursuits. *Philosophers* was felt to be too wide and lofty a term, and was very properly forbidden them by Mr. Coleridge, both in his capacity of philologer and metaphysician; *savans* was rather assuming, besides being French instead of English; some ingenious gentleman proposed that, by analogy with *artist*, they might form *scientist*, and added that there could be no scruple in making free with this termination when we have such words as *sciolist*, *economist*, and *atheist* — but this was not generally palatable; others attempted to translate the term by which the members of similar associations in Germany have described themselves, but it was not found easy to discover an English equivalent for *naturforscher*. The process of examination which it implies might suggest such undignified compounds as *nature-poker*, or *nature-peeper*, for these *naturæ curiosi*; but these were indignantly rejected.

The inconveniences of this division of the soil of science into infinitely small allotments have been often felt and complained of. It was one object, we believe, of the British Association, to remedy these inconveniences by bringing together the cultivators of different departments. To remove the evil in another way is one object of Mrs. Somerville's book. If we apprehend her purpose rightly, this is to be done by showing how detached branches have, in the history of science, united by the discovery of general principles.

Source: William Whewell, 'On the Connexion of the Physical Sciences. By Mrs. Somerville', Quarterly Review 51 (1834), pp. 54–68.

THE FIRST COMPUTER

Although mathematical instruments such as the abacus date back thousands of years, automatic calculating machines are a far more recent phenomenon. In 1642 the French mathematician Blaise Pascal (1623–62) devised a key-driven arithmetical machine that featured an ingenious automated 'carry' function (see fig. 11), but it is Charles Babbage's design for an 'Analytical Engine', which he began in 1835 but never completed, that counts as the world's first programmable computer, the punch-card programmes for which were designed by the mathematician Ada Lovelace, daughter of the poet Lord Byron.

Pascal's Arithmetic Machine

This first extract is from Pascal's own description of his arithmetic machine, which he constructed in 1642. It was the first machine that could perform the four fundamental operations of arithmetic: addition, subtraction, multiplication and division. Though the dials were turned by hand with a stylus in order to add or subtract, the 'carry' function was automatic, through which tens were carried over into the next column by a simple ratchet that shifted the sequence of dials one unit up at the moment that their lower value neighbours made the transition from '9' to '0'. It's curious how Pascal's defence of his portable machine strikes the familiar tone of a proud inventor promoting his brainchild:

Dear reader, this notice will serve to inform you that I submit to the public a small machine of my invention, by means of which you alone may, without any effort, perform all the operations of arithmetic, and may be relieved of the work which has often times fatigued your spirit, when you have worked with the counters or with the pen . . .

As for the simplicity of movement of the operations, I have so devised it that, although the operations of arithmetic are in a way opposed the one to the other, – as addition to subtraction, and multiplication to division, – nevertheless they are all performed on this machine by a single unique movement.

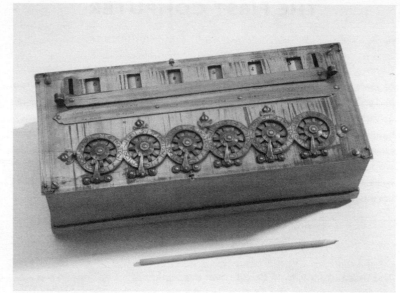

11. This modern replica of Blaise Pascal's portable calculating machine of 1642 shows the dials that were shifted automatically by the ingenious 'carry' function whenever one of their lower value neighbours made the transition from '9' to '0'.

The facility of this movement of operation is very evident since it is just as easy to move one thousand or ten thousand dials, all at one time, if one desires to make a single dial move, although all accomplish the movement perfectly. (I do not know if there remains another principle in nature such as the one upon which I have based this ease of operation.) In addition to the facility of movement in the operation, if you wish to appreciate it, you may compare it with the methods of counters and with the pen. You know that, in operating with counters, the calculator (especially when he lacks practice) is often obliged, for fear of making an error, to make a long series and extension of counters, being afterward compelled to gather up and retake those which are found to be extended unnecessarily, in which you see these two useless tasks, with the double loss of time. This machine facilitates the work and eliminates all unnecessary features. The most ignorant find as many advantages as the most experienced. The instrument makes up for ignorance and for lack of practice, and even without any effort of the operator, it makes possible shortcuts by itself, whenever the numbers are set down. In the same way you know that in operating with the pen one is obliged to retain or to borrow the necessary numbers, and that errors slip in, in these retentions and borrowings, except through very long practice, and in spite of a profound attention which soon fatigues the mind. This machine frees the operator from that vexation; he is relieved from the failing of memory; and without any retaining or borrowing, it does by itself what he wishes, without any thinking on his part. There are a hundred other advantages which practice will reveal, the details of which it would be wearisome to mention.

As to the amount of movement, it is sufficient to say that it is imperceptible, going from left to right and following our method of common writing except that it proceeds in a circle.

And, finally, its speed is evident at once in comparing it with the other two methods of the counters and the pen. If you still wish a more particular explanation of its rapidity, I shall tell you that it is equal to the agility of the hand of the operator. This speed is based, not only on the facility of the movements which have no resistance, but also on the smallness of the dials which are moved with the hand. The result is that, the key board being very short, the movement can be performed in a short time. Thus, the machine is small and hence is easily handled and carried.

And as to the lasting and wearing qualities of the instrument, the durability of the metal of which it is made should be a sufficient warrant: I have been able to give entire assurance to others only after having had the experience of carrying the instrument over more than two hundred and fifty leagues of road, without its showing any damage.

Therefore, dear reader, I ask you again not to consider it an imperfection for this machine to be composed of so many parts, because without these I could not give to it all the qualities which I have explained, and which are absolutely necessary. In this you may notice a kind of paradox, that to render the movement of operation more simple, it is necessary that the machine should be constructed of a movement more complex.

Babbage's Analytical Engine

Two centuries later, in 1835, the English mathematician Charles Babbage (1792–1871) designed the first programmable computer, which he termed the Analytical Engine. It was an improvement on two earlier, unfinished models that he had named Difference Engines, the main development being the use of punch-cards with which to give complex instructions to the machine. Babbage was joined in this work by the mathematician Augusta Ada Lovelace (1815–52), who made a number of improvements to the design as well as writing the programming instructions, which were based on an existing punch-card technology in use in Joseph Marie Jacquard's textile looms in France. As Lovelace observed, 'the Analytical Engine weaves algebraic patterns just as the Jacquard-loom weaves flowers and leaves', a comparison that turned out to be one of the most productive analogies in the history of technology, for when electronic computers first appeared in the twentieth century, programming instructions were fed in on punch-cards, just as Lovelace had suggested. In 1979, the programming language 'Ada' was named in honour of the world's first computer programmer.

The following three extracts are taken from an account of the Analytical Engine written by Luigi Federico Menabrea (1809–96), a military engineer and mathematician from Turin. At Babbage's suggestion, Lovelace translated the piece, adding voluminous notes and corrections that ended up being three times the length of the original article. The first extract is from the translation of Menabrea's original text, while the second two are a selection from Lovelace's copious notes:

Now, admitting that such an engine can be constructed, it may be inquired: what will be its utility? To recapitulate; it will afford the following advantages: – First, rigid accuracy. We know that numerical calculations are generally the stumbling-block to the solution of problems, since errors easily creep into them, and it is by no means always easy to detect these errors. Now the engine, by the very nature of its mode of acting, which requires no human intervention during the course of its operations, presents every species of security under the head of correctness; besides, it carries with it its own check; for at the end of every operation it prints off, not only the results, but likewise the numerical data of the question; so that it is easy to verify whether the question has been correctly proposed. Secondly, economy of time: to convince ourselves of this, we need only recollect that the multiplication of two numbers, consisting each of twenty figures, requires at the very utmost three minutes. Likewise, when a long series of identical computations is to be performed, such as those required for the formation of numerical tables, the machine can be brought into play so as to give several results at the same time, which will greatly abridge the whole amount of the process.

Thirdly, economy of intelligence: a simple arithmetical computation requires to be performed by a person possessing some capacity; and when we pass to more complicated calculations, and wish to use algebraical formulæ in particular cases, knowledge must be possessed which pre-supposes preliminary mathematical studies of some extent. Now the engine, from its capability of performing by itself all these purely material operations, spares intellectual labour, which may be more profitably employed. Thus the engine may be considered as a real manufactory of figures, which will lend its aid to those many useful sciences and arts that depend on numbers. Again, who can foresee the consequences of such an invention? In truth, how many precious observations remain practically barren for the progress of the sciences, because there are not powers sufficient for computing the results! And what discouragement does the perspective of a long and arid computation cast into the mind of a man of genius, who demands time exclusively for meditation, and who beholds it snatched from him by the material routine of operations! Yet it is by the laborious route of analysis that he must reach truth; but he cannot pursue this unless guided by numbers; for without numbers it is not given us to raise the veil which envelopes the mysteries of nature. Thus the idea of constructing an apparatus capable of aiding human weakness in such researches, is a conception which, being realized, would mark a glorious epoch in the history of the sciences.

The following is from one of Ada Lovelace's translator's notes, in which the sense of joint proprietorship with Babbage is evident in her use of 'we'; in it, she introduces the famous comparison between the first computer and the Jacquard loom:

Those who view mathematical science not merely as a vast body of abstract and immutable truths, whose intrinsic beauty, symmetry and logical completeness, when regarded in their connexion together

as a whole, entitle them to a prominent place in the interest of all profound and logical minds, but as possessing a yet deeper interest for
the human race, when it is remembered that this science constitutes
the language through which alone we can adequately express the
great facts of the material world, and those unceasing changes of
mutual relationship which, visibly or invisibly, consciously or uncon-
sciously to our immediate physical perceptions, are interminably
going on in the agencies of the creation we live amidst: those who
thus think on mathematical truth as the instrument through which
the weak mind of man can most effectually read his Creator's works,
will regard with especial interest all that can tend to facilitate the
translation of its principles into explicit practical forms.

The distinctive characteristic of the Analytical Engine, and that
which has rendered it possible to endow mechanism with such extensive faculties as bid fair to make this engine the executive right-hand
of abstract algebra, is the introduction into it of the principle which
Jacquard devised for regulating, by means of punched cards, the
most complicated patterns in the fabrication of brocaded stuffs. It is
in this that the distinction between the two engines lies. Nothing of
the sort exists in the Difference Engine. We may say most aptly, that
the Analytical Engine *weaves algebraic patterns* just as the Jacquard-
loom weaves flowers and leaves. Here, it seems to us, resides much
more of originality than the Difference Engine can be fairly entitled
to claim. We do not wish to deny to this latter all such claims. We
believe that it is the only proposal or attempt ever made to construct
a calculating machine *founded on the principle of successive orders
of differences*, and capable of *printing off its own results*; and that
this engine surpasses its predecessors, both in the extent of the calculations which it can perform, in the facility, certainty and accuracy
with which it can effect them, and in the absence of all necessity for
the intervention of human intelligence *during the performance of its
calculations*. Its nature is, however, limited to the strictly arithmetical, and it is far from being the first or only scheme for constructing
arithmetical calculating machines with more or less of success.

The bounds of *arithmetic* were however outstepped the moment
the idea of applying the cards had occurred; and the Analytical
Engine does not occupy common ground with mere "calculating
machines." It holds a position wholly its own; and the considerations
it suggests are most interesting in their nature. In enabling mecha-

nism to combine together *general* symbols in successions of unlimited variety and extent, a uniting link is established between
the operations of matter and the abstract mental processes of the
most abstract branch of mathematical science. A new, a vast, and a
powerful language is developed for the future use of analysis, in
which to wield its truths so that these may become of more speedy
and accurate practical application for the purposes of mankind than
the means hitherto in our possession have rendered possible. Thus
not only the mental and the material, but the theoretical and the
practical in the mathematical world, are brought into more intimate
and effective connexion with each other. We are not aware of its
being on record that anything partaking in the nature of what is so
well designated the *Analytical* Engine has been hitherto proposed, or
even thought of, as a practical possibility, any more than the idea of
a thinking or of a reasoning machine. A.A.L.

In this second translator's note Lovelace cautions against the suggestion –
made in a moment of exuberance by a carried-away Menabrea – that the
computer possessed any kind of independent intelligence or understanding.
In insisting that a computer can only do what it has been instructed to do
by its programmer, Lovelace formulated an early version of what became
known as 'the Turing test' (after the mathematician and cryptographer Alan
Turing), which determines whether a computer can really be thought to
imitate human intelligence:

It is desirable to guard against the possibility of exaggerated ideas
that might arise as to the powers of the Analytical Engine. In considering any new subject, there is frequently a tendency, first, to *overrate*
what we find to be already interesting or remarkable; and, secondly,
by a sort of natural reaction, to *undervalue* the true state of the case,
when we do discover that our notions have surpassed those that were
really tenable.

The Analytical Engine has no pretensions whatever to *originate*
any thing. It can do whatever we *know how to order it* to perform.
It can *follow* analysis; but it has no power of *anticipating* any
analytical relations or truths. Its province is to assist us in making
available what we are already acquainted with. This it is calculated
to effect primarily and chiefly, of course, through its executive faculties; but it is likely to exert an *indirect* and reciprocal influence on

science itself in another manner. For, in so distributing and combining the truths and the formulæ of analysis, that they may become most easily and rapidly amenable to the mechanical combinations of the engine, the relations and the nature of many subjects in that science are necessarily thrown into new lights, and more profoundly investigated. This is a decidedly indirect, and a somewhat *speculative*, consequence of such an invention. A.A.L.

Two More Contemporary Descriptions

This description of one of Babbage's Difference Engines comes from the diary of the American scholar and traveller George Ticknor (1791–1871). It offers a glimpse into the strength of contemporary wonder at the capacity of one of the first (though at the time only half-built) computers. Ticknor's visit to Babbage was made on Sunday 12 July 1835:

From church we went, by his special invitation, to see Babbage's calculating machine; and I must say, that during an explanation which lasted between two and three hours, given by himself with great spirit, the wonder at its incomprehensible powers grew upon us every moment. The first thing that struck me was its small size, being only about two feet wide, two feet deep, and two and a half high. The second very striking circumstance was the fact that the inventor himself does not profess to know all the powers of the machine; that he has sometimes been quite surprised at some of its capabilities; and that without previous calculation he cannot always tell whether it will, or will not work out a given table. The third was that he can set it to do a certain regular operation, as, for instance, counting 1,2,3,4; and then determine that, at any given number, say the 10,000th, it shall *change* and take a different ratio, like triangular numbers, 1,3,6,9,12 etc.; and afterwards at any other given point, say, 10,550, change again to another ratio. The whole, of course, seems incomprehensible, without the exercise of volition and thought. . . . But he is a very interesting man, ardent, eager, and of almost indefinite intellectual activity, bold and frank in expressing all his opinions and feelings.

This second description, of Babbage himself rather than any of his inventions, is one of the many sharply observed character sketches that populate

Charles Darwin's autobiography; it includes the well-known story of the wayfarers' pump – an excellent example of rational scepticism in action:

I used to call pretty often on Babbage and regularly attended his famous evening parties. He was always worth listening to, but he was a disappointed and discontented man; and his expression was often or generally morose. I do not believe that he was half as sullen as he pretended to be. One day he told me that he had invented a plan by which all fires could be effectively stopped, but added, – "I shan't publish it – damn them all, let all their houses be burnt." The all were the inhabitants of London. Another day he told me that he had seen a pump on a roadside in Italy, with a pious inscription on it to the effect that the owner had erected the pump for the love of God and his country, that the tired wayfarer might drink. This led Babbage to examine the pump closely and he soon discovered that every time that a wayfarer pumped some water for himself, he pumped a larger quantity into the owner's house. Babbage then added – "There is only one thing which I hate more than piety, and that is patriotism." But I believe that his bark was much worse than his bite.

Sources: A Source Book in Mathematics, ed. David E. Smith (New York: McGraw-Hill, 1929), p. 165; L. F. Menabrea, Sketch of the Analytical Engine Invented by Charles Babbage, Esq., with Notes by the Translator, trans. Augusta Ada Lovelace (London: R. & J. Taylor, 1843), pp. 669–731; Life, Letters and Journals of George Ticknor, ed. G. S. Hillard, 2 vols (London: Sampson Low, 1876), I, pp. 407–8; The Autobiography of Charles Darwin, 1809–1882; With original omissions restored, ed. Nora Barlow (London: Collins, 1958), p. 108.

THE SILURIAN SYSTEM

The Silurian System is a geological division that follows on from the Ordovician extinction of around 440 million years ago, ending some 30 million years later with the laying down of the Old Red Sandstone at the start of the Devonian period (c. 416 million years ago). It was named by the Scottish geologist Sir Roderick Impey Murchison (1792–1871), who in the early 1830s set about investigating the greywacke sandstones that underlie the Old Red Sandstone of Herefordshire and Wales. The result was the establishment of the Silurian System, a remarkable series of formations, replete with distinctive remains of early life forms that differ from those of any other rocks in Britain. Murchison's biographer, Sir Archibald Geikie (1835–1924) – himself a geologist of note, who became President of the Royal Society in 1908 – was keen to romanticize this important episode, casting Murchison as a heroic figure who prised out the secrets of the ancient rocks and brought them to geological order:

Murchison was thirty-two years old before he showed any interest in science. But his ardent and active temperament spurred him on. His enthusiasm being thoroughly aroused, his progress became rapid. He joined the Geological Society, and having gained the goodwill of Buckland, went down to Oxford for his first geological excursions under the guidance of that genial professor. He then discovered what field-geology meant, and learnt how the several parts of a landscape depend for their position and form upon the nature of the rocks underneath. He returned to London with his zeal aflame, burning to put into practice the principles of observation he had now been taught. He began among the Cretaceous formations around his father-in-law's home in Sussex, but soon extended his explorations into Scotland, France and the Alps, bringing back with him at the end of each season a bundle of well-filled note-books from which to prepare communications for the Geological Society. These early papers, meritorious though they were, do not call for any special notice here, since they marked no new departure in geological research, nor added any important province to the geological domain.

During six years of constant activity in the field, Murchison, together with Sedgwick, worked out the structure of parts of the west and north of Scotland, and toiled hard in disentangling the complicated structure of the eastern Alps; he also rambled with Lyell over the volcanic areas of Central and Southern France. Thereafter he determined to try whether the "interminable greywacke," as he called it, could not be reduced to order and made to yield a stratigraphical sequence, like that which had been so successfully obtained among younger formations. At the time when he began, that is, in the summer of 1831, absolutely nothing was known of the succession of rocks below the Old Red Sandstone. It was an unknown land, a pathless desert, where no previous traveller had been able to detect any trace of a practicable track towards order, or any clue to a system of arrangement that would enable the older fossiliferous rocks of one country to be paralleled, save in the broadest and most general way, with those of another.

Starting with his "wife and maid, two good grey nags and a little carriage, saddles being strapped behind for occasional equestrian use," Murchison made his way into South Wales. In that region, as was well known, the stratigraphical series could be followed down into the Old Red Sandstone, and within the frame or border of that formation, greywacke was believed to extend over all the rest of the Principality. Let me quote a few sentences in which Murchison describes his first entry into the domain with which his fame is now so inseparably linked. "Travelling from Brecon to Builth by the Herefordshire road, the gorge in which the Wye flows first developed what I had not till then seen. Low terrace-shaped ridges of grey rock, dipping slightly to the south-east, appeared on the opposite bank of the Wye, and seemed to rise quite conformably from beneath the Old Red Sandstone of Herefordshire. Boating across the river at Cavansham Ferry, I rushed up to these ridges, and, to my inexpressible joy, found them replete with Transition fossils, afterwards identified with those at Ludlow. Here then was a key, and if I could only follow this out on the strike of the beds to the north-east, the case would be good."

With unerring instinct Murchison had realised that if the story of old Greywacke was ever to be fully told, a beginning must be made from some known and recognisable horizon. It would have been well-nigh useless to dive into the heart of the Transition hills, and try to work out their complicated structure, for even if a sequence could

then have been determined, there would have been no means of connecting it with the already ascertained stratigraphical series, unless it
could be followed outwards to the Old Red Sandstone. But by commencing at the known base of that series, every fresh stage conquered
was at once a definite platform added to what had already been
established.

The explorer kept along the track of the rocks for many miles
to the north. No hunter could have followed the scent of the fox better than he did the outcrop of the fossiliferous strata, which he saw
to come out regularly from under the lowest members of the Old
Red Sandstone. Directed to the Wye by Buckland, he had the good-
fortune to come at once upon some of the few natural sections where
the order of the higher Transition rocks of Britain, and their relations
to the overlying formations, can be distinctly seen. He pursued the
chase northwards until he lost the old rocks under the Triassic plains
of Cheshire. "For a first survey," he writes, "I had got the upper
grauwacke, so called, into my hands, for I had seen it in several situations far from each other, all along the South Welsh frontier, and in
Shropshire and Herefordshire, rising out gradually and conformably
from beneath the lowest member of the Old Red Sandstone. Moreover, I had ascertained that its different beds were characterized by
peculiar fossils . . . a new step in British geology. In summing up what
I saw and realised in about four months of travelling, I may say that
it was the most fruitful year of my life, for in it I laid the foundation
of my Silurian system. I was then thirty-nine years old, and few could
excel me in bodily and mental activity."

Not only did the work of these four momentous months mark a
new step in British geology. It began the lifting of the veil from the
Transition rocks of the whole globe. It was the first successful foray
into these hitherto intractable masses, and prepared the way for all
that has since been done in deciphering the history of the most
ancient fossiliferous formations, alike in the Old World and in the
New . . . By the summer of 1835, at the instigation of Élie de Beaumont and other geological friends, he had made up his mind as to the
name that should be given to this remarkable assemblage or system
of formations which he had disinterred from out of the chaos of
Greywacke. Following the good rule that stratigraphical terms are
most fitly formed on a geographical basis with reference to the
regions wherein the rocks are most typically developed, he had

looked about for some appropriate and euphonious term that would comprise his various formations and connect them with that borderland of England and Wales where they are so copiously displayed. This territory was in Roman times inhabited by the tribe of the Silures, and so he chose the term Silurian — a word that is now familiar to the geologists of every country.

Murchison communicated his findings in a series of papers published by the Geological Society, but it was clear that there was a growing readership for works of popular science; so he spent some time preparing what Geikie called 'his great work', *The Silurian System*, an 800-page illustrated volume that outlined the full scope of the Silurian formation:

The publication of this splendid monograph forms a notable epoch in the history of modern geology, and well entitles its author to be enrolled among the founders of the science. For the first time, the succession of fossiliferous formations below the Old Red Sandstone was shown in detail. Their fossils were enumerated, described and figured. It was now possible to carry the vision across a vast series of ages, of which hitherto no definite knowledge existed, to mark the succession of their organisms, and thus to trace backward, far farther than had ever before been possible, the history of organised existence on this globe.

Source: Archibald Geikie, *The Founders of Geology*, 2nd edn (London: Macmillan, 1905), pp. 413–20.

THE HUMBOLDT CURRENT

The eminent German naturalist and explorer Alexander von Humboldt (1769–1859) was one of the leading scientific figures of his age, whose books, especially the ambitious, multi-volume *Kosmos* (1845–58), from which the following pages come, were hugely successful throughout Europe and North America. His fame rested on his five-year journey through Latin America from 1799 to 1804, during which he made an array of scientific observations and discoveries, including the first description of the Peruvian current (later renamed the Humboldt current), a cold-water ocean current that flows north along the coasts of Chile and Peru, exerting a significant cooling influence on their climates. Humboldt's approach to natural science is well illustrated by this finely written passage, in which the interrelations between geographic, oceanographic and climatic variables are woven into a masterful account of the movements of the major ocean currents – a good example of the Humboldtian concept of the 'unity of nature' in action:

The currents of the ocean present a remarkable spectacle; maintaining a nearly constant breadth, they cross the sea in different directions, like rivers of which the adjacent undisturbed masses of water form the banks. The line of demarcation between the parts in motion, and those in repose, is most strikingly shewn in places where long bands of sea-weed, borne onward by the current, enable us to estimate its velocity. Analogous phænomena are sometimes presented to our notice in the lower strata of the atmosphere, when, after a violent storm, the path of a limited aerial current may be traced through the forest by long lanes of overthrown trees, whilst those on either side remain unscathed.

The general movement of the sea from east to west between the tropics, known by the name of the equatorial or rotation current, is regarded as the joint effort of the trade winds, and of the progressive propagation of the tide-wave. Its direction is modified by the resistance which it experiences from the eastern coasts of continents. The velocity of this current, computed by Daussy from data supplied by bottles purposely thrown overboard and subsequently picked up in different localities, agrees within one eighteenth part with the velocity of ten French nautical miles in twenty-four hours, which I had previously deduced from comparing the experience of different navigators. Columbus was aware of the existence of this current at the period of his third voyage, (the first in which he sought to enter the tropics in the meridian of the Canary Islands), since we find in his journal the following passage: – "I regard it as proved that the waters of the sea move from east to west as do the heavens" (or the apparent motion of the sun, moon, and stars) . . .

Of the narrow currents, or true oceanic rivers, of which we have spoken, some carry warm water into higher, and others cold water into lower latitudes. To the first class belongs the celebrated gulf stream, the existence of which was recognised by Anghiera, and more particularly by Sir Humphry Gilbert, as early as the sixteenth century, and of which the first origin and impulse is to be sought to the south of the Cape of Good Hope. After a wide circuit, it pours itself from the Caribbean Sea and the Mexican Gulf through the channel of the Bahamas, and following a direction from S.S.W. to N.N.E., deviates more and more from the coast of the United States, until, deflected still further to the east by the banks of Newfoundland, it crosses the Atlantic, and casts an abundance of tropical seeds (Mimosa scandens, Guilandina bonduc, Dolichos urens) on the coasts of Ireland, of the Hebrides, and of Norway. Its north-eastern-most prolongation mitigates the cold of the ocean, and exercises a beneficent influence on the climate of the northernmost point of Scandinavia. At the point where the stream is deflected to the east by the banks of Newfoundland, it sends off an arm toward the south, not far from the Azores. This is the situation of the Sargasso Sea, or that great sea of weed, or bank of fucus, which made so lively an impression on the imagination of Columbus, and which Ovieda calls sea-weed meadows, – "praderias de yerva." These evergreen masses of Fucus natans (one of the most widely distributed of the social sea plants), driven gently to and fro by mild and warm breezes, are the habitation of a countless number of small marine mammals.

This great current, which, in the Atlantic valley, between Africa, America, and Europe, belongs almost entirely to the northern hemisphere, has its counterpart in the Southern Pacific Ocean, in a current the effect of whose low temperature on the climate of the adjacent coasts was first brought into notice by myself in the autumn of 1802. This current brings the cold water of the high south latitudes to the coast of Chile, and follows its shores and those of Peru northward to the bay of Arica, and thence north-westerly to the neighbourhood of Payta, where the most westerly projection of the American coast deflects the stream, and causes it suddenly to quit the shore, taking a due west direction. Here the boundary is so sharply marked, that a ship sailing northwards finds itself passing suddenly from cold to warm water. At certain seasons of the year this cold current brings into the tropics water of 15.6° Cent. (60° Fah.), the temperature of the undisturbed masses of water in the vicinity being from 27.5° to 28.7° Cent. (81.5° to 83.7° Fah.)

We do not know the depth to which oceanic currents (warm or cold) extend: the deflection of the South African current by the Lagullas bank, which is from sixty to seventy fathoms deep, would indicate it to be considerable. The presence of sand banks and shoals, not situated in any line of current, may be recognised by the low temperature of the water over them; and the knowledge of this fact may often conduce to the safety of navigation. It was first discovered by the justly celebrated Benjamin Franklin, who may be said to have thereby transformed the thermometer into a sounding line; and its explanation appears to be, that when in the general movement of the water forming the current, the deeper situated and colder particles strike upon a bank, their motion is inclined upwards, and they mingle with and chill the upper stratum of water. My late illustrious friend, Sir Humphry Davy, attributed the phænomenon rather to the descent of the surface particles, cooled by nocturnal radiation, and to their being prevented from sinking deeper by the shoal, which thus retained them in closer proximity to the surface. Mists are frequently met with over shoals, from the influence of the cooled water in condensing the vapour in the atmosphere. I have seen such mists to the south of Jamaica, and in the Pacific, which have shewn the outline of the shoals beneath so well defined, as to be distinctly recognised from a distance; thus forming to the eye aerial images reflecting the form of the bottom of the ocean. A still more remarkable effect of the cooling influence of shoals is shewn sometimes in the higher regions of the atmosphere. At sea and in very clear weather, clouds are often seen suspended above the site of sand banks or shoals, as well as over low coral or sandy islands. Their bearings may be taken from a distant ship by the compass, precisely as that of a high mountain or a solitary peak.

Source: Alexander von Humboldt, *Cosmos: Sketch of a Physical Description of the Universe*, trans. Elizabeth Sabine, 4 vols (London: Longman, Brown, Green, & Longmans, 1846–58), I, pp. 299–303.

THE TEMPLE OF SERAPIS

Many nineteenth-century geologists were preoccupied by a single as-yet-unanswered question: what was the age of the earth? If, as the majority believed, it was only a few thousand years old, then the transformations it had clearly undergone must have been rapid and extremely violent. If, on the other hand, it was millions, even billions of years old, that allowed plenty of time for such changes to have taken place gradually, just as they do today. These two outlooks, known respectively as catastrophism and uniformitarianism, divided geological opinion until Charles Lyell (1797–1875) made a convincing case for the uniformitarian view in his landmark three-volume survey, *The Principles of Geology* (1830–33). It remains the most important book in the history of geology, and its central argument, that understanding present-day geophysical processes is the key to unlocking the earth's deep history, greatly influenced the young Charles Darwin, who took the first volume with him on the *Beagle* voyage in 1831.

The following extract, from that first volume, contains one of Lyell's most celebrated images: the pillars of the temple of Serapis at Puzzuoli, in southern Italy. The temple had originally been erected at ground level, but it had evidently been submerged at some point in its two-thousand-year history, as the pillars are disfigured by bands of holes drilled by a species of rock-boring clam; some time later the pillars had been lifted again to the level at which they stand today. Considering that such observable changes occurred during the period of recorded history, argued Lyell, think what vast but gradual changes could have happened over millions of years:

Temple of Jupiter Serapis. – This celebrated monument of antiquity affords, in itself alone, unequivocal evidence, that the relative level of land and sea has changed twice at Puzzuoli, since the Christian era, and each movement both of elevation and subsidence has exceeded twenty feet. Before examining these proofs we may observe that a geological examination of the coast of the Bay of Baiæ, both on the north and south of Puzzuoli, establishes in the most satisfactory manner an elevation at no remote period, of more than twenty feet, and the evidence of this change would have been complete even if the temple had to this day remained undiscovered. If we coast along the shore from Naples to Puzzuoli we find, on approaching the latter place, that the lofty and precipitous cliffs of indurated tuff, resembling that of which Naples is built, retire slightly from the sea, and that a low level tract of fertile land, of a very different aspect, intervenes between the present sea-beach, and what was evidently the ancient line of coast. The inland cliff is in many parts eighty feet high near Puzzuoli, and as perpendicular as if it was still undermined by the waves. At its base, the new deposit attains a height of about twenty feet above the sea, and as it consists of regular sedimentary deposits, containing marine shells, its position proves that since its formation there has been a change of more than twenty feet in the relative level of land and sea.

The sea encroaches on these new incoherent strata, and as the soil is valuable, a wall has been built for its protection; but when I visited the spot in 1828, the waves had swept away part of this rampart, and exposed to view a regular series of strata of tuff, more or less argillaceous, alternating with beds of pumice and lapilli, and containing great abundance of marine shells, of species now common on this coast, and amongst them *Cardium rusticum*, *Ostrea edulis*, *Donax trunculus*, and others. The strata vary from about a foot to a foot and a half in thickness, and one of them contains abundantly remains of works of art, tiles, squares of mosaic pavement of different colours, and small sculptured ornaments, perfectly uninjured. Intermixed with these I collected some teeth of the pig and ox. These fragments of building occur below as well as above strata containing marine shells . . . But there are no tides in the Mediterranean; and to suppose that sea to have sunk generally from twenty to twenty-five feet since the shores of Campania were covered with sumptuous buildings, is an hypothesis obviously untenable. The observations, indeed, made during modern surveys on the moles and cothons (docks) constructed by the ancients in various ports of the Mediterranean, have proved that there has been no sensible variation of level in that sea during the last two thousand years. A very slight change would have been perceptible; and had any been ascertained to have taken place, and had it amounted only to a difference of a few feet, it would not have appeared very extraordinary, since the equilibrium of the Mediterranean is only restored by a powerful current from the Atlantic.

Thus we arrive, without the aid of the celebrated temple, at the conclusion that the recent marine deposit at Puzzuoli was upraised in

T. Bradley. Sc.

12. This engraving of the Temple of Serapis at Puzzuoli formed the frontispiece to volume 1 of Lyell's *Principles of Geology* (1830). Note the dark bands on the pillars, caused by rock-drilling molluscs during a period of sea-level change some time during the last 2,000 years.

modern times above the level of the sea, and that not only this change of position, but the accumulation of the modern strata, was posterior to the destruction of many edifices, of which they contain the imbedded remains. If we now examine the evidence afforded by the temple itself, it appears, from the most authentic accounts, that the three pillars now standing erect, continued, down to the middle of the last century, half buried in the new marine strata before described. The upper part of the columns, being concealed by bushes, had not attracted the notice of antiquaries; but, when the soil was removed in 1750, they were seen to form part of the remains of a splendid edifice, the pavement of which was still preserved, and upon it lay a number of columns of African breccia and of granite . . .

The pillars are forty-two feet in height; their surface is smooth and uninjured to the height of about twelve feet above their pedestals. Above this, is a zone, twelve feet in height, where the marble has been pierced by a species of marine perforating bivalve – *Lithodomus*, Cuv. The holes of these animals are pear-shaped, the external opening being minute, and gradually increasing downwards. At the bottom of the cavities, many shells are still found, notwithstanding the great numbers that have been taken out by visitors. The perforations are so considerable in depth and size, that they manifest a long continued abode of the Lithodomi in the columns; for, as the inhabitant grows older and increases in size, it bores a larger cavity, to correspond with the increasing magnitude of its shell. We must, consequently, infer a long continued immersion of the pillars in sea-water, at a time when the lower part was covered up and protected by strata of tuff and the rubbish of buildings, the highest part at the same time projecting above the waters, and being consequently weathered, but not materially injured. On the pavement of the temple, lie some columns of marble, which are perforated in the same manner in certain parts, one, for example, to the length of eight feet, while, for the length of four feet, it is uninjured. Several of these broken columns are eaten into, not only on the exterior, but on the cross fracture, and, on some of them, other marine animals have fixed themselves. All the granite pillars are untouched by Lithodomi. The platform of the Temple is at present about one foot below high-water mark, (for there are small tides in the Bay of Naples,) and the sea, which is only one hundred feet distant, soaks through the intervening soil. The upper part of the perforations then are at least twenty-three feet above high-water mark, and it is clear, that the columns must have continued for a long time in an erect position, immersed in salt-water. After remaining for many years submerged, they must have been upraised to the height of about twenty-three feet above the level of the sea.

So far the information derived from the Temple corroborates that before obtained from the new strata in the plain of La Starza, and proves nothing more. But as the temple could not have been built originally at the bottom of the sea, it must have first sunk down below the waves, and afterwards have been elevated . . .

In 1828 excavations were made below the marble pavement of the Temple of Serapis, and another costly pavement of mosaic was found, at the depth of five feet or more below the other. The existence of these two pavements at different levels seems clearly to imply some subsidence previously to all the changes already alluded to, which had rendered it necessary to construct a new floor at a higher level . . . In concluding this subject, we may observe, that the interminable controversies to which the phenomenon of the Bay of Baiæ gave rise, have sprung from an extreme reluctance to admit that the land rather than the sea is subject alternately to rise and fall. Had it been assumed that the level of the ocean was invariable, on the ground that no fluctuations have as yet been clearly established, and that, on the other hand, the continents are inconstant in their level, as has been demonstrated by the most unequivocal proofs again and again, from the time of Strabo to our own times, the appearances of the temple at Puzzuoli could never have been regarded as enigmatical. Even if contemporary accounts had not distinctly attested the upraising of the coast, this explanation should have been proposed in the first instance as the most natural, instead of being now adopted unwillingly when all others have failed. To the strong prejudices still existing in regard to the mobility of the land, we may attribute the rarity of such discoveries as have been recently brought to light in the Bay of Baiæ and the Bay of Conception. A false theory it is well known may render us blind to facts, which are opposed to our prepossessions, or may conceal from us their true import when we behold them. But it is time that the geologist should in some degree overcome those first and natural impressions which induced the poets of old to select the rock as the emblem of firmness – the sea as the image of inconstancy. Our modern poet, in a more philosophical spirit, saw in the latter 'The image of Eternity', and has finely contrasted the fleeting existence of the successive empires which have flourished and fallen, on the borders of the ocean, with its own unchanged stability.

> — Their decay
> Has dried up realms to deserts: – not so thou,
> Unchangeable, save to thy wild waves' play:
> Time writes no wrinkle on thine azure brow;
> Such as creation's dawn beheld, thou rollest now.
> CHILDE HAROLD, Canto iv.

Source: Charles Lyell, *Principles of Geology: Being an Attempt to Explain the Former Changes of the Earth's Surface, by Reference to Causes Now In Operation*, 3 vols (London: John Murray, 1830–33), I, pp. 449–59.

DARWIN'S DANGEROUS IDEA

Charles Darwin's theory of natural selection – 'the single best idea that any-one has ever had', as the philosopher Daniel Dennett described it – disrupted the Victorian view of nature as a benignly governed kingdom, every corner of which revealed the wisdom of the divine creator. Darwin instead recast nature as a scene of pitiless struggle, a battlefield over which the species were condemned to fight for their very existence. The social and theological impact of Darwin's ideas can hardly be overstated; writing *The Origin of Species* (1859) was, as its author observed, 'like confessing a murder', and Victorian culture was both unnerved and energized by Darwin's revelations.

Darwin (1809–82) was a powerful writer as well as a great scientist, and the second of the two extracts, below, from the final pages of *The Origin of Species*, reveals the narrative skills of a thinker who had agonized over how best to express his views. Note how careful he is to cushion 'the religious feelings' of his readers, particularly those of his wife Emma, who suffered genuine distress at the thought of her husband's loss of his Christian faith. But I thought I would preface the *Origin*'s finale with the following passage from Darwin's autobiography, in which he describes the dawning of his dangerous idea, as well as the long and eventful process of turning it into a book:

In October 1838, that is, fifteen months after I had begun my systematic enquiry, I happened to read for amusement Malthus on *Population*, and being well prepared to appreciate the struggle for existence which everywhere goes on from long-continued observation of the habits of animals and plants, it at once struck me that under these circumstances favourable variations would tend to be preserved, and unfavourable ones to be destroyed. The result of this would be the formation of new species. Here, then, I had at last got a theory by which to work; but I was so anxious to avoid prejudice, that I determined not for some time to write even the briefest sketch of it. In June 1842 I first allowed myself the satisfaction of writing a very brief abstract of my theory in pencil in 35 pages; and this was enlarged during the summer of 1844 into one of 230 pages, which I had fairly copied out and still possess.

But at the same time I overlooked one problem of great importance; and it is astonishing to me, except on the principle of

Columbus and his egg, how I could have overlooked it and its solution. This problem is the tendency in organic beings descended from the same stock to diverge in character as they become modified. That they have diverged greatly is obvious from the manner in which species of all kinds can be classed under genera, genera under families, families under sub orders, and so forth; and I can remember the very spot in the road, whilst in my carriage, when to my joy the solution occurred to me; and this was long after I had come to Down. The solution, as I believe, is that the modified offspring of all dominant and increasing forms tend to become adapted to many and highly diversified places in the economy of nature.

Early in 1856 Lyell advised me to write out my views pretty fully, and I began at once to do so on a scale three or four times as extensive as that which was afterwards followed in my *Origin of Species*; yet it was only an abstract of the materials which I had collected, and I got through about half the work on this scale. But my plans were overthrown, for early in the summer of 1858 Mr Wallace, who was then in the Malay archipelago, sent me an essay *On the Tendency of Varieties to depart indefinitely from the Original Type*; and this essay contained exactly the same theory as mine. Mr Wallace expressed the wish that if I thought well of his essay, I should send it to Lyell for perusal.

The circumstances under which I consented at the request of Lyell and Hooker to allow for an extract of my MS., together with a letter to Asa Gray, dated September 5, 1857, to be published at the same time with Wallace's Essay, are given in the *Journal of the Proceedings of the Linnean Society*, 1858, p. 45. I was at first very unwilling to consent, as I thought Mr Wallace might consider my doing so unjustifiable, for I did not then know how generous and noble was his disposition. The extract from my MS. and the letter to Asa Gray had neither been intended for publication, and were badly written. Mr Wallace's essay, on the other hand, was admirably expressed and quite clear. Nevertheless, our joint productions excited very little attention, and the only published notice of them which I can remember was by Professor Haughton of Dublin, whose verdict was that all that was new in them was false, and what was true was old. This shows how necessary it is that any new view should be explained at considerable length in order to arouse public attention.

In September 1858 I set to work by the strong advice of Lyell and
Hooker to prepare a volume on the transmutation of species, but was
often interrupted by ill-health, and short visits to Dr. Lane's delightful hydropathic establishment at Moor Park. I abstracted the MS.
begun on a much larger scale in 1856, and completed the volume on
the same reduced scale. It cost me thirteen months and ten days' hard
labour. It was published under the title of *The Origin of Species*, in
November 1859. Though considerably added to and corrected in the
later editions, it has remained substantially the same book.

It is no doubt the chief work of my life. It was from the first
highly successful. The first small edition of 1250 copies was sold on
the day of publication, and a second edition of 3000 copies soon
afterwards. Sixteen thousand copies have now (1876) been sold in
England and considering how stiff a book it is, this is a large sale.

Part of the success of *The Origin of Species* was down to Darwin's skills as a
writer, as evidenced by the final few paragraphs of one of the most famous
and influential books ever published:

I have now recapitulated the chief facts and considerations which
have thoroughly convinced me that species have been modified, during a long course of descent, by the preservation or the natural
selection of many successive slight favourable variations. I cannot
believe that a false theory would explain, as it seems to me that the
theory of natural selection does explain, the several large classes of
facts above specified. It is no valid objection that science as yet
throws no light on the far higher problem of the essence or origin of
life. Who can explain what is the essence of the attraction of gravity?
No one now objects to following out the results consequent on this
unknown element of attraction; notwithstanding that Leibnitz formerly accused Newton of introducing "occult qualities and miracles
into philosophy."

I see no good reason why the views given in this volume should
shock the religious feelings of any one. It is satisfactory, as showing
how transient such impressions are, to remember that the greatest
discovery ever made by man, namely, the law of the attraction of
gravity, was also attacked by Leibnitz, "as subversive of natural and
inferentially of revealed religion." A celebrated author and divine
has written to me that "he has gradually learnt to see that it is just as

noble a conception of the Deity to believe that He created a few original forms capable of self-development into other and needful forms, as to believe that He required a fresh act of creation to supply the voids caused by the action of His laws." . . .

When the views entertained in this volume on the origin of species, or when analogous views are generally admitted, we can dimly foresee that there will be a considerable revolution in natural history. Systematists will be able to pursue their labours as at present; but they will not be incessantly haunted by the shadowy doubt whether this or that form be in essence a species. This I feel sure, and I speak after experience, will be no slight relief. The endless disputes whether or not some fifty species of British brambles are true species will cease. Systematists will have only to decide (not that this will be easy) whether any form be sufficiently constant and distinct from other forms, to be capable of definition; and if definable, whether the differences be sufficiently important to deserve a specific name. This latter point will become a far more essential consideration than it is at present; for differences, however slight, between any two forms, if not blended by intermediate gradations, are looked at by most naturalists as sufficient to raise both forms to the rank of species. Hereafter we shall be compelled to acknowledge that the only distinction between species and well-marked varieties is, that the latter are known, or believed, to be connected at the present day by intermediate gradations, whereas species were formerly thus connected. Hence, without quite rejecting the consideration of the present existence of intermediate gradations between any two forms, we shall be led to weigh more carefully and to value higher the actual amount of difference between them. It is quite possible that forms now generally acknowledged to be merely varieties may hereafter be thought worthy of specific names, as with the primrose and cowslip; and in this case scientific and common language will come into accordance. In short, we shall have to treat species in the same manner as those naturalists treat genera, who admit that genera are merely artificial combinations made for convenience. This may not be a cheering prospect; but we shall at least be freed from the vain search for the undiscovered and undiscoverable essence of the term species . . .

In the distant future I see open fields for far more important researches. Psychology will be based on a new foundation, that of the necessary acquirement of each mental power and capacity by gradation. Light will be thrown on the origin of man and his history.

Authors of the highest eminence seem to be fully satisfied with the view that each species has been independently created. To my mind it accords better with what we know of the laws impressed on matter by the Creator, that the production and extinction of the past and present inhabitants of the world should have been due to secondary causes, like those determining the birth and death of the individual. When I view all beings not as special creations, but as the lineal descendants of some few beings which lived long before the first bed of the Silurian system was deposited, they seem to me to become ennobled. Judging from the past, we may safely infer that not one living species will transmit its unaltered likeness to a distant futurity. And of the species now living very few will transmit progeny of any kind to a far distant futurity; for the manner in which all organic beings are grouped, shows that the greater number of species of each genus, and all the species of many genera, have left no descendants, but have become utterly extinct. We can so far take a prophetic glance into futurity as to foretell that it will be the common and widely-spread species, belonging to the larger and dominant groups, which will ultimately prevail and procreate new and dominant species. As all the living forms of life are the lineal descendants of those which lived long before the Silurian epoch, we may feel certain that the ordinary succession by generation has never once been broken, and that no cataclysm has desolated the whole world. Hence we may look with some confidence to a secure future of equally inappreciable length. And as natural selection works solely by and for the good of each being, all corporeal and mental endowments will tend to progress towards perfection.

It is interesting to contemplate an entangled bank, clothed with many plants of many kinds, with birds singing on the bushes, with various insects flitting about, and with worms crawling through the damp earth, and to reflect that these elaborately constructed forms, so different from each other, and dependent on each other in so complex a manner, have all been produced by laws acting around us. These laws, taken in the largest sense, being Growth with Reproduction; Inheritance which is almost implied by reproduction; Variability from the indirect and direct action of the external conditions of life, and from use and disuse; a Ratio of Increase so high as to lead to a Struggle for Life, and as a consequence to Natural Selection, entailing Divergence of Character and the Extinction of less-improved forms. Thus, from the war of nature, from famine and death, the most exalted object which we are capable of conceiving, namely, the production of the higher animals, directly follows. There is grandeur in this view of life, with its several powers, having been originally breathed into a few forms or into one; and that, whilst this planet has gone cycling on according to the fixed law of gravity, from so simple a beginning endless forms most beautiful and most wonderful have been, and are being, evolved.

13. Following the publication of *The Descent of Man*, Darwin grew used to seeing cartoons that depicted him as an ape. This example appeared in the *Hornet* magazine in 1871.

There was, at first, a great deal of opposition to Darwin's ideas, particularly when *The Descent of Man* was published in 1871, for it was in that volume that Darwin introduced the idea – obliquely hinted at in *The Origin of Species* – that mankind was also part of the great evolutionary story, and that we are likely to have shared a common ancestor with other higher primates. The claim that 'man is descended from a hairy, tailed quadruped, probably arboreal in its habits' proved a godsend to cartoonists, and Darwin grew wearily familiar with seeing himself depicted as a bearded monkey up a tree (see fig. 13). The following verses, from a long and vigorously argued anti-evolutional poem entitled *Our Modern Philosophers* (1884) by an anonymous author writing under the pen-name 'Psychosis', adopt a more subtle form of attack by summing up the argument in a mock-heroic style:

Miasma dense spread o'er the earth;
    The sun's fierce rays pierced stagnant pools,
Which by degrees brought forth the birth
    Of life from molecules.
Low marshes and deep quagmires dank
    Were filled with every loathsome thing,
Which wriggled up each slimy bank.
    Amazed with life, and wondering
Whence they all came, and what they were,
They licked their dirty sides, nor asked what brought them there.

"Natural Selection" then began.
    Now mark the process, as I trace
The wondrous origin of man
    And descent of the human race.
The vertebrated genus then
    (Thou hast been shown its slimy source)
Was the first generant of men,
    And centrifugal lineal force
That procreated, stage by stage,
And improved mammal life in each succeeding age.

It took two million years before
    Mankind became a perfect breed.
My science says it took e'en more –
    On this wise men are all agreed.

The human germ was in a trout,
    A toad, a tadpole, or an eel;
And as the germ was changed about
    Natural selection fixed its seal;
Through vertebrated life it ran,
And after many years the germ became a man.

And thus by graduation came
    The creatures that now rule the earth.
Mammalia held the germ the same,
    Thou oft she changed in form at birth.
Arrived at monkey-hood, 'twas then
    Nature had little more to do –
She moulded monkeys into men,
    Which by analogy is true.
No scientist will ever doubt it;
Nor would the simple bard, if he knew more about it.

Sources: *The Autobiography of Charles Darwin, 1809–1882; With original omissions restored*, ed. Nora Barlow (London: Collins, 1958), pp. 120–22; Charles Darwin, *On the Origin of Species by Means of Natural Selection, or the Preservation of Favoured Races in the Struggle for Life* (London: John Murray, 1859), pp. 514–25; 'Psychosis', *Our Modern Philosophers: Darwin, Bain and Spencer; or, The Descent of Man, Mind and Body: A Rhyme with Reasons, Essays, Notes and Quotations* (London: T. Fisher Unwin, 1884), pp. 15–16.

BUTTERFLY HUNTING IN MALAYA

Although Alfred Russel Wallace (1823–1913) is best known today for having independently proposed a theory of natural selection in an article that prompted the reluctant Charles Darwin to hurry up and publish his own delayed version (see p. 243), he was famous in his lifetime as a naturalist and explorer, and his book *The Malay Archipelago* (1869), from which the following extract comes, was one of the nineteenth century's biggest-selling travel books. Wallace spent nearly eight years exploring the islands of the East Indies (modern-day Malaysia and Indonesia), returning to England in 1862 to write up his travels, as well as to sort out the vast collection of more than 125,000 natural-history specimens he had amassed. From these he derived a theory of animal distribution, in which distinct species inhabit either side of an ecological boundary, known as the Wallace Line, that runs between the islands of Bali and Lombok. The following is Wallace's account of his discovery of the Croesus butterfly (*Ornithoptera croesus*) on the island of Batchian (now Bacan), on the eastern side of the line. His feelings of delight, which he describes with great candour, are in sharp contrast to his determination to collect as many dead specimens as he could of what he knew was an extremely rare creature:

During my very first walk into the forest at Batchian, I had seen sitting on a leaf out of reach, an immense butterfly of a dark colour marked with white and yellow spots. I could not capture it as it flew away high up into the forest, but I at once saw that it was a female of a new species of Ornithoptera or "bird-winged butterfly," the pride of the Eastern tropics. I was very anxious to get it and to find the male, which in this genus is always of extreme beauty. During the two succeeding months I only saw it once again, and shortly afterwards I saw the male flying high in the air at the mining village. I had begun to despair of ever getting a specimen, as it seemed so rare and wild; till one day, about the beginning of January, I found a beautiful shrub with large white leafy bracts and yellow flowers, a species of Mussænda, and saw one of those noble insects hovering over it, but it was too quick for me, and flew away. The next day I went again to the same shrub and succeeded in catching a female, and the day after a fine male. I found it to be as I had expected, a perfectly new and most magnificent species, and one of the most gorgeously coloured butterflies in the world. Fine specimens of the male are more than seven inches across the wings, which are velvety black and fiery orange, the latter colour replacing the green of the allied species. The beauty and brilliancy of this insect are indescribable, and none but a naturalist can understand the intense excitement I experienced when I at length captured it. On taking it out of my net and opening the glorious wings, my heart began to beat violently, the blood rushed to my head, and I felt much more like fainting than I have done when in apprehension of immediate death. I had a headache the rest of the day, so great was the excitement produced by what will appear to most people a very inadequate cause.

I had decided to return to Ternate in a week or two more, but this grand capture determined me to stay on till I obtained a good series of the new butterfly, which I have since named Ornithoptera crœsus. The Mussænda bush was an admirable place, which I could visit every day on my way to the forest; and as it was situated in a dense thicket of shrubs and creepers, I set my man Lahi to clear a space all round it, so that I could easily get at any insect that might visit it. Afterwards, finding that it was often necessary to wait some time there, I had a little seat put up under a tree by the side of it, where I came every day to eat my lunch, and thus had half an hour's watching about noon, besides a chance as I passed it in the morning. In this way I obtained on an average one specimen a day for a long time, but more than half of these were females, and more than half the remainder worn or broken specimens, so that I should not have obtained many perfect males had I not found another station for them.

As soon as I had seen them come to flowers, I sent my man Lahi with a net on purpose to search for them, as they had also been seen at some flowering trees on the beach, and I promised him half a day's wages extra for every good specimen he could catch. After a day or two he brought me two very fair specimens, and told me he had caught them in the bed of a large rocky stream that descends from the mountains to the sea about a mile below the village. They flew down this river, settling occasionally on stones and rocks in the water, and he was obliged to wade up it or jump from rock to rock to get at them. I went with him one day, but found that the stream was far too rapid and the stones too slippery for me to do anything, so I left it entirely to him, and all the rest of the time we stayed in

Batchian he used to be out all day, generally bringing me one, and on good days two or three specimens. I was thus able to bring away with me more than a hundred of both sexes, including perhaps twenty very fine males, though not more than five or six that were absolutely perfect.

Source: Alfred Russel Wallace, *The Malay Archipelago: The Land of the Orang-utan and the Bird of Paradise: A Narrative of Travel, with Studies of Man and Nature*, 2 vols (London: Macmillan, 1869), II, pp. 23–5.

DROPS OF WATER

Ever since Antoni van Leeuwenhoek wrote his celebrated letter describing 'little animals in water' in 1676, generations of biologists have devoted themselves to the study of microorganisms, which are, it transpires, some of the oldest and most abundant life forms on the planet. The diatoms described below by the popular-science writer Agnes Catlow (c. 1807–89) show up in Jurassic-era fossils, and are present today in every body of water, from dewponds to oceans, where they play a vital primary production role at the base of the oceanic food chain.

Catlow begins her discussion with a suggestive comparison between astronomy and microscopy, both of which were dependent on the increasing refinement of their instruments. But while the first enables us to judge our planetary place in the great system of the universe, the microscope – 'a wonderful brazen tunnel, with crystal doors at the entrance' – instead reveals to us a teeming cosmos present in every drop of water. (The squeamish are then reassured that it is perfectly safe to drink):

The reflective mind experiences great delight in the investigation of those minute objects, which to the unassisted eye are invisible, but which, by the aid of a good microscope, may be studied at ease; our curiosity, however, is never satisfied, for, though by magnifying an object we find wonders revealed which before were hidden, we know that if our glasses were of a still higher power, we should discover more of the mechanism, and find out the use of many parts, that without this increased aid would remain in uncertainty. As our microscopes will probably never be made sufficiently powerful to show clearly all the minute creatures contained in water, we shall still remain ignorant of some of them. And this fact forms a parallel case to that relating to the stars; for astronomers have informed us, that, by the increased power of their telescopes, myriads of stars have been discovered, beyond those seen without this powerful aid, and that every year, as the glasses are improved, more and more stars appear: we may, therefore, in viewing the series the other way, imagine with some degree of probability, that water may still teem with life even beyond the reach of the highest powers of our glasses, and we may never be able to say with perfect truth of any drop of water, that it is free from animal or vegetable life. The thought is as overpowering in the one case as in the other, and we should be thankful that, by the aid of science, such systems are laid open for our investigation and study. Dr. Chalmers, in speaking of these two wonderful instruments, says very impressively – "While the telescope enables us to see a system in every star, the microscope unfolds to us a world in every atom. The one instructs us that this mighty globe, with the whole burthen of its people and its countries, is but a grain of sand in the vast field of immensity; the other, that every atom may harbour the tribes and families of a busy population."

After reassuring any of her readers who might be 'disgusted after viewing water through a microscope, and suppose that all water abounds in living creatures – this is an error: There are none, or very few, in spring water', Catlow then introduces the first of her four 'drops', each one illustrated by a beautiful hand-coloured plate (fig. 14). Having described some of its larger and brighter contents, she moves on to the smaller, more delicate organisms of the *Naviculacea* genus:

I shall not again repeat the opinions of naturalists, or the disputes as to the nature of this branch of the family *Bacillaria*, but merely name the most interesting of the genera. *Navicula* is so called from its shape, all the individuals being more or less in the form of a little boat, – and their quiet gliding motion increases the resemblance. The bright colours, or delicate transparent appearance of the lorica, make these little objects also very attractive; and I have watched a number of them moving in different directions with great pleasure, for their pace is so slow and gentle that all their habits may be noted. They appear to me to have decidedly a will of their own, and to avoid dangers of obstacles in their path, just as many of those animalcules do which are decidedly considered as belonging to the animal kingdom; and yet, as Dr. Meyen says, they are by no means so free and active as the spores of the Algæ, we are again puzzled. When watched with attention, they are found to glide slowly along at the bottom of the water, in a straight line, though occasionally a little slightly zigzag, as if to avoid the roughness of the glass; then, when anything obstructs their path, they do not go round it, but immediately turn back on almost the same track, without turning the body, and this motion I have seen a specimen keep up for some time: a little tap on the glass will arrest their progress, and then they immediately reverse their motion. When, however, they meet with a substance which may afford them food, they stop, and either glide under or about it. They appear to have cilia at the sides, for I have often observed small substances propelled along the body of the creature, and even running backwards and forwards several times, as if the *Navicula* might be extracting food from it. I do not know a prettier sight for the observer, than a drop of water containing several of these curious little creatures of different colours and shapes, gliding in various directions, and all actively engaged in their pursuits. Ehrenberg describes about forty species, some found in salt water, some in fresh, and others fossil. They are seen single, and in pairs, but never in chains, or bands, like the rest of this division of the family. They are generally broader in the middle than at the ends: some have the two ends very sharp, others are more blunt, and one or two are more round than long; a few have the ends curved different ways. Some are green, some brown or red, and others almost transparent; a few, also, are of a golden yellow. *N. viridis* (fig. 14, no. 14), *N. amphisbæna* (no. 15), and *N. acus* (no. 16), are frequently seen.

The genus *Bacillaria*, from which the family derives its name, is composed of bodies of singular construction. They seem originally to form connected lines, but, when mature, separate generally only in part, forming zigzag chains: in this state they move slightly, and one marine species, *B. paradoxa*, when separated from its companions, moves quickly like a *Navicula*. *B. vulgaris* (fig. 14, no. 17) is by many botanists looked upon as a vegetable, and named *Diatoma flocculosum*: it is found both in fresh and salt water, and has a straight lorica three or four times as long as broad; and when seen sideways it is in the form of a spindle. *B. cuneata* (fig. 14, no. 18) is of wonderful construction – each individual is wedge-shaped, and they still preserve the straight ribbon or chain-like form, by being placed alternately the broad and narrow end together. These chains of the *Bacillaria* are free and floating; as well as those of the next genus, *Fragilaria*, specimens of which are frequently seen to rise in the water, and turn round; the individuals also, when separated, move forward gently. They may be distinguished from *Navicula* by being square at the ends, instead of pointed, and having two openings instead of one. In *F. grandis* there are as many as thirty individuals found linked

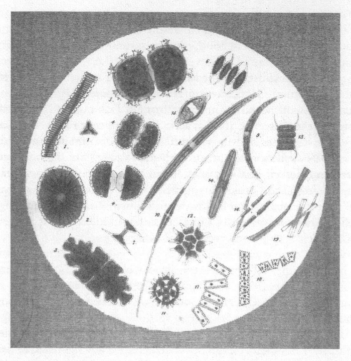

14. Drop I from Agnes Catlow's *Drops of Water*, showing a number of species of diatom, including those of the *Naviculacea* genus, as seen through a microscope.

together. *F. rhabdosoma* (fig. 14, no. 19) is very slender, and the bands drop to pieces very easily, and then the individuals creep about. The last genus of this division is *Meridion*, the separate individuals of which are wedge-shaped, but, unlike the *Bacillaria cuneata* mentioned above, they are arranged with all the small ends placed together, so that the band is not straight, but forms a spiral, if many are linked together; if there are but few, then a circle, or part of a circle, may be formed: *M. vernale* is the species usually seen. How marvellous are all these variations, when we consider the minute size of the objects!

Source: Agnes Catlow, *Drops of Water: Their Marvellous and Beautiful Inhabitants Displayed by the Microscope* (London: Reeve and Benham, 1851), pp. 3–5; 70–76.

THE PHYSICS OF A WET TOWEL

The Irish-born physicist John Tyndall (1820–93) was one of the nineteenth century's foremost popularizers of science and technology. An evening lecture that he delivered at London's Royal Institution in 1853 was such a success that later that year he was appointed Professor of Natural Philosophy at the Institution, where he would remain for the rest of his career. Though he became a noted research scientist, with a particular interest in radiant heat, he was best known as an inspiring speaker with an ability to communicate difficult ideas to large general audiences. The following extract, from the transcript of a talk he gave at the Royal Institution in the spring of 1854, is a typical Tyndall performance, in which the subject – in this case, the physics of light – is approached from a surprising angle:

At an agricultural college in Hampshire, with which I was connected for some time, and which is now converted into a school for the general education of youth, a Society was formed among the boys, who met weekly for the purpose of reading reports and papers upon various subjects. The Society had its president treasurer; and abstracts of its proceedings were published in a little monthly periodical issuing from the school press. One of the most remarkable features of these weekly meetings was, that after the general business had been concluded, each member enjoyed the right of asking questions on any subject on which he desired information. The questions were either written out previously in a book, or, if a question happened to suggest itself during the meeting, it was written upon a slip of paper and handed in to the Secretary, who afterwards read all the questions aloud . . .

To the mind which loves to detect in the tendencies of the young the instincts of humanity generally, such questions are not without a certain philosophic interest, and I have therefore thought it not derogatory to copy a few of them, and to introduce them here. They run as follows:—

What are the duties of the Astronomer Royal?

What is frost?

Why are thunder and lightning more frequent in summer than in winter?

What occasions falling stars?

What is the cause of the sensation called 'pins and needles'?

What is the cause of waterspouts?

What is the cause of hiccup?

If a towel be wetted with water, why does the wet portion become darker than before?

What is meant by Lancashire Witches?

Does the dew rise or fall?

What is the principle of the hydraulic press?

Is there more oxygen in the air in summer than in winter?

What are those rings we see round the gas and sun?

What is thunder?

How is it that a black hat can be moved by forming round it a magnetic circle, while a white hat remains stationary?

What is the cause of perspiration?

Is it true that men were once monkeys?

What is the difference between the *soul* and the *mind*?

Is it contrary to the rules of Vegetarianism to eat eggs?

In looking over these questions, which were wholly unprompted, and have been copied almost at random from the book alluded to, we see that many of them are suggested directly by natural objects, and are not such as had an interest conferred on them by previous culture . . . Take, for example, the case of the wetted towel, which at first sight appears to be one of the most unpromising questions in the list. Shall we tell the proposer to repress his curiosity, as the subject is improper for him to know, and thus interpose our wisdom to rescue the boy from the consequences of a wish which acts to his prejudice? Or, recognising the propriety of the question, how shall we answer it? It is impossible to answer it without reference to the laws of optics – without making the boy to some extent a natural philosopher. You may say that the effect is due to the reflection of light at the common surface of two media of different refractive indices. But this answer presupposes on the part of the boy a knowledge of what reflection and refraction are, or reduces you to the necessity of explaining them.

On looking more closely into the matter, we find that our wet towel belongs to a class of phenomena which have long excited the interest of philosophers. The towel is white for the same reason that snow is white, that foam is white, that pounded granite or glass is white, and that the salt we use at table is white. On quitting one medium and entering another, a portion of light is always reflected, but on this condition – the media must possess different refractive indices. Thus, when we immerse a bit of glass in water, light is reflected from the common surface of both, and it is this light which enables us to see the glass. But when a transparent solid is immersed in a liquid of the same refractive index as itself, it immediately disappears. I remember once dropping the eyeball of an ox into water; it vanished as if by magic, with the exception of the crystalline lens, and the surprise was so great as to cause a bystander to suppose that the vitreous humour had been instantly dissolved. This, however, was not the case, and a comparison of the refractive index of the humour with that of water cleared up the whole matter. The indices were identical, and hence the light pursued its way through both as if they formed one continuous mass.

In the case of snow, powdered quartz, or salt, we have a transparent solid mixed with air. At every transition from solid to air, or from air to solid, a portion of light is reflected, and this takes place so often that the light is wholly intercepted. Thus from the mixture of two transparent bodies we obtain an opaque one. Now the case of the towel is precisely similar. The tissue is composed of semi-transparent vegetable fibres, with the interstices between them filled with air; repeated reflection takes place at the limiting surfaces of air and fibre, and hence the towel becomes opaque like snow or salt. But if we fill the interstices with water, we diminish the reflection; a portion of the light is transmitted, and the darkness of the towel is due to its increased transparency. Thus the deportment of various minerals, such as hydrophane and tabasheer, the transparency of tracing paper used by engineers, and many other considerations of the highest scientific interest, are involved in the simple enquiry of this unsuspecting little boy.

Source: John Tyndall, 'On the Study of Physics', in *Fragments of Science for Unscientific People: A Series of Detached Essays, Lectures, and Reviews*, 6th edn, 2 vols (London: Longmans, Green, 1892), I, pp. 285–9.

PASTEURIZATION

Most of us will be familiar with the pasteurization of milk – the elimination of harmful bacteria by rapid heating and cooling – but how many of us knew that the process was originally developed as a means of prolonging the shelf-life of beer? The French chemist and microbiologist Louis Pasteur (1822–95) spent some years researching yeast fermentation before he turned to the more practical issue of the apparently spontaneous spoiling of beer and wine. Having studied the low fermentation techniques in use among the French and German breweries, he travelled to London in 1871 to investigate the high fermentation techniques that produced the characteristically dark varieties of British ale and porter. As Pasteur's son-in-law René Vallery-Radot (1853–1933) recalled in his biography of the great man, the visit to the Whitbread brewery caused a certain amount of consternation, especially when Pasteur produced his travelling microscope and treated the assembled managers to an impromptu microbiology lesson:

The great French *savant* was most courteously received by the managers of one of the most important breweries in London, who offered to show him round the works where 250 men were employed. But Pasteur asked for a little of the barm of the porter which was flowing into a trough from the cask. He examined that yeast with a microscope, and soon recognized a noxious ferment which he drew on a piece of paper and showed to the bystanders, saying, "This porter must leave much to be desired," to the astonished managers, who had not expected this sudden criticism. Pasteur added that surely the defect must have been betrayed by a bad taste, perhaps already complained of by some customers. Thereupon the managers owned that that very morning some fresh yeast had had to be procured from another brewery. Pasteur asked to see the new yeast, and found it incomparably purer, but such was not the case with the barm of the other products then in fermentation – *ale* and *pale ale*.

By degrees, samples of every kind of beer on the premises were brought to Pasteur and put under the microscope. He detected marked beginnings of disease in some, in others merely a trace, but a threatening one. The various foremen were sent for; this scientific visit seemed like a police inquiry. The owner of the brewery, who had

been fetched, was obliged to register, one after the other, these experimental demonstrations. It was only human to show a little surprise,
perhaps a little impatience of wounded feeling. But it was impossible
to mistake the authority of the French scientist's words: "Every
marked alteration in the quality of the beer coincides with the development of micro-organisms foreign to the nature of true beer yeast."
It would have been interesting to a psychologist to study in the
expression of Pasteur's hearers those shades of curiosity, doubt, and
approbation, which ended in the thoroughly English conclusion that
there was profit to be made out of this object lesson.

Pasteur afterwards remembered with a smile the answers he
received, rather vague at first, then clearer, and finally – interest and
confidence now obtained – the confession that there was in a corner
of the brewery a quantity of spoilt beer, which had gone wrong only
a fortnight after it was made, and was not drinkable. "I examined it
with a microscope," said Pasteur, "and could not at first detect any
ferments of disease; but guessing that it might have become clear
through a long rest, the ferments now inert having dropped to the
bottom of the reservoirs, I examined the deposit at the bottom of
the reservoirs. It was entirely composed of filaments of disease un-
mixed with the least globule of alcoholic yeast. The complementary
fermentation of that beer had therefore been exclusively a morbid
fermentation."

When he visited the same brewery again, a week later, he found
that not only had a microscope been procured immediately, but the
yeast of all the beer then being brewed had been changed.

Pasteur was convinced that micro-organisms from the environment were
responsible for the contamination of beer and wine, but he needed further
proof. Back in France, he continued his researches, which led him to the conclusions, as outlined below, which were the essentials of his germ theory of
disease; not surprisingly, Pasteur was an obsessively fastidious dinner-guest,
as Vallery-Radot recalled:

He was now going through a series of experiments, buying at Bertin's
much praised cafés samples of various famous beers – Strasburg,
Nancy, Vienna, Burton's, etc. After letting the samples rest for
twenty-four hours he decanted them and sowed one drop of the
deposit in vessels full of pure wort, which he placed in a temperature

of 20°C. After fifteen or eighteen days he studied and tasted the yeasts formed in the wort, and found them all to contain ferments of diseases. He sowed some pure yeast in some other vessels, with the same precautions, and all the beers of this series remained pure from strange ferments and free from bad taste; they had merely become *flat.*

He was eagerly seeking the means of judging how his laboratory tests would work in practice. He spent some time at Tantonville, in Lorraine, visiting an immense brewery, of which the owners were the brothers Tourtel. Though very carefully kept, the brewery was yet not quite clean enough to satisfy him. It is true that he was more than difficult to please in that respect; a small detail of his everyday life reveals this constant preoccupation. He never used a plate or a glass without examining them minutely and wiping them carefully; no microscopic speck of dust escaped his short-sighted eyes. Whether at home or with strangers he invariably went through this preliminary exercise, in spite of the anxious astonishment of his hostess, who usually feared that some negligence had occurred, until Pasteur, noticing her slight dismay, assured her that this was but an inveterate scientist's habit. If he carried such minute care into daily life, we can imagine how strict was his examination of scientific things and of brewery tanks.

After those studies at Tantonville with his curator, M. Grenet, Pasteur laid down three great principles –

1. Every alteration either of the wort or of the beer itself depends on the development of micro-organisms which are ferments of diseases.

2. These germs of ferments are brought by the air, by the ingredients, or by the apparatus used in breweries.

3. Whenever beer contains no living germs it is unalterable.

When once these principles were formulated and proved they were to triumph over all professional uncertainties. And in the same way that wines could be preserved from various causes of alteration by heating, bottled beer could escape the development of disease ferments by being brought to a temperature of 50° to 55°. The application of this process gave rise to the new word *"pasteurized"* beer, a neologism which soon became current in technical language.

Pasteur foresaw the distant consequences of these studies, and wrote in his book on beer –

"When we see beer and wine subjected to deep alterations because they have given refuge to micro-organisms invisibly introduced and now swarming within them, it is impossible not to be pursued by the thought that similar facts may, *must*, take place in animals and in man. But if we are inclined to believe that it is so because we think it likely and possible, let us endeavour to remember, before we affirm it, that the greatest disorder in the mind is to allow the will to direct the belief."

This shows us once more the strange duality of this inspired man, who associated in his person the faith of an apostle with the inquiring patience of a scientist.

Source: René Vallery-Radot, *The Life of Pasteur*, trans. Mrs R. L. Devonshire (London: Constable, 1906), pp. 210–14.

STAMMERING

The Oscar-winning film *The King's Speech* (2010) told the story of Lionel Logue (1880–1953), an Australian-born speech therapist who helped King George VI overcome a severe stammer that hindered his efforts at public speaking. A century earlier, J. H. Ayres Poett, a Dublin-born doctor who developed a specialized interest in the causes and treatment of stammering, held a reputation equal to Logue's. He began his career in Dublin in 1827, before moving first to London and then to Paris, where he worked as a hospital surgeon. Upon his return to London in 1856 he published his *Practical Treatise on Stammering*, based on thirty years' experience in the field. It was, he claimed, 'the first practical effort to bring the disease within the legitimate domains of medicine by an analysis of its pathology, an enumeration of its causes, and a description of its most successful mode of cure,' though he was well aware that 'cure' was (and remains) a somewhat optimistic term. In this first extract from his treatise Poett outlines the physiology of the condition:

Stammering may be defined as "a convulsive effort of the organs of speech to pronounce a syllable or word." Whatever may be the variety, or species of stammering, it will invariably be found capable of being included within this definition. To understand fully this theory or explanation of stammering, it will be necessary to hold in recollection that human speech requires two separate sets of volitions, one to the larynx for the production of sounds, the other to the mouth for their temporary obstruction, or, as it is called, pronunciation, and that these sets of volitions follow each other so rapidly as to make it impossible for the finest ear to separate the sound made in the larynx from that same sound pronounced in the mouth, although to our reason it must be evident that the action in the larynx must have been prior to that in the mouth, or in other words, what has been finished in the mouth must have been commenced in the larynx, and as a consequence that there must be two distinct sets of volitions following each other very rapidly to constitute speech; that the first of these volitions must be to produce the sound of the syllable or word, and the second to temporarily obstruct, modify, or pronounce that sound, so as to make it a fit representative of a consonant.

The volition to sound a syllable must therefore always precede the volition to pronounce it, for without sound there can be no pronunciation. The attempt to pronounce without having previously formed the sound to pronounce with would be the mere pantomime of speech. On the perfect accord between these two sets of volitions must depend the dexterity of the muscular actions concerned in pronunciation, and if this accord between them should not exist, one of two things must inevitably take place: – first, simple volition to sound without pronouncing; secondly, simple volition to pronounce without sounding; but sound, or a series of sounds, is mere vocalization, and not speech, and mere pronunciation without sound is an absurd attempt at an impossibility, being an attempt to pronounce sounds without sounds, and like all such efforts, if seriously continued, must end in convulsive action. And I may here observe, that under whatever peculiarity of effort stammering may exhibit itself, it will always be found accompanied or characterized by this want of accord between the organs of sound and the organs of pronunciation, inducing an anticipatory effort to pronounce without a sound.

To sound or vocalize alone is easy enough to every stammerer; if he will only avoid trying to pronounce, he can do so, but on the least attempt at pronunciation the volition to sound will be suspended, and the attempt to pronounce being continued, stammering will be produced, as the stammerer's mind is incapable of effecting the two volitions following each other sufficiently rapidly for the purposes of speech, he can sound alone, if he does not try to pronounce, or he can go through the dumb show of pronunciation alone, if he does not try to sound, but he cannot do both at once; and if he tries to do so he only neglects the first volition to sound, and flies off to the second, which, as I said before, being an impossibility, must, if earnestly persevered in, produce the convulsive efforts and involuntary movements which characterise stammering.

Poett's reputation was based on his track record of successfully treating the condition. 'I by no means wish to insinuate that I have been invariably fortunate in every case I have treated, far from it', he wrote; 'no physician can predicate unvarying success in the treatment of any disease, much less of a one like stammering – but I do contend that my mode of cure will be found effectual in the generality of cases.' Poett advocated a strict course of exercise of the 'muscles of sound and pronunciation' that was intended to build dexterity in the compound actions of speech. The larynx, he observed, was 'a stringed and a wind instrument', which was why learning to play it was so difficult. But, as he outlines in this second extract, there was another, less physical, dimension to the treatment:

Even in young children just commencing to stammer, this disciplining of the muscles of speech will be necessary, as the bad habit must be exchanged for, or supplanted by another, so entirely opposite as to render stammering impossible. I have not much faith in elocutional means taken alone, they must be combined in every case with moral ones, regulating the volitions, and in cases of very young children, combined with medical means for the regulation of the vital powers already alluded to in the enumeration of the predisposing and proximate causes . . .

The regulating of the irregular volitions will be the second indication of cure which is entirely disciplinetic, while by restoring self-confidence and consciousness of power by demonstrating to the reason of the patient that his stammer is a misplaced convulsive effort of his own, and which he can avoid if he only chooses to take the right way to do so, in fact that it is a voluntary convulsion and not an involuntary one, at least in the commencement; all this makes such a moral appeal as to entitle it to the third and most important place in the indications of cure. He must be taught over and over again to analyse, as it were, his stammer, and so convince himself of its being his own voluntary act, and not a necessary one which he cannot avoid; the youngest children are, strange to say, the soonest convinced of this fact. The impressing it on the mind of the stammerer is one of the most essential parts of the treatment, and for this purpose I daily demonstrate on a dried preparation of the larynx, the rhyma glottidis, and chordæ vocales where the sounds are made, the temporary stopping of which, in the mouth, moulds them into consonants, arranged afterwards into syllables, words, and phrases. I show how the stammerer, unlike other people, has not the power to make the sound perfectly, and stop it successfully at the same moment, his defective volition not giving him this dexterity. I show him that he consequently endeavours to effect with one set of actions what requires two, and that, overlooking the organs of sound, he tries to pronounce without it, which is an absurdity, being an attempt to produce sound without sound. I show him that he can sound alone, or go through the pantomime of pronunciation alone, easily enough, provided he does not attempt to do the two things at once, and that, as unfortunately, he cannot do that, he must begin with what he can do, which is to vocalise alone, without any attempt whatsoever at pronunciation, which establishes a habit of withdrawing the volition to the larynx, and thus counteracting, as it were, the morbid tendency to constantly anticipate the sound of the syllable, in the violent and unreasonable effort to stop or pronounce it. I then, by gradual advances, allow him to foreshadow, as it were, his future pronunciatory efforts in a peculiar mode of speech, christened by my patients the rule, as it is one from which, when under my influence or that of my family, the patients are not allowed to deviate from. This rule implies a peculiar, monotonous, drawling mode of speech, in which all the syllables have equal time, so that the volition can more easily direct itself to the preserving a continued stream of sound in the rhyma glottidis; it implies, also, a very imperfect pronunciation, barely sufficient to make the sounds intelligible, so as to restrain that determined effort to pronounce which characterises stammering; on this rule it is impossible to stammer, and yet stammerers can read and speak on it so as to be perfectly intelligible; it is, however, peculiarly monotonous and disagreeable, and almost impossible to practise except with the professor, and amongst stammerers themselves . . . But the rule is not always so unnatural and repulsive, every day lets them down, as it were, one step nearer to the usual modes of pronunciation, until at last every peculiarity of it fades away. As the professor himself practises it, and as all his patients have to do so in various grades, no unpleasant feeling is excited when amongst themselves; its practice out of doors is more mortifying, but use, example, and necessity, finally triumph, and once the stammerer can overcome the shame of speaking on the rule, from that moment his certain cure can be dated.

Source: J. H. Ayres Poett, *A Practical Treatise on Stammering; its Pathology, Predisposing, Exciting, and Proximate Causes, and its most Successful Mode of Cure, Scientifically Explained,* 2nd edn (London: John Churchill, 1858), pp. 14–16; 24–8.

DINOSAUR TEETH

Dinosaurs were the talismanic creatures of nineteenth-century public science, their bones rising from the earth to haunt the Victorian imagination as it grappled with new ideas about creation, extinction, and the age of the planet. One of the pioneers of dinosaur science was the English country doctor Gideon Mantell (1790–1852), who with his wife Mary Ann (née Woodhouse), unearthed dozens of fossil bones and teeth from the Cretaceous chalk downlands of Kent and Sussex. It was the Mantells' patient reconstruction of an enormous extinct reptile, which they named the Iguanodon, that attracted the greatest attention, particularly as the shape of its surviving teeth indicated that the creature had been herbivorous. Gideon Mantell's account of the episode appears below, followed by a short extract from the anatomist Richard Owen's landmark paper of 1842, in which he first proposed the use of the word 'dinosaur':

Gideon Mantell reconstructs the Iguanodon

It is now several years, since the discovery of a mutilated fragment of a tooth, led me to suspect the existence of a gigantic herbivorous animal in the strata of Tilgate Forest, which subsequent researches confirmed. This is the fragment: it is part of the crown of a tooth, resembling in its prismatic form the incisor of one of the herbivorous mammalia, worn by use. The enamel is thick in front and thin behind, and by this disposition a sharp cutting edge is maintained in every stage. Here, then, is a character, which if we bear in mind the principles of comparative anatomy . . . will afford us certain indications as to the nature of the animal to which it belonged. The structure of the tooth, and its worn surface, prove that it is referable to a species that fed on vegetables; the absence of a fang, and the appearance of the base, not broken, but *indented*, show that the shank has been absorbed from the pressure of a new tooth, which has grown up and supplanted the old one; a process too familiar to require explanation.

In the teeth before us we trace every gradation of this change, from the perfect form (fig. 15, no. 2,) – the partially worn specimen, (nos. 4, 5,) to the mere stump, (nos. 1, 3,) in which the crown is worn flat, and the absorption of the fang complete. The teeth, when perfect, are of a prismatic form, and remarkable for the prominent ridges which extend down the front, and the serrated margins of the crown (no. 6, the serrated edge magnified).

But although by this mode of induction the grand division of the animal kingdom to which the original belonged was determined, a

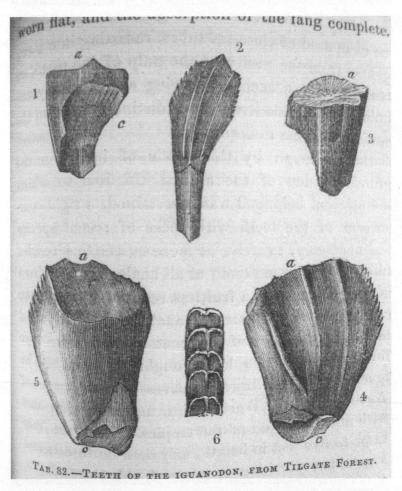

TAB. 82.—TEETH OF THE IGUANODON, FROM TILGATE FOREST.

15. Teeth of the iguanodon, from Gideon Mantell, *The Wonders of Geology* (1838).

rigid comparison of the teeth with those of recent species was necessary, to arrive at more satisfactory results. In a fossil state, no teeth
at all analogous had been noticed; and after a fruitless research
through the collections of comparative anatomy in London, I found,
in the jaws of a recent Iguana, the type for which I had so long sought
in vain. The Iguanas are land lizards, natives of many parts of America and the West Indies, and are rarely met with north or south of the
tropics. They are from three to five feet in length, and feed on insects
and vegetables, climbing trees, and chipping off the tender shoots.
They nestle in the hollows of rocks, and deposit their eggs, which are
like those of turtles, in the sands or banks of rivers. The Iguana is furnished with a row of very small, closely-set, pointed teeth, with
serrated edges, which are attached at the base, and by the outer surfaces of the fangs, to the jaw, the alveolar process forming an
external parapet: there is no internal bony covering. The new teeth
arise at the base of the old, and supplant them by occasioning the
absorption of the fangs. The teeth of the Iguana closely resemble the
perfect tooth, fig. 15, no. 2, except in size; those of the recent animal
scarcely exceeding in magnitude the teeth of the common mouse . . .
But apart from this discrepancy, the teeth and mode of dentition of
the fossil animal are so perfectly analogous to those of the Iguana,
that I have named the original the IGUANODON, signifying an animal having teeth like the Iguana. In the course of the present summer
a portion of jaw has been discovered, which confirms all the inferences that many years since I ventured to deduce from the teeth
alone.

From the gigantic size of the fossil teeth, as compared with the
recent, I was led to infer that many of the colossal bones, collected
from time to time, in Tilgate Forest, belonged to the same kind of
animal. By comparing the bones with the skeleton of the Iguana,
(presented me by Baron Cuvier,) I succeeded in determining many
parts of the skeleton; and at length was enabled to restore, as it were,
the form of the Iguanodon, and ascertain its proportions; the correctness of my inferences was shortly to be put to the test, by a discovery
in a neighbouring county.

In May, 1834, some workmen employed in a stone-quarry, in the
occupation of Mr. W. H. Bensted, of Maidstone, observed in a mass
of rock which they had blasted, several portions of what they supposed to be petrified wood; they preserved the largest piece for the

inspection of Mr. Bensted, who at once perceived that it was a portion of bone belonging to some gigantic animal. He therefore gave directions that every fragment should be collected, and after much labour and research, succeeded in obtaining those pieces, which are now united, and form a specimen of the highest interest; he also cleared away part of the surrounding stone, so as to expose the bones, which I have since completely developed and joined together.

The specimen consists of a considerable number of the bones, composing the inferior portion of the skeleton of an Iguanodon, which, when living, must have been upwards of 60 feet in length. The bones are imbedded in the stone in a very confused manner, few of them being in their natural order of juxta-position, and all more or less flattened and distorted . . .

The geological position of this specimen forms an exception to what has been previously remarked of the fossils of the Wealden; for while the bones in the latter are found associated with terrestrial and fluviatile remains only, the Maidstone specimen is imbedded in a marine deposit. This discrepancy, however, in no wise affects the arguments previously advanced, as to the fluviatile origin of the strata of the Wealden; it merely shows that part of the delta had subsided, and was covered by the chalk ocean, whilst the country of the Iguanodon was still in existence. The body of an Iguanodon was then drifted out to sea, and became imbedded in the sand of the ocean; in like manner, as at the present day, bones of land quadrupeds may not only be engulfed in deltas, but also in the deposits of the adjacent sea.

This specimen possesses a high interest, because it proves that the separate bones found in the strata of Tilgate Forest, and which I had assigned to the Iguanodon, solely from analogy, have been correctly appropriated; and we obtain also a knowledge of many interesting facts relating to the structure and economy of the original. I can notice but one of these inductions. As the Iguana lives chiefly upon vegetables, it is furnished with long slender feet, by which it is enabled to climb trees with facility, in search of food. But no tree could have borne the weight of the colossal Iguanodon, – its movements must have been confined to the land and water, and it is evident that its enormous bulk must have required limbs of great strength. Accordingly we find, that the hind feet, as in the Hippopotamus, Rhinoceros, and other large mammalia, were composed of strong, short, massy bones, furnished with claws, not hooked as in the Iguana, but compressed as in the land tortoises; thus forming a powerful support for the enormous leg and thigh. But the bones of the hands, or fore feet, are analogous to those of the Iguana, – long, slender, flexible, and armed with curved claws the exact counterpart of the nail-bones of the recent animal; thus furnishing prehensile instruments fitted to seize the palms, arborescent ferns, and dragon-blood plants, which probably constituted the food of the original. Here we have another interesting example of that admirable adaptation of structure to the necessities and conditions of every form of existence, which is alike manifest, whether our investigations be directed to the beings around us, or to the structure of those which have long since passed away.

Richard Owen (1804–92) names the Dinosaurs

[The discovery of the bones of] creatures far surpassing in size the largest of existing reptiles, will, it is presumed, be deemed sufficient ground for establishing a distinct tribe or sub-order of Saurian Reptiles, for which I would propose the name of *Dinosauria*: Gr. δεινός, fearfully great; σαύρος, a lizard.

Of this tribe the principal and best established genera are the *Megalosaurus*, the *Hylæosaurus*, and the *Iguanodon*; the gigantic Crocodile-lizards of the dry land, the peculiarities of the osteological structure of which distinguish them as clearly from the modern terrestrial and amphibious *Sauria*, as the opposite modifications for an aquatic life characterize the extinct *Enaliosauria*, or Marine Lizards.

Sources: Gideon Mantell, *The Wonders of Geology; or, A Familiar Exposition of Geological Phenomena*, 2 vols (London: Relfe & Fletcher, 1838), I, pp. 355–63; Richard Owen, 'Report on British Fossil Reptiles: Part II', *Report of the Eleventh Meeting of the British Association for the Advancement of Science, held at Plymouth in July 1841* (London: John Murray, 1842), p. 103.

CARNIVOROUS PLANTS

'You know none of the botanists agree with you', wrote Asa Gray to Mary Treat (1830–1923), when she first communicated her observations of the feeding habits of insectivorous plants. 'I cannot help it', she replied, 'for I have now seen it for myself, and know *it is* so.' And Mary Treat was right, bladderworts and pitcher plants, along with sundews and butterworts, kill and slowly digest their insect prey using a variety of means at their disposal.

Carnivorous plants have been objects of eerie fascination ever since the Venus flytrap was first described in 1763 by Arthur Dobbs (1689–1765), Governor of North Carolina. Since then, many other meat-eating plants have been discovered by botanists, and more than six hundred species have now been identified. The work of Mary Treat, a New Jersey naturalist best known for her long-running newspaper column 'Home Studies in Nature' (published in book form in 1885), was much admired by Charles Darwin, whose own book *Insectivorous Plants* (1875) contains numerous references to her observations. Darwin learned a great deal from Treat's ideas, and was fulsome in his acknowledgements to her work, yet, because Darwin's book appeared in the same year as Treat's landmark paper, 'Plants that Eat Animals', Treat was evidently worried that her findings were about to be eclipsed: note her attempt towards the end of her paper to claim respectful priority over the great English evolutionist:

About a year ago (in December 1873) a young man now at Cornell University and myself, on placing some of the bladders under the microscope, noticed animalcules – dead Entomostraca, &c., apparently imprisoned therein. But our attention was not sufficiently aroused to follow up the subject very closely; we laughingly called it "our new carnivorous plant." But as the bladders always seemed to be open, the significance of the fact of the imprisoned animal was not very apparent. We thought it could hardly be for the purpose of feeding the plant, but a kind of wanton cruelty. Still, my curiosity was aroused. I soon found larger animals in the bladders – dead larvæ of some aquatic insect – large enough to be seen distinctly with the naked eye. But I was not aroused to earnest work until I watched the movements of an imprisoned living larva, and saw its struggles and final death. This was in October, 1874. I now visited the ponds and procured abundant material.

The plant that I experimented mostly with was the one known to botanists as Utricularia clandestina. I soon became so deeply interested that I scarcely took note of time, and the small hours of the morning frequently found me absorbed in the work.

My observations were now more specially directed to the structure of the little bladder, which is quite complicated and very beautiful. It seems to be composed of irregular cells, and evenly distributed over the inner surface are clusters of star-like points, always four in number, arranged very regularly. The office of these stellate points I am as yet unable to determine. I have thought they might be to prevent the living animal from making too much exertion.

My next work was to see what prevented the escape of the animal from the bladder, and to this end I directed all my attention for several days. The animal that I found most commonly entrapped was a snake-like larva, about the length of the mosquito larva, but more slender and of lighter colour. Under the microscope it appears quite formidable, with fierce-looking jaws, and a pair of telescopic horns which it thrusts in and out at pleasure. Near the head are two beautiful brush-like telescopic feet, and at the other extremity are two more feet, which the animal thrusts out as a sort of propeller while swimming. I worked with this larva for several days, determined, if possible, to see him walk into the trap. I repeatedly took individuals from the water and placed them in the live-box with a spray of plant containing bladders; but it was of no use, the obstinate things would not accommodate me. The light or unnatural position, or both combined, made them fairly frantic, and they dashed about, paying no attention to the bladders. But I entrapped them another way. I put growing stems of the plant in a small dish of water with several larvæ, and set it aside. In a few hours thereafter I would find the living larvæ imprisoned. This served for another purpose, but not for the object I was aiming at. Forced to give up this plan of seeing the larvæ enter the bladder, I now directed my attention to the smaller ones – animalcules proper. I placed the bladders in water inhabited by numerous tiny creatures, and soon had the satisfaction of seeing the *modus operandi* by which the victim was caught.

The entrance into the bladder has the appearance of a tunnel-net, always open at the large end, but closed at the other extremity. The little animals seemed to be attracted into this inviting retreat. They would sometimes dally about the open entrance for a short time, but would sooner or later venture in, and easily open or push apart the closed entrance at the other extremity. As soon as the animal was fairly in, the forced entrance closed, making it a secure prisoner.

I was very much amused in watching a water-bear (Tardigrada) entrapped. It went slowly walking around the bladder, as if reconnoitering – very much like its larger namesake; finally it ventured in at the entrance, and easily opened the inner door and walked in. The bladder was transparent and quite empty, so that I could see the movements of the little animal very distinctly, and it seemed to look around as if surprised to find itself in so elegant a chamber; but it was soon quiet, and on the morning following it was entirely motionless, with its little feet and claws standing out as if stiff and rigid. The wicked plant had killed it very much quicker than it kills the snake-like larva . . .

So these points were settled to my satisfaction – that the animals were entrapped, and killed, and slowly macerated. But how was I to know that these animals were made subservient to the plant? If I could only prove that the contents of the bladders were carried directly into the circulation, my point was gained. This now was my sole work for several days, to examine closely the contents of the bladders. I found the fluid contents to vary considerably, from a dark muddy to a very light transparent colour. Hundreds of these bladders, one after another, were put to the test under the microscope, and I found that to a greater or lesser extent I could trace the same colour that I found in the bladder into the stem on which the bladder grew, though the observation was not so clear and satisfactory as I could wish. After more critical examination I arrived at the conclusion that the cells themselves, and not their contents, change to a red colour; the stems also take on this colour, so as to make it appear as if a red fluid was carried from the bladders into the main stem, which is not specifically the fact so far as the observations yet made determine, though the main point, that the contents of the bladders are carried into the circulation, does not seem open to question.

The next step was to see how many of the bladders contained animals, and I found almost every one that was well developed contained one or more of their remains, in various stages of digestion. The snake-like larva above mentioned was the largest and most constant animal found. On some of the stems that I examined fully nine out of every ten of the bladders contained this larva or its remains. When first caught it was fierce, thrusting out its horns and feet and drawing them back, but otherwise it seemed partly paralysed, moving its body but very little; even small larvæ of this species, that had plenty of room to swim about, were soon very quiet, although they showed signs of life from twenty-four to thirty-six hours after they were imprisoned. In about twelve hours, as nearly as I could make out, they lost the power of drawing their feet back, and could only move the brush-like appendages. There was some variation with different bladders as to the time when maceration or digestion began to take place, but usually, on a growing spray, in less than two days after a large larva was captured the fluid contents of the bladders began to assume a cloudy or muddy appearance, and often became so dense that the outline of the animal was lost to view.

Nothing yet in the history of carnivorous plants comes so near to the animal as this. I was forced to the conclusion that these little bladders are in truth like so many stomachs, digesting and assimilating animal food.

What it is that attracts this particular larva into the bladders is left for further investigation. But here is the fact, that animals are found there, and in large numbers, and who can deny that the plant feeds directly upon them? The why and wherefore is no more inexplicable than many another fact in Nature. And it only goes to show that the two great kingdoms of Nature are more intimately blended than we had heretofore supposed, and, with Dr. Hooker, we may be compelled to say, "our brother organisms – plants."

About December 1, after I had made the most of my observations, I wrote to Dr. Asa Gray and to Mr. Darwin, both on the same day, telling them of my discovery. Dr. Gray then informed me that Mr. Darwin had been engaged in the same work on Utricularia, and also sent me a note from him, bearing date August 5. From this note it would appear that at that date he had not worked the matter up as far as I had – at least had not found so many imprisoned animals – but with his superior faculties, he may have far outstripped me.

Ten years later, on Christmas Day, 1885, Treat treated her butterwort plant to its gruesome Christmas dinner:

On December 25th I placed tiny bits of raw fresh beef on ten leaves of P. pumila. In six hours the secretion was so copious that the spoon-tipped ends of seven leaves were filled. The secretion had mingled with the juice of the beef and looked bloody, but the meat itself was white and tender. In a little less than twelve hours the fluid had changed colour; it now looked clear, and remained so until it was gradually absorbed.

Sources: Mary Treat, 'Plants That Eat Animals', *American Naturalist* 9 (1875), pp. 658–62; Mary Treat, *Home Studies in Nature* (New York: American Book Company, 1885), p. 169.

LUNAR COSMORAMA

Ever since Galileo published the first detailed sketches of its cratered surface, the moon has been an object of allure for generations of scientists and observers; but of all the pre-Apollo-era lunar enthusiasts, James Nasmyth (1808–90) was surely the most devoted. An Edinburgh-born mechanical engineer, Nasmyth made a fortune from his invention of the steam hammer in 1839, which was specifically made to forge the paddle shaft for the SS *Great Britain*. The success of his invention allowed Nasmyth to retire from business in 1856, free to pursue his other interests, namely astronomy and photography, which he brought together in the form of a magnificently illustrated book on the moon, co-written with James Carpenter, from which the following extract and accompanying photograph are taken.

Though Nasmyth designed and built his own reflecting telescopes in order to gain ever-more detailed lunar views, his photographs were not of the moon itself, but of a series of plaster models he constructed, based on his own observations (it was not yet possible to take high-resolution photographs through a telescope). What follows is one of my favourite passages from Nasmyth's book, a rhapsody of informed speculation, in the shape of an imagined lunar travelogue:

Let us, in imagination, take our stand high upon the eastern side of the rampart of one of the great craters. Height, it must be remarked, is more essential on the moon to command extent of view than upon the earth, for on account of the comparative smallness of the lunar sphere the dip of the horizon is very rapid. Such height, however, would be attained without great exercise of muscular power, since equal amounts of climbing energy would, from the smallness of lunar gravity, take a man six times as high on the moon as on the earth. Let us choose, for instance, the hill-side of Copernicus. The day begins by a sudden transition. The faint looming of objects under the united illumination of the half-full earth and the zodiacal light is the lunar precursor of day-break. Suddenly the highest mountain peaks receive the direct rays of a portion of the sun's disc as it emerges from below the horizon. The brilliant lighting of these summits serves but to increase, by contrast, the prevailing darkness, for they seem to float like islands of light in a sea of gloom. At a rate of motion twenty-eight times slower than we are accustomed to, the light tardily creeps down the mountain-sides, and in the course of about twelve hours the whole of the circular rampart of the great crater below us, and towards the east, shines out in brilliant light, unsoftened by a trace of mountain-mist. But on the opposite side, looking into the crater, nothing but blackness is to be seen. As hour succeeds hour, the sun-beams reach peak after peak of the circular rampart in slow succession, till at length the circle is complete and the vast crater-rim, 50 miles in diameter, glistens like a silver-margined abyss of dark-ness. By-and-by, in the centre, appears a group of bright peaks or bosses. These are the now illuminated summits of the central cones, and the development of the great mountain cluster they form hence-forth becomes an imposing feature of the scene. From our high standpoint, and looking backwards to the sunny side of our cos-morama, we glance over a vast region of the wildest volcanic desolation. Craters from five miles diameter downwards crowd together in countless numbers, so that the surface, as far as the eye can reach, looks veritably frothed over with them. Nearer the base of the rampart on which we stand, extensive mountain chains run to north and to south, casting long shadows towards us; and away to southward run several great chasms a mile wide and of appalling blackness and depth. Nearer still, almost beneath us, crag rises on crag and precipice upon precipice, mingled with craters and yawning pits, towering pinnacles of rock and piles of scoria and volcanic *débris*. But we behold no sign of existing or vestige of past organic life. No heaths or mosses soften the sharp edges and hard surfaces: no tints of cryptogamous or lichenous vegetation give a complexion of life to the hard fire-worn countenance of the scene. The whole landscape, as far as the eye can reach, is a realization of a fearful dream of desolation and lifelessness — not a dream of death, for that implies evidence of pre-existing life, but a vision of a world upon which the light of life has never dawned.

Looking again, after some hours' interval, into the great crateral amphitheatre, we see that the rays of the morning sun have crept down the distant side of the rampart, opposite to that on which we stand, and lighted up its vast landslipped terraces into a series of seeming hill-circles with all the rude and rugged features of a terres-trial mountain view, and none of the beauties save those of desolate grandeur. The plateau of the crater is half in shadow 10,000 feet below, with its grand group of cones, now fully in sight, rising from its centre. Although these last are twenty miles away and the base of the opposite rampart fully double that distance, we have no means of judging their remoteness, for in the absence of an atmosphere there can be no aërial perspective, and distant objects appear as brilliant and distinct as those which are close to the observer. Not the brightness only, but the various colours also of the distant objects are preserved in their full intensity; for colour we may fairly assume there must be. Mineral chlorates and sublimates will give vivid tints to certain parts of the landscape surface, and there must be all the

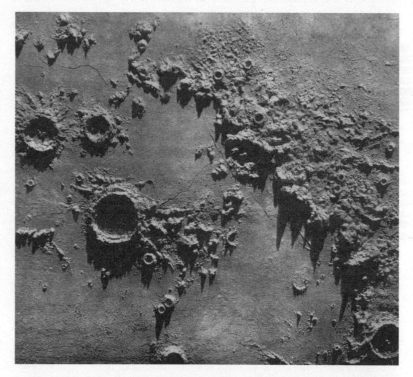

16. This atmospheric photograph of the lunar crater Archimedes is not exactly what it seems. James Nasmyth built plaster models of the lunar surface, based on his careful telescopic observations; he then took skilfully lit photographs of his dioramas, achieving better results than were available at the time from direct lunar photography. The image is one of many that featured in his pioneering book *The Moon* (1874).

more sombre colours which are common to mineral matters that have been subjected to fiery influence. All these tints will shine and glow with their greater or less intrinsic lustres, since they have not been deteriorated by atmospheric agencies, and far and near they will appear clear alike, since there is no aërial medium to veil them or tarnish their pristine brightness.

In the lunar landscape, in the line of sight, there are no means of estimating distances; only from an eminence, where the intervening ground can be seen, is it possible to realize *magnitude* in a lunar cosmorama and comprehend the dimensions of the objects it includes.

And with no air there can be no diffusion of light. As a consequence, no illumination reaches those parts of the scene which do not receive the direct solar rays, save the feeble amount reflected from contiguous illuminated objects, and a small quantity shed by the crescent earth. The shadows have an awful blackness. As we stand upon our chosen point of observation, we see on the lighted side of the rampart almost dazzling brightness, while beneath us, on the side away from the sun, there is a region many miles in area impenetrable to the sight, for there is no object within it receiving sufficient light to render it discernible; and all around us, far and near, there is the violent contrast between intense brightness of isolated parts and deep gloom of those in equally intense shadow. The black though starlit sky helps the violence of this contrast, for the bright mountains in the distance around us stand forth upon a background formed by the darkness of interplanetary space. The visible effects of these conditions must be in every sense unearthly and truly terrible. The hard, harsh glowing light and pitchy shadows; the absence of all the conditions that give tenderness to an earthly landscape; the black noonday sky, with the glaring sun ghastly in its brightness; the entire absence of vestiges of any life save that of the long since expired volcanoes — all these conspire to make up a scene of dreary, desolate grandeur that is scarcely conceivable by an earthly habitant, and that the description we have attempted but insufficiently portrays.

Source: James Nasmyth and James Carpenter, *The Moon: Considered as a Planet, a World, and a Satellite* (London: John Murray, 1874), pp. 164–6.

KRAKATOA DUST

The eruption of Mt Krakatoa in August 1883 was one of the most significant events in the history of meteorology; for while volcanologists had to make their way to Indonesia's Sunda Strait in order to make on-site observations, everyone else needed only to look up to see its full atmospheric impact. This was because the eruption had jettisoned millions of tonnes of volcanic particulates into the stratosphere, which then drifted over much of the earth as a semi-opaque band of dust and gas; as it threaded slowly westwards, the dust-veil caused luridly coloured sunsets and afterglows to advance across the evening skies. Apart from their beauty, these twilight phenomena offered a valuable resource for understanding the behaviour of atmospheric circulation; it was as though a kind of dye had been added to the earth's upper atmosphere, rendering its usually unseen movements suddenly and thrillingly visible. For the two and a half years that it remained aloft, the Krakatoa dust remained an object of close observation for scientists all over the world.

Joseph Wharton

In suburban Philadelphia, for example, a metallurgist named Joseph Wharton (1826–1909) had been entranced by 'the splendid roseate glows' that descended every evening through the hazy sky, and he devised his own experiment for testing whether they really had been caused by the dust from Krakatoa. After extracting samples of the dust that had fallen to earth in a snowdrift near his house, he set about comparing them with ground-up pumice from the Sunda Strait that had been collected by the crew of a Philadelphia cargo steamer that had returned to port in early 1884. As a control measure, he compared his snow-caught sample with dust collected from other sources, such as nearby steel works and blast furnaces, in order to rule out the possibility of industrial contamination. Wharton later wrote up his experiment in the *Proceedings of the American Philosophical Society*, and it reads as an excellent example of inductive science in action, as well as a reminder of the truly global impact of the Krakatoa eruption:

The splendid roseate glows which in the winter of 1883–4 were visible in the western sky after sunset and in the eastern sky before sunrise, gave rise to many conjectures, but apparently to almost no experiments. A few persons believed these glows to be sunlight reflected from the under surface of a stratum of fine solid particles suspended at a great height in the atmosphere; some thought with me that those particles might be volcanic dust which had floated to us from the eruption at Krakatoa, but, as no one offered any proof of this, I attempted on the morning of January 20, 1884, to demonstrate it. Six miles northward from the centre of Philadelphia, where I reside, a light and fine snow was then gently falling in an almost calm atmosphere, presumably from a high altitude. Of that snow, while it was yet falling, I collected about a gallon by skimming it carefully with my hands from a considerable surface in a field a hundred yards to windward of my house and a quarter-mile from the nearest windward building.

This very clean new-fallen snow I melted under cover in the porcelain bowl it was gathered in, and was at first unable to detect any sediment; after maintaining for several minutes a gentle rotatory movement of the bowl in order to bring into its deepest part any solid matter which might be present, I poured off most of the water and evaporated the remainder. A minute quantity of fine dust was then discerned by the tiny vitreous reflections which it gave in the sunlight. My practice in chemical analysis, and therefore in weighing small quantities, affords some justification for the estimate that the total weight of this dust was less than one-hundredth of a grain.

Under the microscope, where it was immediately placed, this dust showed the characteristics of volcanic glass; it consisted in part of irregular, flattish, blobby fragments, mostly transparent and showing no trace of crystalline structure, in part of transparent filaments more or less contorted, sometimes attached together in wisps, and mostly sprinkled with minute glass particles. The filaments of glass had about the same diameter as single filaments of silk placed on the microscope slide for comparison with them.

Having microscopically examined the dust again and again, I ignited it upon platinum to destroy any organic matter which might be present, and thereafter found the filaments, the flat plates, and the amorphous accretions of glass quite unchanged.

No pyroxene, augite, or magnetite, such as have elsewhere been observed in volcanic dust, was present; it may be assumed that, if at first mingled with the glass, those heavier minerals had been dropped during the long voyage of more than ten thousand miles of space and more than four months of time . . .

Early in February, 1884, the ship *J. E. Ridgeway* arrived at Philadelphia from Manila by the Strait of Sunda. On February 12, I visited that ship, and read on her log-book that at 10 P.M., October 27, 1883, in south latitude 7° 57' and east longitude 100° 54' (about five hundred miles W. S. W. from Krakatoa), she encountered a vast field of floating pumice, through which she sailed until 7 A.M., October 29. So abundant was this pumice that the ship's speed was reduced from nine knots when she entered it to two knots at 6 P.M., October 28; several hours after that time her speed gradually increased, as the pumice became less dense, from two knots to eight, and finally, when she cleared it, to her normal nine knots. No volcanic ash had fallen upon the ship, as she arrived too late upon the scene.

Some of this pumice I took directly from the hands of the mate and steward, who had collected it from the sea and had kept it in their private lockers. It can scarcely be doubted that this pumice was ejected from Krakatoa.

Now, on placing under the microscope small crumbs of that pumice and filaments picked out from its cavities, I recognized just such transparent flattish scraps and ragged accretions as were among the dust found in the snow-fall of January 20, while the filaments, though less varied and interesting than those then collected, were quite similar in character, even to the tiny glass particles sprinkled upon them.

A minor point of resemblance was that the yellow color of one little vesicular mass in the dust caught on January 20, was fairly matched by a slight streak of similar color in the pumice.

In March, 1884, I collected dust from the steel works at South Bethlehem, Pa., and also dust from a blast furnace there, in order to compare them with the dust found in the snow and with the filaments and crumbs of pumice from the ship *J. E. Ridgeway*.

After separating from these dusts the large proportion which was attracted by the magnet, the remnant showed in each case many vitreous particles; that from the iron furnace largely spheroidal or globular, with a few filaments; that from the steel works partly minute rounded particles, but containing many filaments of great tenuity. Neither contained such clear vitreous plates and aggregations as abounded in the snow-dust, while the filaments in both were of dark color, and smooth, straight form, distinctly different from the colorless and frequently contorted filaments of the snow-dust.

It is difficult to resist the conclusions (1) that the vitreous dust found in the snow-fall of January 20, 1884, was not derived from iron or steel furnaces, (2) that it was of similar origin to the floating pumice found by the ship *J. E. Ridgeway*, (3) that it was ejected by the huge volcanic explosion of Krakatoa.

Gerard Manley Hopkins

The extraordinary sunsets that were seen all over the world in the months following the eruption attracted a great deal of comment and speculation. In January 1884 the poet Gerard Manley Hopkins (1844–89), then a classics tutor at Stonyhurst College in Lancashire, sent a lyrical description of the volcanic skyscapes over the Ribble Valley to the journal *Nature*; as can be seen from the following extract, the letter was both a vivid prose poem and a well-observed piece of meteorological reportage:

I subjoin an account of the sunset of the 16th [December 1883], which was here very remarkable, from my own observations and those of one of the observatory staff.

A bright glow had been round the sun all day and became more remarkable towards sunset. It then had a silvery or steely look, with soft radiating streamers and little colour; its shape was mainly elliptical, the slightly longer axis being vertical; the size about 20° from the sun each way. There was a pale gold colour, brightening and fading by turns for ten minutes as the sun went down. After the sunset the horizon was, by 4.10, lined a long way by a glowing tawny light, not very pure in colour and distinctly textured in hummocks, bodies like a shoal of dolphins, or on what are called gadroons, or as the Japanese conventionally represent waves. The glowing vapour above this was as yet colourless; then this took a beautiful olive or celadon green, not so vivid as the previous day's, and delicately fluted; the green belt was broader than the orange, and pressed down on and contracted it. Above the green in turn appeared a red glow, broader and burlier in make; it was softly brindled, and in the ribs or bars the colour was rosier, in the channels where the blue of the sky shone through it was a mallow colour. Above this was a vague lilac. The red was first noticed 45° above the horizon, and spokes or beams could be seen in it, compared by one beholder to a man's open hand. By 4.45 the red had driven out the green, and, fusing with the remains of the orange, reached the horizon. By that time the east, which had a rose tinge, became of a duller red, compared to sand: according to my observation, the ground of the sky in the east was green or else tawny, and the crimson only in the clouds. A great sheet of heavy dark cloud, with a reefed or puckered make, drew off the west in the course of the pageant: the edge of this and the smaller pellets of cloud that filed across the bright field of the sundown caught a livid green. At 5 the red in the west was fainter, at 5.20 it became notably rosier and livelier; but it was never of a pure rose. A faint dusky blush was left as late as 5.30, or later. While these changes were going on in the sky, the landscape of Ribblesdale glowed with a frowning brown.

The two following observations seem to have to do with the same phenomena and their causes. For some weeks past on fine bright days, when the sun has been behind a big cloud and has sent up (perspectively speaking) the dark crown or paling of beams of shadow in such cases commonly to be seen, I have remarked, upon the ground of the sky, sometimes an amber, sometimes a soft rose colour, instead of the usual darkening of the blue. Also on moonlight nights, and particularly on December 14, a sort of brown or muddy cast, never before witnessed, has been seen by more than one observer, in the sky.

Sources: Joseph Wharton, 'Dust from the Krakatoa Eruption of 1883', *Proceedings of the American Philosophical Society* 32 (1894), pp. 343–5; Gerard Manley Hopkins, 'The Remarkable Sunsets', *Nature* 29 (3 January 1884), pp. 222–3.

EDISON'S CARBON TELEPHONE

Although Alexander Graham Bell (1847–1922) is often credited with the invention of the telephone – he was awarded the first patent for it in 1876 – its development was more of a collective, international endeavour than popular history suggests. One of the many scientists and inventors who contributed to the evolving technology was the former telegraph operator Thomas Edison (1847–1931), best known today for his work on the electric light bulb. In 1877–8, at his research lab in Menlo Park, New Jersey, Edison developed the carbon microphone, which greatly enhanced the fidelity of the electric telephone; it was subsequently used as a transmitter in conjunction with the Bell receiver until the early 1980s. The high-fidelity wonders of the new carbon telephone were demonstrated to an enthusiastic London audience at the Royal Institution, on Albemarle Street, in 1879, as described in this exuberant account by the Irish physicist and mountaineer John Tyndall (who we met before, on p. 257):

In 1879 the performance of the Edison telephone was illustrated in the theatre of the Royal Institution. Through the kindness of Lord John Manners and of the Post Office authorities, a wire, passing through the air from Albemarle Street to Piccadilly Circus, was placed at my service. The two ends of this wire being connected with the public water-pipes at the respective stations, a circuit was established through which a voltaic current could flow. In the circuit, at each end of the wire, was placed an ordinary carbon telephone (to be referred to immediately), into which the messages were spoken. But while the receiver, at the Circus, of the messages sent from the Royal Institution, was a Bell's magnetic telephone held to the ear, the receiver at the Royal Institution was Edison's loudspeaking telephone. The nephew of Mr. Edison, who bore his name, was stationed at the Circus, while Mr. Adams operated with the new instrument in Albemarle Street. Passages from Shakespeare, Scott, Tennyson, Macaulay, and Burns, spoken by me through the carbon telephone, were received at the Circus, there repeated by Mr. Edison, and returned with an accuracy and loudness which enabled them to be heard throughout the theatre. Not only were selected phrases thus heard, but a poem of Emerson's was read out here from beginning to

end, and sent back line by line with extraordinary fidelity and distinctness. Various expressions, moreover, following the quotations,
such as 'Excellent!' 'Perfectly satisfactory!' 'Exceedingly good!' were
promptly returned and heard with amusing intensity by the audience. Perhaps the most striking illustration of the pliant power of the
instrument was its capability to reproduce a whistled tune. Mr. Edison's whistling at the Circus was heard in Albemarle Street almost as
distinctly as if it had been produced upon the spot. After the lecture
I quitted the theatre for a time, during which some members of the
audience took my place. On my return I resumed the carbon telephone, and spoke into it. Mr. Edison immediately detected this
difference in tone, and, on being asked who it was that now spoke,
answered correctly. By this new instrument, therefore, the varying
qualities of the human voice are in a remarkable degree reproduced.

These extraordinary effects were obtained with an apparatus so
simple, and apparently so rude, that, without hearing the instrument,
its alleged performance could hardly have been believed . . . In the
Edison telephone there is a small rotating cylinder, and a flat strip of
metal, one end of which is pressed down upon the cylinder by a
spring. The other end of this metal rubber is attached to the centre of
a thin circular plate of mica about four inches in diameter. The cylinder is formed of powdered chalk, with which are mixed a little
hydrate of potash and acetate of mercury, the powder being squeezed
to hardness by pressure in a cylindrical mould. Through the centre
of the cylinder passes a metallic axis. This is connected with one end
of the secondary wire of a very small induction coil, the other end of
the wire being joined to the flat strip of metal above referred to.
When moistened, the cylinder becomes to all intents and purposes
an electrolyte, every passage of a current producing an amount of
decomposition exactly proportionate to the current's strength and
duration.

From a very small voltaic battery, a current was sent through the
primary wire of the small induction coil, thence through the carbon
telephone held by me, onward to Piccadilly Circus, from which it
returned under the earth to Albemarle Street. As long as this current
flowed without any variation of strength, no effect whatever was
produced upon the Edison telephone. A hand turning the crank of
the chalk cylinder experienced a uniform resistance, the mica plate
being drawn inwards with a constant force. In the carbon telephones

employed in these experiments, a thin cake of fine petroleum lamp-black was held between two thin plates of platinum, on one of which the voice impinged. The alternate compression and relaxation of the lampblack, by varying the resistance, produced variations in the voltaic current corresponding to the vibrations of the voice. Every variation thus introduced into the primary current started an induced current in the small secondary coil, while every such induced current produced its due amount of electro-chemical decomposition at the common surface of chalk cylinder and metallic rubber. By this decomposition a lubricant was liberated underneath the rubber, which immediately yielded, by slipping, to the tension of the mica plate. Each slip was of momentary duration, being followed by a frictional 'bite' which drew the mica diaphragm inwards as before. Thus the vibrations of the voice — of its tones and overtones — were in the first place impressed upon the primary current, every variation of the latter being followed by a proportionate discharge of secondary currents through the induction coil. By their electrolytic action these induced discharges produced and controlled the slipping of the metal rubber, causing it to vibrate longitudinally in accordance with the vocal vibrations. These were finally transferred, with their qualities to a surprising extent intact, to the plate of mica, and thence to the surrounding air. The mica plate might, therefore, be regarded as a magnified tympanic membrane, the latter, like the plate, being drawn inwards by the bones of the ear. It may seem amazing that the mica should be able to take up and reproduce with such intensity and distinctness the manifold vibrations involved in whistling and speaking; but the wonder was anticipated by an artificer more ancient than Mr. Edison, in the construction of the tympanum itself.

Source: John Tyndall, *Sound*, 4th edn (London: Longmans, Green, 1883), pp. 201–5.

FIDGETING

Sir Francis Galton (1822–1911), coiner of the phrase 'nature versus nurture', was a true Victorian polymath; the breadth of his interests was extraordinary – among his many achievements was the first printed weather map in 1875; the first cartographic survey of South-West Africa (present-day Namibia), for which he was awarded the Gold Medal of the Royal Geographical Society; and founding the discipline of psychometric testing – though he is probably best known today for his enthusiasm for the now discredited field of eugenics (a term he also coined), the aim of which was to 'improve' the human gene pool through a programme of selective breeding – 'the self direction of human evolution', as its supporters took to describing it. Galton's association with eugenics has cost him his reputation as one of the nineteenth century's most outstanding thinkers, a sage who sought to apply statistical methodologies to a variety of human problems, including weather, criminality, and – in this instance – boredom.

The Royal Geographical Society may have given Galton its highest honour, but he didn't always enjoy his visits there; frequently bored by the lengthy presentations, he began to wonder whether a mental state such as boredom could usefully be measured among an audience. 'For instance', he recalled in his memoirs, 'by counting the number of their Fidgets. I not infrequently tried this at the Royal Geographical Society, for even there dull memoirs are occasionally read . . . I have often amused myself with noticing the increase in that number as the audience becomes tired. The use of a watch attracts attention, so I reckon time by the number of my breathings, of which there are fifteen in a minute. The counting is reserved for the fidgets. These observations should be confined to persons of middle age. Children are rarely still, while elderly philosophers will sometimes remain rigid for minutes together.' Galton wrote up his preliminary findings in a short, witty paper that he sent to the journal *Nature* in 1885. Note the way he treats the lecture hall as a calibrated instrument, its columns acting as a virtual grid with which to divide the audience into convenient sectors for study – a fine example of scientific observation applied to an everyday circumstance:

Latterly – no matter where – I was present at a crowded and expectant meeting. The communication proved tedious, and I could not hear much of it, so from my position at the back of the platform I studied the expressions and gestures of the bored audience.

The feature that an instantaneous photograph, taken at any moment, would have most prominently displayed was the unequal horizontal interspace between head and head. When the audience is intent each person forgets his muscular weariness and skin discomfort, and he holds himself rigidly in the best position for seeing and hearing. As this is practically identical for persons who sit side by side, their bodies are parallel, and again, as they sit at much the same distances apart, their heads are correspondingly equidistant. But when the audience is bored the several individuals cease to forget themselves and they begin to pay much attention to the discomforts attendant on sitting long in the same position. They sway from side to side, each in his own way, and the intervals between their faces, which lie at the free end of the radius formed by their bodies, with their seat as the centre of rotation varies greatly. I endeavoured to give numerical expression for this variability of distance, but for the present have failed. I was, however, perfectly successful in respect to another sign of mutiny against constraint, inasmuch as I found myself able to estimate the frequency of fidget with much precision. It happened that the hall was semicircularly disposed and that small columns under the gallery were convenient as points of reference. From where I sat, 50 persons were included in each sector of which my eye formed the apex and any adjacent pair of columns the boundaries. I watched most of these sections in turn, some of them repeatedly, and counted the number of distinct movements among the persons they severally contained. It was curiously uniform, and about 45 per minute. As the sectors were rather too long for the eye to surely cover at a glance, I undoubtedly missed some movements on every occasion. Partly on this account and partly for the convenience of using round numbers I will accept 50 movements per minute among 50 persons, or an average of 1 movement per minute in each person, as nearly representing the true state of the case. The audience was mostly elderly; the young would have been more mobile. Circumstances now and then occurred that roused the audience to temporary attention, and the effect was twofold. First, the frequency of fidget diminished rather more than half; second, the amplitude and period of each movement were notably reduced. The swayings of head, trunk, and arms had before been wide and sluggish, and when rolling from side to side the individuals seemed to "yaw"; that is to say, they lingered in extreme positions. Whenever they became intent this peculiarity disappeared, and they performed their fidgets smartly. Let me suggest to observant philosophers when the meetings they attend may prove dull, to occupy themselves in estimating the frequency, amplitude and duration of their fellow-sufferers. They must do so during periods both of intentness and of indifference, so as to eliminate what may be styled "natural fidget," and then I think they may acquire the new art of giving numerical expression to the amount of boredom expressed by the audience generally during the reading of any particular memoir.

No doubt the boring lecture will always be with us, although, as the science historian Steven Shapin has argued, it doesn't have to be this way. In a passionately argued essay entitled 'Hyperprofessionalism and the Crisis of Readership in the History of Science', published in the journal *Isis* in 2005, he lamented the fact that academics feel almost duty-bound to write and speak incomprehensibly, as though it was some hard-won badge of professionalism: 'they resist using the vernacular because it doesn't sound smart enough; they infer from obscurity to profundity.' This is bad enough on paper (where at least you can stop reading), but in its spoken form it is lethally dull, even for fellow academics. 'My guess is that at most [academic] colloquia more than half the audience is not "there"', claims Shapin; 'if they are civilly minded, they are desperately trying to arrange their faces to simulate rapt attention; if they are not, they are correcting proofs or reading a newspaper tactically positioned on the floor. I've even heard of a wireless senior academic who does his e-mail during colloquia, but I can't make myself believe that's really so.'

Sources: Francis Galton, 'The Measure of Fidget', *Nature* 32 (1885), pp. 174–5; Steven Shapin, 'Hyperprofessionalism and the Crisis of Readership in the History of Science', *Isis* 96 (2005), pp. 238–43.

SURFACE TENSION

'I had a passionate interest in natural science, especially physics, and would have loved to study it,' recalled the seventy-year-old Agnes Pockels in 1932; but German universities did not admit women in the 1880s and 1890s, so she taught herself at home using textbooks supplied by her younger brother, Friedrich, who went on to earn a PhD in physics from the University of Göttingen. As a full-time carer for her infirm parents, Pockels (1862–1935) spent much of her life at the family home in Brunswick, where she attended to her scientific interests whenever time permitted. She became fascinated by the complex behaviour of dirty washing-up water, and devised a simple piece of apparatus – now known as the Pockels trough – with which she was able to demonstrate that the surface tension of a liquid was greatly reduced by the presence of organic impurities; the greater the concentration of impurities, the greater the measurable reduction in surface tension (the molecular forces acting at the surface of a liquid or solid).

Unable to find any mention of this in her brother's textbooks, she tried without success to interest his tutors in her work. But in 1888 she learned that the British physicist Lord Rayleigh (John William Strutt, 3rd Baron Rayleigh (1842–1919), Professor of Natural Philosophy at Cambridge), had begun a similar line of enquiry, so she wrote him a long letter describing the results of her kitchen-sink experiments. Rayleigh recognized the originality of Pockels's work and, to his credit, arranged for the letter to be translated and published in the journal Nature in 1891. Two extracts from this letter, as well as Rayleigh's covering note, appear below. Pockels went on to publish several more articles on the subject, and although she never held a scientific post, she was widely recognized as a pioneer of surface science. In 1931 she shared the Laura Leonard Prize for her work, and in 1932 was awarded an honorary doctorate from the Carolina-Wilhelmina University of Brunswick: the same year that Irving Langmuir received a Nobel Prize in Chemistry for work on the chemistry of surfaces that he conducted with a modified version of Pockels's tin trough:

I shall be obliged if you can find space for the accompanying translation of an interesting letter which I have received from a German lady, who with very homely appliances has arrived at valuable results respecting the behaviour of contaminated water surfaces. The earlier part of Miss Pockels' letter covers nearly the same ground as some of my own recent work, and in the main harmonizes with it. The later sections seem to me very suggestive, raising, if they do not fully answer, many important questions. I hope soon to find opportunity for repeating some of Miss Pockels' experiments.

RAYLEIGH. March 2.

Brunswick, January 10.

My Lord, — Will you kindly excuse my venturing to trouble you with a German letter on a scientific subject? Having heard of the fruitful researches carried on by you last year on the hitherto little understood properties of water surfaces, I thought it might interest you to know of my own observations on the subject. For various reasons I am not in a position to publish them in scientific periodicals, and I therefore adopt this means of communicating to you the most important of them.

First, I will describe a simple method, which I have employed for several years, for increasing or diminishing the surface of a liquid in any proportion, by which its purity may be altered at pleasure.

A rectangular tin trough, 70 cm. long, 5 cm. wide, 2 cm. high, is filled with water to the brim, and a strip of tin about 1 1/2 cm. wide laid across it perpendicular to its length, so that the under side of the strip is in contact with the surface of the water, and divides it into two halves. By shifting this partition to the right or the left, the surface on either side can be lengthened or shortened in any proportion, and the amount of the displacement may be read off on a scale held along the front of the trough.

No doubt this apparatus suffers, as I shall point out presently, from a certain imperfection, for the partition never completely shuts off the two separate surfaces from each other. If there is a great difference of tension between the two sides, a return current often breaks through between the partition and the edge of the trough (particularly at the time of shifting). The apparatus, however, answers for attaining any condition of tension which is at all possible, and in experiments with very clean surfaces there is little to be feared in the way of currents breaking through.

I always measured the surface tension in any part of the trough by the weight necessary to separate from it a small disk (6 mm. in diameter), for which I used a light balance, with unequal arms and a sliding weight.

I will now put together the most important results obtained with this apparatus, most of which, though perhaps not all, must be known to you.

I. *Behaviour of the surface tension of water.* — The surface tension of a strongly contaminated water surface is *variable* — that is, it varies with the size of the surface. The minimum of the separating weight attained by diminishing the surface is to the maximum, according to my balance, in the ratio of 52 : 100.

If the surface is further extended, after the maximum tension is attained, the separating weight remains *constant*, as with oil, spirits of wine, and other normal liquids. It begins, however, to diminish again, directly the partition is pushed back to the point of the scale at which the increase of tension ceased.

The water surface can thus exist in two sharply contrasted conditions; the *normal* condition, in which the displacement of the partition makes no impression on the tension, and the *anomalous* condition, in which every increase or decrease alters the tension.

II. *Mobility.* — Upon the purity of the surface depends its mobility, and in consequence the persistence of a wave once set in motion. So long, however, as the water surface is in its anomalous condition, the damping of the waves is constant and just at the degree of purity at which *the tension ceases to alter the decrease of the damping begins.*

If the balance is loaded with just the maximum weight which the surface tension can hold, and the normal surface is contracted till the weight breaks away, a measure is obtained of the relative amount of contamination by the ratio of the length of the surface before and after contraction; for, the purer the surface, the smaller must be the fraction to which it is reduced before it begins to enter the anomalous state. By counting, with different relative contaminations, how often a wave excited by a small rod at the end of the trough passed along the surface adjusted to a length of 30 cm. before it ceased to be visible, I obtained approximately the following values for the number of the passages: —

Relative contamination	0	5	10	15	20	25	30
Number of visible wave passages	17	17	17	17	12	08	03

The numbers of the upper row indicate the length at which the surface becomes anomalous in 30ths of its whole length; those of the second row are, as may be imagined, rather uncertain, particularly the greater ones, although they are the mean of many observations.

A *perfectly clean* surface, whose tension remains constant, even under the greatest contraction, can be approximately produced with the adjustable trough, by placing the partition quite at the end, and pushing it from thence to the middle. The surface on one side is thus formed entirely afresh, from the interior of the liquid.

III. *Effect on a water surface of contact with solid bodies.* — Every solid body, however clean, which is brought in contact with a newly formed surface, contaminates it more or less decidedly, according to the substance of which the body consists. With many substances, such as camphor or flour, this effect is so strong that the tension of the surface is lowered to a definite value; with others (glass, metals) it is only shown by the increase of relative contamination. The contaminating current which goes out from the circumference of a body — for example, of a floating fragment of tinfoil — is easily made visible by dusting the water with Lycopodium or flowers of sulphur. I will call it, for the sake of brevity, "the solution current."

The solution current of a body which is introduced into a perfectly clean water surface lasts until the relative contamination produced by it has attained a definite value, which is *different for every substance*.

Pockels went on to describe a series of other experiments, in which she contaminated the test solution with floating pieces of metal, glass, tinfoil, wax, and wood, as well as dissolved substances such as sugar and bicarbonate of soda. Having noted the variable effects on the nature of the surface tension in each case, she concluded that 'the effect of soluble matter on the surface tension has absolutely nothing to do with the change which the cohesion of the water undergoes, through matter dissolved in the body of the liquid, for both sugar and soda solutions have a higher maximum tension than pure water, and yet these same substances introduced into the surface produce a fall in the separating weight.' As Rayleigh pointed out, these were suggestive ideas, which asked more questions than they answered; fittingly, her final paragraph contained a remarkable blend of deference and self-confidence:

What I have further observed regarding solutions in the surface and the like, seems to me less remarkable, and part of it still very uncertain. I therefore confine myself to these short indications, but I believe that much might be discovered in this field if it were thoroughly investigated. I thought I ought not to withhold from you these facts which I have observed, although I am not a professional physicist; and again begging you to excuse my boldness, I remain, with sincere respect,

Yours faithfully,
(Signed) AGNES POCKELS.

Source: Agnes Pockels, 'Surface Tension', *Nature* 43 (1891), pp. 437–9.

THE HOUSEHOLD LABORATORY

'Every house is a laboratory', wrote the environmental chemist Ellen Henrietta Richards (1842–1911) in her most popular work, *The Chemistry of Cooking and Cleaning* (3rd edn, 1907), a book that sought to lessen the drudgery of housework through the appliance of chemical knowledge. Agnes Pockels would have known exactly what she meant (see previous entry). Having worked as a cleaner to pay her own college fees – she graduated from Vassar in 1870 – Ellen Swallow, as she was known before her marriage, attended the Massachusetts Institute of Technology, where she became the first American woman to earn a chemistry degree.* She taught at MIT for a number of years, although she never received a salary (in fact she donated an annual bursary to fund science studentships for women); she later worked as an environmental chemist at the Lawrence Experiment Station, the leading American research institute for the chemistry of water sanitation. In 1892 Richards came across a new term, 'Oecologie', which had been coined by the German biologist Ernst Haeckel (1834–1919), and in it she recognized a one-word summary of her principal scientific interests; with Haeckel's blessing, she translated the word as 'ecology', thereby introducing a powerful new concept to the Anglophone scientific world.

Richards was also a pioneer of what she called 'the Chemistry of Common Life', the forerunner of what became known as home economics (in 1908 she was appointed the first President of the American Home Economics Association). The following pages are from a chapter of *The Chemistry of Cooking and Cleaning* entitled 'Chemicals and their Use in the Household':

Every woman, whether she knows it or not, is every day performing simple experiments in chemistry. Every match that is lighted, every use of soap on the body, the clothes or the utensils, depends upon chemical laws for the reactions which take place.

There is no process of cooking or cleaning that does not rest upon a foundation of chemical or physical law. Therefore every house is a laboratory. Each is presided over by a director of greater or less intelligence.

* Ernest Rutherford's famous comment, 'If you can't explain your research to the cleaning lady, it isn't worth doing', always reminds me of both Ellen Swallow Richards and Agnes Pockels. — R.H.

When intellectual interest and manual dexterity unite, 'drudgery' is eliminated.

There may be, too, at all times an attitude of discovery. Many of the most important chemical processes have been found out, it is said, accidentally.

In most persons an experiment awakens interest. The housewife should be a cautious experimenter.

An understanding of simple chemical reactions tends also to economy in household management.

The thrifty housewife may not only save many dollars by restoring tarnished furniture and stained fabrics, but may also keep her belongings fresh and 'as good as new' by the judicious use of a few chemical substances always ready at her hand.

It is essential, however, that she know their properties and the effect they are likely to have on the materials to be treated, lest more harm than good result from their use. A good example is the instant disappearance of all red iron-rust stains when treated with a drop of hydrochloric acid. If, however, the acid is not completely washed out, the fabric will become eaten, and holes will appear, which, in the housekeeper's eye, are worse than the stains. This danger may be entirely removed by adding ammonia to the final rinsing water, which neutralizes any remaining acid, and the stained tray-cloth or sheet is perfectly whitened.

It is well that the household laboratory should be supplied with the following substances . . .

I. Alkalies — substances with a soapy feeling and which turn red litmus blue. In solutions they neutralize the effects of acids. When the neutralization is complete the solution is said to be neutral, and it will not change the color of litmus. This neutral substance is a chemical salt.

Alkalies, except ammonia, injure wool fibres, hardening, roughening and shrinking them, while the caustic alkalies dissolve them. Only weak alkalies should be used on linen or silk.

Potassium hydroxide, KOH, caustic potash. This can be bought in solid form to use in drains for removing grease. It forms a soap, which must be washed out of the pipes by thorough flushing. It is one of the strongest alkalies and must be used with caution. It is dissolved in water with the evolution of great heat, causing rapid boiling. Spatters from it on fabrics are liable to burn holes, on wood will darken the color, and on the flesh may cause deep burns . . .

Sodium hydroxide, NaOH, caustic soda. This compound resembles caustic potash, is effective in slightly less degree for the same purposes; and, being cheaper, is much more extensively used. The element sodium is common in all marine and seashore plants, and is the metal in common salt, from which it is now prepared.

Sodium carbonate, Na_2CO_3, soda-ash (originally from the ashes of seaweed). When a hot solution of soda-ash is cooled, a crystalline form is left known as *sal-soda, washing soda, soda crystals*. The crystals lose water when exposed to the air and crumble to powder. This powder is, therefore, stronger than the crystals.

Sal-soda is used most commonly in softening hard water, for keeping the plumbing pipes free from grease and to remove grease and hardened food from cooking utensils. A convenient way to prepare sal-soda for general use is to put one pound of the ash in one quart of water. Let this boil until the soda is dissolved. Bottle when cold. It is the second strongest alkaline cleaner, cheaper and more safely used than potash.

Sodium bicarbonate, $NaHCO_3$, cooking soda, carbonated soda-ash. In cooking it is used to neutralize acids, to aerate dough and to produce effervescence in acid solutions by liberating carbon dioxide. This is the saleratus of today. The true saleratus or "pearlash" used seventy-five years ago was the corresponding potassium salt, $KHCO_3$. This was often obtained by burning corncobs, mixing the ashes with water and allowing the solution to evaporate to dryness . . .

Borax, $Na_2B_4O_7$. A weakly alkaline substance, most useful with hard water, as a bleacher and as an antiseptic. It is much more expensive than sal-soda, but is less liable to injure fabrics and to irritate the skin. Its action on colors is less than that of ammonia.

Ammonia. The gas NH_3 is dissolved in water in varying proportions, forming ammonium hydroxide, or aqua ammonia, NH_4OH. It is the only volatile alkali. It is a useful substance in nearly all cleaning processes, and to neutralize acids.

"Household ammonia" is subject to impurities due to processes of manufacture. These often fade colors or cause white materials to turn yellow. It is safer and cheaper to buy the concentrated ammonia from a druggist or a dealer in chemicals and add the water at home.

This concentrated ammonia may be diluted one-half to one-sixth and yet be sufficiently strong for most uses . . .

II. Acids — substances with a sour taste and which change blue litmus red. An acid will also liberate carbon dioxide from cooking or washing soda. In general, weak acids lighten the color of wood, while strong acids burn it. Acids act injuriously on all metals if allowed to remain in contact with them. The metallic salts that are formed are often very poisonous.

When used with caution acids are effectual in removing iron and fruit stains. They remove color from some fabrics.

Acetic acid, $H(C_2H_3O_2)$, is the acid of vinegar and for many purposes it may be used in this form. For delicate processes, however, the other substances present may stain or interfere with the action, so that the pure acetic acid, much diluted, is better. It is volatile, therefore any excess is not likely to injure the fabric by concentration in it on drying, as will hydrochloric and oxalic acids. While acids clean copper and brass quickly by combining with the tarnishing salts, thus exposing a fresh surface, this new surface soon tarnishes again, and the process must be repeated. If any acid remains, as it is likely to do in seams and grooves, metallic salts will be formed. Copper acetate, which is formed when brass or copper is treated with vinegar, is very dangerous.

Sour milk contains lactic acid; lemons, citric acid, and this is perhaps the best natural acid to use for cleaning purposes. It should be used with caution, however, as it is strong enough to affect some colors and reacts with metals, as copper, brass and iron. Rhubarb, tomatoes, sorrel, etc., contain acid principles. They can be used in emergencies and are less liable to "eat" fabrics.

Oxalic acid, $H_2(C_2O_4)$, is found naturally in some plants, as oxalis and sorrel. It is bought in crystals, which are quickly soluble in hot and more slowly in cold water. It is very useful in removing stains from white fabrics and may be used on some colors (any acid to be used on colored fabrics should be tested first on a piece of the goods or on some hidden part, as a seam).

Hot solutions of oxalic acid are more effective than cold. When very strong it makes the finger nails brittle and may irritate the skin temporarily. It must be labeled "Poison," and should be kept out of the reach of children.

Note. — The strong acids destroy the coats of the stomach and therefore are poisonous in a general sense, although not in the strict sense in which strychnine is.

Tartaric acid, $H_2(C_4H_4O_6)$, the acid of cream of tartar is one of the safest acid agents. The Rochelle salt of Seidlitz powders is a sodium potassium tartrate. In "soda powders" one paper contains tartaric acid, the other sodium bicarbonate.

The crude tartar or argol is formed as a hard crust or deposit on the bottom and sides of vessels in which wine is manufactured.

Hydrochloric or *muriatic acid*, HCl, most valuable for removing iron stains from fabrics and other materials. It sometimes injures silk and must be used with caution on colored goods. A twenty per cent solution is effective, but it will lose its strength unless very tightly corked with glass or rubber. The fumes escaping around the stopper will rust metals and "eat" fabrics even at some distance.

Whenever this acid is used there should be thorough rinsing of the fabric in water, preferably warm, and then neutralization in ammonia.

III. Bleachers.

Hydrogen peroxide, H_2O_2, may be purchased as a five per cent liquid. This is a powerful oxidizing agent. It loses the extra atom of oxygen readily and should be kept in a dark place, preferably closed with a rubber stopper.

It may be safely used with all fibres, being especially good for wool. The bleaching action is permanent.

It is an excellent disinfectant for wounds, sore throat, etc.

Sulphur dioxide, SO_2, is made by burning sulphur in the air. With moisture the dioxide forms sulphurous acid, H_2SO_3.

This is effective on moist silk, wool, straw and paper. The fumes should not be breathed. The country housewife attaches the wet, yellowed straw hat to the bottom of a barrel, which is then inverted over a kettle of coals and sulphur.

This is much less destructive to the fibre than chloride of lime, but the color often returns. The fumes from a burning match held under the wet hand will remove the purple stains left from black kid gloves. They are also very effective for blueberry and blackberry stains.

The common sulphur candle is a convenient means of obtaining sulphur fumes. They may be readily concentrated by inverting over the candle a paper, cardboard or other funnel.

Chloride of lime is also called bleaching powder. Its composition is not definitely known, but approximately is $CaOCl_2$. When dissolved in water it shows the presence of both calcium hypochlorite and calcium chloride. A similar compound is sometimes called "chlorinated lime." (Chlorinated soda is also on the market.)

Sodium thiosulphite — called also hyposulphite, $Na_2S_2O_3$ + $5H_2O$, the "hypo" of the photographer — is effective in removing the marks of indelible ink containing silver nitrate.

IV. Solvents. For grease there are naphtha, benzene, gasoline, ether, chloroform, extremely volatile; kerosene, turpentine, carbon tetrachloride, coal tar benzene or benzole C_6H_6, alcohol, less volatile.

The vapors of all these substances are heavier than air, therefore sink. They should be used out-of-doors or by an open window, and *never* where there is *any fire*. There should be a current of air near the floor to ensure quick removal of the vapor.

Source: Ellen H. Richards and S. Maria Elliott, *The Chemistry of Cooking and Cleaning: A Manual for Housekeepers*, 3rd edn (Boston: Whitcomb & Barrows, 1907), pp. 144–58.

IN SEARCH OF KRYPTON

For the ancient Greeks, there were four fundamental elements – earth, air, fire and water – to which Aristotle added ether, the quintessence (or 'fifth element'). By the mid-nineteenth century, however, scientists had identified more than sixty chemical elements, but had found no clear way to make sense of them. In 1829 the German chemist Johann Döbereiner (1780–1849) noticed a repeating trend in the properties of certain elements, in which one element had an atomic weight equal to the average of two of its neighbours, an idea that was known as 'Döbereiner's Triads'. In 1864 the British chemist John Newlands (1837–98) noted similarities between every eighth element – if arranged by atomic weight – which he called the 'law of octaves'. Building on these earlier insights, the Russian chemist Dmitri Mendeleev (1834–1907) came up with a way of ordering the sixty-six known elements in a grid formation, grouping them into families according to their 'periodicity': the pattern of recurring trends in their chemical properties. 'I began to look about and write down the elements with their atomic weights and typical properties on separate cards', he recalled, 'and this soon convinced me that the properties of elements are in periodic dependence upon their atomic weights.' It was Mendeleev's idea to leave gaps in the arrangement in order to accommodate future discoveries: a simple but brilliant innovation that allowed chemists to predict the precise atomic properties of as-yet undiscovered elements, and when gallium, scandium and germanium were isolated in the 1870s and 1880s they fitted exactly into the empty spaces that Mendeleev had reserved for them. It was like completing a game of patience.

Chemists continued to search for new elements, and in a letter to Lord Rayleigh in May 1894 the Scottish chemist William Ramsay (1852–1916) asked a simple but insightful question: 'Has it occurred to you that there is room for gaseous elements at the end of the first column of the periodic table?' The answer was yes, and by the end of that year the pair had isolated a new gas that they named *argon* (from the Greek for 'inactive'), and realized that there must be others in the same group, all with predictable atomic properties. The following extract is from the transcript of a talk given by Ramsay in September 1897 at a scientific conference in Toronto, Canada. In it, he foretells the discovery of a new elementary gas with an atomic weight of 20. The gas certainly exists, he says; it just hasn't been found yet:

The subject of my remarks today is a new gas. I shall describe to you later its curious properties; but it would be unfair not to put you at once in possession of the knowledge of its most remarkable property – it has not yet been discovered. As it is still unborn, it has not yet been named. The naming of a new element is no easy matter. For there are only twenty-six letters in our alphabet, and there are already over seventy elements. To select a name expressible by a symbol which has not already been claimed for one of the known elements is difficult, and the difficulty is enhanced when it is at the same time required to select a name which shall be descriptive of the properties (or want of properties) of the element.

It is now my task to bring before you the evidence for the existence of this undiscovered element.

It was noticed by Döbereiner, as long ago as 1817, that certain elements could be arranged in groups of three. The choice of the elements selected to form these triads was made on account of their analogous properties, and on the sequence of their atomic weights, which had at that time only recently been discovered. Thus calcium, strontium and barium formed such a group; their oxides, lime, strontia and baryta, are all easily slaked, combining with water to form soluble lime-water, strontia-water and baryta-water. Their sulphates are all sparingly soluble and resemblance has been noticed between their respective chlorides and between their nitrates. Regularity was also displayed by their atomic weights. The numbers then accepted were 20, 42.5 and 65; and the atomic weight of strontium, 42.5, is the arithmetical mean of those of the other two elements . . .

The upshot of these efforts to discover regularity was that, in 1864, Mr. John Newlands, having arranged the elements in eight groups, found that when placed in the order of their atomic weights, 'the eighth element, starting from a given one, is a kind of repetition of the first, like the eighth note of an octavo in music.' To this regularity he gave the name 'The Law of Octaves.'

The development of this idea, as all chemists know, was due to the late Professor Lothar Meyer, of Tübingen, and to Professor Mendeléeff, of St. Petersburg. It is generally known as the 'Periodic Law.' One of the simplest methods of showing this arrangement is by means of a cylinder divided into eight segments by lines drawn parallel to its axis; a spiral line is then traced round the cylinder, which will, of course, be cut by these lines eight times at each revolution.

Holding the cylinder vertically, the name and atomic weight of an element is written at each intersection of the spiral with a vertical line, following the numerical order of the atomic weights. It will be found, according to Lothar Meyer and Mendeléeff, that the elements grouped down each of the vertical lines form a natural class; they possess similar properties, form similar compounds, and exhibit a graded relationship between their densities, melting points, and many of their other properties. One of these vertical columns, however, differs from the others, inasmuch as on it there are three groups, each consisting of three elements with approximately equal atomic weights. The elements in question are iron, cobalt and nickel; palladium, rhodium and ruthenium; and platinum, iridium and osmium. There is apparently room for a fourth group of three elements in this column and it may be a fifth. And the discovery of such a group is not unlikely, for when this table was first drawn up Professor Mendeléeff drew attention to certain gaps, which have since been filled up by the discovery of gallium, germanium and others.

The discovery of argon at once raised the curiosity of Lord Rayleigh and myself as to its position in this table. With a density of nearly 20, if a diatomic gas, like oxygen and nitrogen, it would follow fluorine in the periodic table; and our first idea was that argon was probably a mixture of three gases, all of which possessed nearly the same atomic weights, like iron, cobalt and nickel. Indeed, their names were suggested, on this supposition, with patriotic bias, as Anglium, Scotium and Hibernium! But when the ratio of its specific heats had, at least in our opinion, unmistakeably shown that it was molecularly monatomic, and not diatomic, as at first conjectured, it was necessary to believe that its atomic weight was 40, and not 20, and that it followed chlorine in the atomic table and not fluorine. But here arises a difficulty. The atomic weight of chlorine is 35.5, and that of potassium, the next element in order in the table, is 39.1; and that of argon, 40, follows, and does not precede, that of potassium, as it might be expected to do. It still remains possible that argon, instead of consisting wholly of monatomic molecules, may contain a small percentage of diatomic molecules; but the evidence in favor of this supposition is, in my opinion, far from strong . . .

The discovery of helium has thrown a new light on this subject. Helium, it will be remembered, is evolved on heating certain minerals, notably those containing uranium; although it appears to be contained in others in which uranium is not present, except in traces. Among these minerals are cleveite, monazite, fergusonite, and a host of similar complex mixtures, all containing rare elements, such as niobium, tantalum, yttrium, cerium, etc. The spectrum of helium is characterized by a remarkably brilliant yellow line, which had been observed as long ago as 1868 by Professors Frankland and Lockyer in the spectrum of the sun's chromosphere, and named 'helium' at that early date.

The density of helium proved to be very close to 2.0, and, like argon, the ratio of its specific heat showed that it, too, was a monatomic gas. Its atomic weight, therefore, is identical with its molecular weight, viz., 4.0, and its place in the periodic table is between hydrogen and lithium, the atomic weight of which is 7.0.

Ramsay went on to demonstrate that there was a gap in the periodic table between helium and argon, for an as yet unknown element with an atomic weight 16 units higher than that of helium, and 20 units lower than argon, namely 20; Ramsay and his assistant, Morris Travers (1872–1961), travelled Europe in search of this 'needle in a haystack', as he described it, venturing to Iceland in order to collect gases from boiling springs, and to the Pyrenees the following summer to collect gases from the mineral springs – all to no avail:

It by no means follows that the gas does not exist; the only conclusion to be drawn is that we have not yet stumbled on the material which contains it. In fact, the haystack is too large and the needle too inconspicuous. Reference to the periodic table will show that between the elements aluminium and indium there occurs gallium, a substance occurring only in the minutest amount on the earth's surface; and following silicon, and preceding tin, appears the element germanium, a body which has as yet been recognized only in one of the rarest of minerals, argyrodite. Now, the amount of helium in fergusonite, one of the minerals which yields it in reasonable quantity, is only 33 parts by weight to 100,000 of the mineral; and it is not improbable that some other mineral may contain the new gas in even more minute proportion. If, however, it is accompanied in its still undiscovered source by argon and helium, it will be a work of extreme difficulty to effect a separation from these gases.

It may have been difficult work, but less than a year later, in June 1898,

Ramsay and Travers announced that they had finally tracked down this elusive gas in a sample of liquid air. They named it *krypton*, from the Greek for
'hidden'. A fortnight later they discovered *neon* ('new'), and shortly after that
xenon ('the stranger'). The new family of noble gases was almost complete
(radioactive radon was added in 1900). In 1902, Mendeleev added the noble
gases to the periodic table, and in 1904 Ramsay and Rayleigh received Nobel
Prizes for their work.

As can be seen from the periodic table, below, there are now ninety-four
natural elements, plus a further twenty-four synthetic ones, arranged by
periods (the horizontal rows) and groups (the vertical columns): the noble
gases can be seen in the column on the far right.

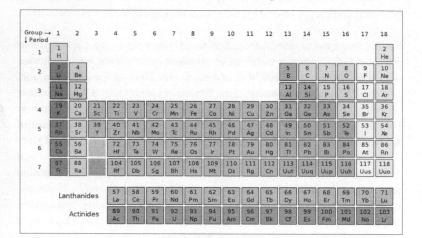

17. The periodic table.

Source: William Ramsay, 'An Undiscovered Gas', *Science* 6 (1897), pp. 493–502.

RADIOACTIVITY

Marie Curie (1867–1934) was born Maria Skłodowska in Warsaw, Poland, but moved to Paris in 1891 to study chemistry and physics at the Sorbonne. In 1895, while still a postgraduate student, she married a physics teacher, Pierre Curie (1859–1906), before embarking on the research that would change the course of science. The work began in 1896, when the French physicist Henri Becquerel (1852–1908) accidentally left a specimen of the mineral ore pitchblende on a covered photographic plate; when he developed the plate the next day he discovered a photographic image of the pitchblende container, and realized that something in the ore must have emitted a form of invisible radiation. Becquerel's discovery attracted Marie Curie's attention, and she and her husband set about investigating the properties of pitchblende in a dilapidated shed in a corner of the Paris School of Physics. What followed was four years of arduous work, during which the husband-and-wife team sifted through six tonnes of muddy pitchblende, subjecting it to painstaking chemical and physical analysis that eventually led to the discovery of two new radioactive elements: polonium and radium (the former named in honour of Marie Curie's country of birth). In 1903 the Curies shared the Nobel Prize in Physics with Henri Becquerel; eight years later the widowed Marie became the first person ever to win a second Nobel Prize, although as a woman she remained disbarred from the Académie des Sciences.

The first of the following extracts from Marie Curie's autobiography describes the background to her work on radioactivity (she coined the word in 1898), while the second paints a vivid picture of her time in the filthy shed:

Our attention was caught by a curious phenomenon discovered in 1896 by Henri Becquerel. The discovery of the X-ray by Roentgen had excited the imagination, and many physicians were trying to discover if similar rays were not emitted by fluorescent bodies under the action of light. With this question in mind Henri Becquerel was studying uranium salts, and, as sometimes occurs, came upon a phenomenon different from that he was looking for: the spontaneous emission by uranium salts of rays of a peculiar character. This was the discovery of radioactivity.

The particular phenomenon discovered by Becquerel was as follows: uranium compound placed upon a photographic plate covered with black paper produces on that plate an impression analogous to that which light would make. The impression is due to uranium rays that traverse the paper. These same rays can, like X-rays, discharge an electroscope, by making the air which surrounds it a conductor.

Henri Becquerel assured himself that these properties do not depend on a preliminary isolation, and that they persist when the uranium compound is kept in darkness during several months. The next step was to ask whence came this energy, of minute quantity, it is true, but constantly given off by uranium compounds under the form of radiations.

The study of this phenomenon seemed to us very attractive and all the more so because the question was entirely new and nothing yet had been written upon it. I decided to undertake an investigation of it.

It was necessary to find a place in which to conduct the experiments. My husband obtained from the director of the School the authorization to use a glassed-in study on the ground floor which was then being used as a storeroom and machine shop.

In order to go beyond the results reached by Becquerel, it was necessary to employ a precise quantitative method. The phenomenon that best lent itself to measurement was the conductibility produced in the air by uranium rays. This phenomenon, which is called *ionization*, is produced also by X-rays and investigation of it in connection with them had made known its principal characteristics.

For measuring the very feeble currents that one can make pass through air ionized by uranium rays, I had at my disposition an excellent method developed and applied by Pierre and Jacques Curie. This method consists in counterbalancing on a sensitive electrometer the quantity of electricity carried by the current with that which a piezo-electric quartz can furnish. The installation therefore required a Curie electrometer, a piezo-electric quartz, and a chamber of ionization, which last was formed by a plate condenser whose higher plate was joined to the electrometer, while the lower plate, charged with a known potential, was covered with a thin layer of the substance to be examined. Needless to say, the place for such an electrometric installation was hardly the crowded and damp little room in which I had to set it up.

My experiments proved that the radiation of uranium compounds can be measured with precision under determined conditions, and that this radiation is an atomic property of the element of uranium. Its intensity is proportional to the quantity of uranium contained in the compound, and depends neither on conditions of chemical combination, nor on external circumstances, such as light or temperature.

I undertook next to discover if there were other elements possessing the same property, and with this aim I examined all the elements then known, either in their pure state or in compounds. I found that among these bodies, thorium compounds are the only ones which emit rays similar to those of uranium. The radiation of thorium has an intensity of the same order as that of uranium, and is, as in the case of uranium, an atomic property of the element.

It was necessary at this point to find a new term to define this new property of matter manifested by the elements of uranium and thorium. I proposed the word radioactivity which has since become generally adopted; the radioactive elements have been called radio elements.

During the course of my research, I had had occasion to examine not only simple compounds, salts and oxides, but also a great number of minerals. Certain ones proved radioactive; these were those containing uranium and thorium; but their radioactivity seemed abnormal, for it was much greater than the amount I had found in uranium and thorium had led me to expect.

This abnormality greatly surprised us. When I had assured myself that it was not due to an error in the experiment, it became necessary to find an explanation. I then made the hypothesis that the ores uranium and thorium contain in small quantity a substance much more strongly radioactive than either uranium or thorium. This substance could not be one of the known elements, because these had already been examined; it must, therefore, be a new chemical element.

I had a passionate desire to verify this hypothesis as rapidly as possible. And Pierre Curie, keenly interested in the question, abandoned his work on crystals (provisionally, he thought) to join me in the search for this unknown substance.

We chose, for our work, the ore pitchblende, a uranium ore, which in its pure state is about four times more active than oxide of uranium. Since the composition of this ore was known through very careful chemical analysis, we could expect to find, at a maximum, 1 per cent of new substance. The result of our experiment proved that there were in reality new radioactive elements in pitchblende, but that their proportion did not reach even a millionth per cent!

The method we employed is a *new method in chemical research based on radioactivity*. It consists in inducing separation by the ordinary means of chemical analysis, and of measuring, under suitable conditions, the radioactivity of all the separate products. By this means one can note the chemical character of the radioactive element sought for, for it will become concentrated in those products which will become more and more radioactive as the separation progresses. We soon recognized that the radioactivity was concentrated principally in two different chemical fractions, and we became able to recognize in pitchblende the presence of at least two new radioactive elements: polonium and radium. We announced the existence of polonium in July, 1898, and of radium in December of the same year.

This second extract from Curie's memoirs describes the squalid conditions in which the world-famous discoveries were made:

We were very poorly equipped with facilities for this purpose. It was necessary to subject large quantities of ore to careful chemical treatment. We had no money, no suitable laboratory, no personal help for our great and difficult undertaking. It was like creating something out of nothing, and if my earlier studying years had once been called by my brother-in-law the heroic period of my life, I can say without exaggeration that the period on which my husband and I now entered was truly the heroic one of our common life.

We knew by our experiments that in the treatment of pitchblende at the uranium plant of St. Joachimsthal, radium must have been left in the residues, and, with the permission of the Austrian government, which owned the plant, we succeeded in securing a certain quantity of these residues, then quite valueless, — and used them for extraction of radium. How glad I was when the sacks arrived, with the brown dust mixed with pine needles, and when the activity proved even greater than that of the primitive ore! It was a stroke of luck that the residues had not been thrown far away or disposed of in some way, but left in a heap in the pine wood near the plant. Some time later, the Austrian government, on the proposition of the Academy of Science of Vienna, let us have several tons of similar residues at a low price. With this material was prepared all the radium I had in my laboratory up to the date when I received the precious gift from the American women.

The School of Physics could give us no suitable premises, but for lack of anything better, the Director permitted us to use an abandoned shed which had been in service as a dissecting room of the School of Medicine. Its glass roof did not afford complete shelter against rain; the heat was suffocating in summer, and the bitter cold of winter was only a little lessened by the iron stove, except in its immediate vicinity. There was no question of obtaining the needed proper apparatus in common use by chemists. We simply had some old pine-wood tables with furnaces and gas burners. We had to use the adjoining yard for those of our chemical operations that involved producing irritating gases; even then the gas often filled our shed. With this equipment we entered on our exhausting work.

Yet it was in this miserable old shed that we passed the best and happiest years of our life, devoting our entire days to our work. Often I had to prepare our lunch in the shed, so as not to interrupt some particularly important operation. Sometimes I had to spend a whole day mixing a boiling mass with a heavy iron rod nearly as large as myself. I would be broken with fatigue at the day's end. Other days, on the contrary, the work would be a most minute and delicate fractional crystallization, in the effort to concentrate the radium. I was then annoyed by the floating dust of iron and coal from which I could not protect my precious products. But I shall never be able to express the joy of the untroubled quietness of this atmosphere of research and the excitement of actual progress with the confident hope of still better results. The feeling of discouragement that sometimes came after some unsuccessful toil did not last long and gave way to renewed activity. We had happy moments devoted to a quiet discussion of our work, walking around our shed.

One of our joys was to go into our workroom at night; we then perceived on all sides the feebly luminous silhouettes of the bottles or capsules containing our products. It was really a lovely sight and one always new to us. The glowing tubes looked like faint, fairy lights.

Thus the months passed, and our efforts, hardly interrupted by short vacations, brought forth more and more complete evidence. Our faith grew ever stronger, and our work being more and more known, we found means to get new quantities of raw material and to carry on some of our crude processes in a factory, allowing me to give more time to the delicate finishing treatment.

At this stage I devoted myself especially to the purification of the radium, my husband being absorbed by the study of the physical properties of the rays emitted by the new substances. It was only after treating one ton of pitchblende residues that I could get definite results. Indeed we know to-day that even in the best minerals there are not more than a few decigrammes of radium in a ton of raw material.

At last the time came when the isolated substances showed all the characters of a pure chemical body. This body, the radium, gives a characteristic spectrum, and I was able to determine for it an atomic weight much higher than that of the barium. This was achieved in 1902. I then possessed one decigramme of very pure radium chloride. It had taken me almost four years to produce the kind of evidence which chemical science demands, that radium is truly a new element. One year would probably have been enough for the same purpose, if reasonable means had been at my disposal. The demonstration that cost so much effort was the basis of the new science of radioactivity.

Source: Marie Curie, *Pierre Curie, with Autobiographical Notes*, trans. Charlotte and Vernon Kellogg (New York: Macmillan, 1923), pp. 92–8; 180–8.

N-RAYS

Following one of the greatest scientific achievements, here is one of the most notorious failures: Prosper-René Blondlot's 'discovery' of N-rays in 1903. Blondlot (1849–1930) was an eminent French physicist whose work on radio waves had shown that, like all electromagnetic waves, they travel at the speed of light. But he is fated to be remembered only for his mistaken belief that he had discovered a new form of electromagnetic radiation, analogous to the X-ray, which he named the 'N ray' in honour of his home town of Nancy. Blondlot's discovery was initially hailed as a major scientific achievement, but It soon became clear that no other scientist was able to replicate his experiments. What was going on? The science journal *Nature* commissioned the American physicist Robert W. Wood (1868–1955) to visit Blondlot in his laboratory and observe the results for himself; his subsequent report constituted one of the most thorough debunkings in the history of science:

In the late autumn of 1903, Professor R. Blondlot, head of the Department of Physics at the University of Nancy, member of the French Academy, and widely known as an investigator, announced the discovery of a new ray, which he called N ray, with properties far transcending those of the X rays. Reading of his remarkable experiments with these rays in the *Comptes rendus* of the Academy, the leading scientific journal of France, I attempted to repeat his observations, but failed to confirm them after wasting a whole morning. According to Blondlot, the rays were given off spontaneously by many metals. A piece of paper, very feebly illuminated, could be used as a detector, for, wonder of wonders, when the N rays fell upon the eye they increased its ability to see objects in a nearly dark room . . .

Blondlot next announced that he had constructed a spectroscope with aluminum lenses and a prism of the same metal, and found a spectrum of lines separated by dark intervals, showing that there were N rays of different refrangibility and wave length. He measured the wave lengths. The flame of N-ray research was now a conflagration. Jean Becquerel, son of Henri Becquerel, whose discovery of the rays from uranium had laid the foundation for the discovery of radium by the Curies, claimed that N rays could be transmitted over a wire, just as light can be transmitted along the inside of a bent glass rod by internal reflection. One end of a wire near the faintly luminous detector caused variation of its intensity as the other end, some meters away, was passed over the skull of a living person. If the subject was anesthetized with ether, the N rays from the brain first increased and then decreased as the sleep deepened. He claimed that metals could be anesthetized with ether, chloroform, or alcohol, in which condition they ceased to emit or transmit the rays. Biologists, physiologists, psychologists, chemists, botanists, and geologists climbed on the band wagon. The nerve centers of the spinal cord in their relation to disease and previous surgical operations were studied by the N rays which they emitted. The rays were given off by growing plants, vegetables, and even by a human corpse. A. Charpentier found the sense of hearing and smell were increased by N rays as well as the sense of sight. A tuning fork in vibration gave a powerful N ray. By early summer Blondlot had published twenty papers, Charpentier twenty, and J. Becquerel ten, all describing new properties and sources of the rays.

Nearly one hundred papers on N rays were published in the *Comptes rendus* in the first half of the year 1904. The N ray was polarized, magnetized, hypnotized, and tortured in all of the ways that had forced confessions from light rays — but only Frenchmen could observe the phenomena. Scientists in all other countries were frankly skeptical, in fact ridiculed these fantastic impossibilities. But the French Academy stamped Blondlot's work with its approval by awarding him the Lalande prize of 20,000 francs and its gold medal 'for the discovery of the N rays.'

In August 1904, at the Cambridge meeting of the British Association for the Advancement of Science, Woods was asked to travel to the University of Nancy and make a personal visit to Blondlot in order to test his claims:

So I visited Nancy before rejoining my family in Paris, meeting Blondlot by appointment at his laboratory in the early evening. He spoke no English, and I elected German as our means of communication, as I wanted him to feel free to speak confidentially to his assistant, who was apparently a sort of high-class laboratory janitor.

He first showed me a card on which some circles had been painted in luminous paint. He turned down the gas light and called my attention to their increased luminosity when the N ray was turned on. I said that I saw no change. He said that was because my eyes were not sensitive enough, so that proved nothing. I asked him if I could move an opaque lead screen in and out of the path of the rays while he called out the fluctuations of the screen. He was almost 100 per cent wrong and called out fluctuations when I made no movement at all, and that proved a lot, but I held my tongue. He then showed me the dimly lighted clock, and tried to convince me that he could see the hands when he held a large flat file just above his eyes. I asked if I could hold the file, for I had noticed a flat wooden ruler on his desk, and remembered that wood was one of the few substances that *never* emitted N rays. He agreed to this, and I felt around in the dark for the ruler and held it in front of his face. Oh, yes, he could see the hands perfectly. This also proved something.

But the crucial and most exciting test was now to come. Accompanied by the assistant, who by this time was casting rather hostile glances at me, we went into the room where the spectroscope with the aluminum lenses and prism was installed. In place of an eyepiece, this instrument had a vertical thread, painted with luminous paint, which could be moved along in the region where the N-ray spectrum was supposed to be by turning a wheel having graduation and numerals on its rim. This wheel turned a horizontal screw with a movable nut on which the thread was mounted. Blondlot took a seat in front of the instrument and slowly turned the wheel. The thread was supposed to brighten as it crossed the invisible lines of the N-ray spectrum. He read off the numbers on the graduated scale for a number of the lines, by the light of a small, dark room, red lantern. This experiment had convinced a number of skeptical visitors, as he could repeat his measurements in their presence, always getting the same numbers. He claimed that a movement of the thread of 0.1 mm. was sufficient to change the luminosity, and when I said that seemed impossible, as the slit of the spectroscope was 2 mm. wide, he said that was one of the inexplicable properties of the N rays. I asked him to repeat his measurements, and reached over in the dark and lifted the aluminum prism from the spectroscope. He turned the wheel again, reading off the same numbers as before. I put the prism back before the lights were turned up, and Blondlot told his assistant that his eyes were tired. The assistant had evidently become suspicious, and asked Blondlot to let him repeat the reading for me. Before he turned down the light I had noticed that he placed the prism very exactly on its little round support, with two of its corners exactly on the rim of the metal disk. As soon as the light was lowered, I moved over towards the prism, with audible footsteps, but *I did not touch the prism.* The assistant commenced to turn the wheel, and suddenly said hurriedly to Blondlot in French, "I see nothing; there is no spectrum. I think the American has made some *dérangement.*" Whereupon he immediately turned up the gas and went over and examined the prism carefully. He glared at me, but I gave no indication of my reactions. This ended the séance, and I caught the night train for Paris.

Next morning I sent off a letter to *Nature*, London's scientific weekly, giving a full account of my findings, not, however, mentioning the double-crossing incident at the end of the evening and merely locating the laboratory as 'one in which most of the N-ray experiments had been carried on.'* *La Revue scientifique*, France's weekly semipopular scientific journal, published a translation of my letter and started an *Enquête*, or inquiry, asking French scientists to express their opinions as to the reality of the N rays. About forty letters were published in the succeeding numbers, only a half dozen backing Blondlot. The most scathing was one by Le Bel, who said, 'What a spectacle for French science when one of its distinguished savants measures the position of the spectrum lines, while the prism reposes in the pocket of his American colleague!'

Only two papers on N rays appeared in the *Comptes rendus* after this. They may have been delayed in the mail. The Academy at its annual meeting in December, when the prize and medal were presented, announced the award as given to Blondlot 'for his life work, taken as a whole.'

Source: William B. Seabrook, *Dr Wood, Modern Wizard of the Laboratory* (New York: Harcourt, Brace, 1941), pp. 234–9.

* R. Wood, 'The N Rays', *Nature* 70 (1904), p. 530.

PARANOID SCHIZOPHRENIA

The modern classification of mental illness grew out of the work of the German psychiatrist Emil Kraepelin (1856–1926), who in 1893 divided mental illnesses into two broad categories: 'manic-depressive psychosis' (mood disorders that tended to recur at regular intervals), and 'dementia praecox' (premature dementia), which usually appeared in early adulthood and remained present throughout the patient's life. As Kraepelin's terminology suggests, he regarded the latter condition as a degenerative form of brain disease, a dementia from which no sufferer could hope to recover. In 1908 the Swiss psychiatrist Eugen Bleuler (1857–1939) proposed that 'dementia praecox' should be renamed 'schizophrenia' – a word that translates roughly as 'splitting of the mind' – as it was beginning to be apparent that the condition was not degenerative; indeed patients could and did respond to treatment, some even making a complete recovery.

Although the label 'schizophrenia' has often been used to refer (inaccurately) to a split personality, Bleuler's terminology highlighted the way in which the various facets of mental life, such as perception, thought and emotion, appear to 'split off' from one another as the condition worsens and, in the case of paranoid schizophrenia (the most common variety), hallucinations and delusions begin to dominate the patient's mind. The following case history of a twenty-five-year-old paranoid schizophrenic is taken from one of Kraepelin's clinical lectures, which were written with enormous sympathy and insight, although the contrast between the calm, clinical tone of the psychiatrist and the noisy distress of the patient still makes for slightly uncomfortable reading:

Gentlemen, – The merchant, aged twenty-five, whom you see before you today, has made himself conspicuous by putting leaves and ferns into his buttonhole. He takes a seat with a certain amount of ceremony, and gives positive, concise, and generally relevant answers to our questions. We learn that he was admitted to the hospital a year ago, that he afterwards spent six weeks at home, and that he has now been here again for six months. The patient makes no explicit statements about the nature of the disturbances which appeared, but he admits, when he is asked about it, that he did not speak for some time, he does not know why. But he remembers most of the details of what he has been through. Although he knows where he is, he mistakes the people about him, calls us by wrong names, and takes us for merchants. While he is more or less indifferent at first, taking very little interest in us and looking round with a conceited expression, he gradually becomes rather excited, grows rude, irritable, and threatening, and breaks out into an incoherent flood of words, in which there is a quite senseless play on syllables – "Macbeth – mach'ins Bett," "Irr ich mich nicht – Klinik," "je suis – Jesus," and so on.

At the same time, the patient intimates that he is the German Emperor, and that the Grand Duke is his father-in-law; he has been promised the Grand Duke's daughter since 1871. He is studying astronomy here. He denies that he has hallucinations. He declines to obey orders, but after some persuasion he finally stretches out his hand stiffly to shake hands. The patient is divertible; he often breaks off in his talk, and he intersperses it with curious snorting noises. His mood is changeable, but on the whole very much exalted. Often, more especially when he makes his jesting play on words, the patient bursts into a tittering laugh. His behaviour shows no marked excitement. His deportment is pompous and affected.

The diagnosis will have to rest principally on the peculiar aberrations in the patient's actions – the *mannerisms*, the *play on words*, the signs of *negativism*, and also on his *emotional indifference*, while he is yet quite collected. The patient does not consider himself ill, but stays here without making any resistance, does not worry at all, forms no plans for the future, and expresses no desires. We know this picture well already as a form in which *dementia præcox* appears. What seems strange about the condition is the absence of strong excitement on the one hand, and on the other the delusionary mistaking of people and the ideas of grandeur quite quietly expressed . . .

The patient is said to belong to a healthy family. He was clever at school, was always serious and conscientious, and served as a one-year's conscript. So far back as three years ago he complained of being shaky and excited, and no longer able to work as he had once done. Then, after he had been particularly active and enterprising for some time, marked depression of spirits set in fifteen months ago. The patient became sleepless, and complained of pains in the back of his head. He felt stupid and unfit for work, took no pleasure in his business, played apathetically with his fingers, lay in bed all day, and thought that he had abused his principal's confidence and embezzled.

In this way he came to the hospital. Here he showed a striking conjunction of emotional dulness with good comprehension and perfect lucidity. Very soon he sank into a stupor, became dumb, showed signs of automatic obedience alternating with negativism, and masturbated very much. He was afraid that the French were coming, and that the knife would be whetted and people made away with, heard threatening voices, felt electricity in bed, wished to die, and ate hardly anything. He was perfectly indifferent when he was visited. It was only quite slowly that he became a little more active, got out of bed, and followed the doctor in his shirt, without speaking, or at the most murmured in a low tone to himself, or occasionally uttered irrelevant expressions, as to the meaning of which nothing could be learned from him.

In this condition he was taken home by his relations. There, too, he was almost dumb, and ate little, but jumped into a carriage one day, saying, "I drive a cab." He drove a little way, got out, and remained standing on the same spot for an hour and a quarter. During the next few days he suddenly became excited, clapped his hands, stamped with his feet, threw himself about, spoke quite incoherently in a loud voice about princesses, the Grand Duke, pardon, beheading, and the like, and laughed a great deal to himself. As he then became violent and broke windows, he was brought back to the hospital. Here he was very irritable and disobliging, gave senseless, irrelevant answers to questions, expressed disconnected ideas of grandeur, mistook people, and spoke and behaved affectedly. In speech and writing he plainly showed confusion of speech with playing on words. Thus he wrote in a letter, "2, x 4 = 8, that is the day of the Lord come to pass; good mine to bad man is bad; bad wicked mine d'or to good man better as first recovery; I B flat in A – Saucier, you got – Minister – Mercier." When shown a knife, he answered, "Knife, razor, Barber of Bagdad, Salem aleikum." On seeing a gold piece, "Louis d'or, Napoleon, Empress Eugénie, la France, Spain, thither will we go." He often spoke out of the window, and said he was talking to spirits or acting a play. His sleep was very much disturbed. The physical examination showed no abnormalities worth noticing, but great dermatography and mechanical excitability of the facials. His weight had increased considerably.

The patient was originally supposed to be suffering from maniacal-depressive insanity. His previous history and the alternation of excited and depressed moods seemed greatly to favour this view. But in the further course of the case *katatonic* symptoms came out prominently during both the stuporose and the excited periods – negativism, mannerisms, automatic obedience, and confusion of speech. The diagnosis we derived from the clinical condition is thus fully confirmed by the whole development of the disease until now. It is true that divertibility and quibbling, as we see them here, are generally considered to be symptoms of mania. I may, however, point out that in maniacal cases it is only in the very worst states of excite- ment that the incoherence reaches as high a degree as it did here, where rationality was fully maintained and the excitement was comparatively slight. Under these circumstances we must certainly adhere to the diagnosis of katatonia, and may therefore suppose that considerable improvement in the patient's condition is still possible, but that even in the most favourable event a certain degree of dulness and want of freedom will probably remain in the spheres of emotion and action.*

The First Electroshock Treatment

Although the idea of using electricity as a treatment for mental illness dates back to the eighteenth century, it was not until the 1930s that it was put into clinical practice. Electric-shock therapy, better known today as electro- convulsive therapy (ECT), was introduced in 1938 by the Italian neurologist Ugo Cerletti (1877–1963), of the Department of Mental and Neurological Diseases at the University of Rome. Cerletti, who had been a student of Emil Kraepelin and Alois Alzheimer, believed that electrically induced convul- sions could have therapeutic benefits – a view strengthened during a visit to an abattoir in Rome where the modern technique of electric stunning had recently been introduced. He noted that the animals (pigs, he recalled) were not killed by the electric current, merely rendered unconscious through a carefully induced seizure. Cerletti later described the introduction of ECT as

---

* When the patient had been eight months in the asylum, we were able to discharge him substantially improved. He was not very accessible, but fairly intelligent and free from delusions. Now, six years later, he is actively engaged in his business, but is shy, 'does not get on with people,' shows marked catalepsy, and expresses hypochondriacal delusions.

'neither a discovery nor an invention. We shall call it a deed of courage, the courage of passing from the "possibility" to the practical application', and his account, below, of the first clinical trial is notable for the nervousness of the experimenters, as well as the fear and distress suffered by the unfortunate patient, a thirty-nine-year-old engineer from Milan who had been discovered at Rome railway station wandering about in an agitated state. On 18 April 1938 he was brought to Cerletti's institute, where, due to his repeated claims of being telepathically influenced by malign machines, he was diagnosed with paranoid schizophrenia. His fate was to be chosen as the subject of the first clinical experiment in induced electric convulsions in a human being; first he was strapped down, and then, as Cerletti recalled:

Two large electrodes were applied to the frontoparietal regions, and I decided to start cautiously with a low-intensity current of 80 volts for 0.2 seconds. As soon as the current was introduced, the patient reacted with a jolt and his body muscles stiffened; then he fell back on the bed without loss of consciousness. He started to sing abruptly at the top of his voice, then he quieted down.

Naturally, we, who were conducting the experiment, were under great emotional strain and felt that we had already taken quite a risk. Nevertheless, it was quite evident to all of us that we had been using a too low voltage. It was proposed that we should allow the patient to have some rest and repeat the experiment the next day. All at once, the patient, who evidently had been following our conversation, said clearly and solemnly, without his usual gibberish: "Not another one! It's deadly!"

I confess that such explicit admonition under such circumstances, and so emphatic and commanding, coming from a person whose enigmatic jargon had until then been very difficult to understand, shook my determination to carry on with the experiment. But it was just this fear of yielding to a superstitious notion that caused me to make up my mind. The electrodes were applied again, and a 110-volt discharge was applied for 0.2 seconds.

We observed the same instantaneous, brief, generalized spasm, and soon after, the onset of the classic epileptic convulsion. We were all breathless during the tonic phase of the attack, and really over-whelmed during the apnea as we watched the cadaverous cyanosis of the patient's face; the apnea of the spontaneous epileptic convulsion is always impressive, but at that moment it seemed to all of us painfully endless. Finally, with the first stertorous breathing and the first clonic spasm, the blood flowed better not only in the patient's vessels but also in our own. Thereupon we observed with the most intensely gratifying sensation the characteristic gradual awaking of the patient "by steps." He rose to sitting position and looked at us, calm and smiling, as though to inquire what we wanted of him. We asked: "What happened to you?" He answered: "I don't know. Maybe I was asleep." Thus occurred the first electrically produced convulsion in man, which I at once named "electroshock."

To complete the history of the first patient treated with electroshock, I must note that after eleven applications, the patient wrote the physicians of the Institute on the following May 15 a well-composed letter in which he described his previous disturbances and treatments received and thanked the physicians for what they had done for him. We later received information from the Mombello Psychiatric Hospital (Milan), where he had been precedently treated, with the following report: "This patient was admitted for the second time in our hospital on December 28, 1937, at which time he presented auditory hallucinations and deliriant ideas of being influenced by means of machines, as well of being persecuted, particularly by his wife. The case was diagnosed as paranoiac schizophrenia. He was treated with eight intravenous injections of cardiazol and showed some improvement. Thereupon he presented a condition of acute rheumatic fever which was cured in 18 days, after which he was released in his wife's care."

Sources: Emil Kraepelin, 'Paranoidal Forms of Dementia Præcox', in Lectures on Clinical Psychiatry, trans. Thomas Johnstone, 2nd edn (London: Ballière, Tindall and Cox, 1906), pp. 153–6; Ugo Cerletti, 'Electroshock Therapy', in Arthur M. Sackler et al. (eds), The Great Physiodynamic Therapies in Psychiatry: An Historical Reappraisal (New York: Hoeber-Harper, 1956), pp. 93–4.

RIPPLES IN THE SAND

This landmark paper in wave mechanics made scientific history when, on 16 June 1904, it became the first to be read in person by a female scientist at the Royal Society of London (before then, papers by women had been read aloud by male representatives). Its author, Hertha Ayrton (1854–1923), who won the Royal Society's Hughes Medal in 1906, had already been nominated to the fellowship of the Society, but her application was rejected on the spurious grounds that she was married – a stipulation that had never applied to men. It would be another forty years before the first woman fellow was finally elected (see p. 362).

Hertha Ayrton was born Phoebe Marks in Portsea, Hampshire, but adopted the name 'Hertha' from the title of a poem by Algernon Swinburne. She studied mathematics at Girton College, Cambridge, at a time when women could attend the university but not claim their degrees; London University did award degrees to women, however, so she moved there and graduated BSc in 1881. After tutoring in mathematics for a while, she began her research career alongside her husband, William Ayrton, and published widely on the subject of the electric arc before moving on to the mechanics of ripple marks. She was a lifelong campaigner for women's access to science. 'I do not agree with sex being brought into science at all,' she once said. 'The idea of "woman and science" is completely irrelevant. Either a woman is a good scientist, or she is not; in any case she should be given opportunities, and her work should be studied from the scientific, not the sex, point of view':

To any one who, for the first time, sees a great stretch of sandy shore covered with innumerable ridges and furrows, as if combed with a giant comb, a dozen questions must immediately present themselves. How do these ripples form? Are they made and wiped out with every tide, or do they take a long time to grow, and last for many tides? What is the relation between the ripple and the waves to which they owe their existence? And a host of others too numerous to mention.

The questions to which I particularly directed my attention at first were the following: — (1) How do the ripples first start? (2) What is the relation between the water waves and the ripples?

During the course of this investigation certain fresh facts have come to light, showing how the principles involved in the formation of ripplemark apply to other phenomena of apparently widely different origin. Some of these are included in the present communication, but the discussion of others, less immediately connected with ripples, I have deferred to a future occasion.

1. *Starting of the First Ripple.*—To the first question as to the *origin* of ripplemark – fundamental as it is – I could, for some time, find no satisfactory answer, either in nature or in books. Even the deeply interesting paper in which Prof. George Darwin described the vortices he had discovered in the water oscillating over ripplemark touched but lightly on this point. Prof. Darwin said: 'When a small quantity of sand is sprinkled [in a glass trough] and the rocking begins, the sand dances backwards and forwards on the bottom, the grains rolling as they go.

'Very shortly the sand begins to aggregate into irregular little flocculent masses, the appearance being something like that of curdling milk. *The position of the masses is, I believe, solely determined by the friction of the sand on the bottom.*' . . .

With all the respect I felt for so eminent an observer – one, also, who had thrown so much light on the subject – I could not concur in this opinion. It seemed to me impossible that chance inequalities, having no relation with one another, but scattered here and there entirely without order, should develop into such ripples as are commonly seen on the sea shore – straight as if ruled, all of the same shape, and all at equal distances apart, or at distance varying according to some definite law. I cast about, therefore, for some other solution to the problem – some way of connecting the ripples with one another from the beginning, without the intervention of chance irregularities in the sand. I may say at once that I have been successful in finding such a solution, and that I am about to show how oscillating water can produce ripplemark on sand which is perfectly smooth and level to start with, and free from irregularity of every sort.

My first experiments were carried out at Margate, with the rather coarse brown sand found there. I tried oscillating water of various depths, over different thicknesses of sand, in vessels of all sorts of shapes and sizes, from a soap dish some 4" x 3" x 2" to a tank 44" x 18" x 18". The oscillations were produced by giving the vessel either slight instantaneous horizontal pushes, or a very small rocking motion, in time with the natural swing of the water, and by putting either rollers or cushions under the vessel to ease the jerks. The sand was made quite level at the beginning of each experiment, by being violently stirred up first, and then gently and irregularly shaken while it was settling.

In every vessel ripples appeared in times varying from a few seconds to a few minutes; and in all those in which the water was simply made to rise and fall alternately at each end of the vessel, without the formation of intermediate waves, two things invariably happened: — (1) Ripples formed first across the *middle* of the vessel, a fact first observed by C. de Candolle, and (2) after prolonged oscillation most of the sand had collected there also in a long ripple-marked heap . . .

Since ripples formed first across the middle of the vessel, and as, also, the sand was gradually removed from near the ends to the middle, by prolonged oscillation, it seemed clear that it was the formation of a small ridge across the middle, during the first few oscillations, that caused the ripples to start there first.

In order both to reduce surface friction, and to render observation easy, I repeated the experiment, scattering a mere pinch of sand as evenly as possible over the smooth bottom of a pie dish, some 8 inches long, containing about an inch of water. In this way, each grain, being isolated from the next, could be easily moved by the water, and as readily observed. The result was very striking. After oscillation of the water for less than half a minute, *the whole of the sand was collected in a straight line across the middle of the dish at right angles to the line of motion.*

Scattering the sand again, and watching carefully how the water moved it, I saw that each swing of the water pushed every grain that was being swept *towards* the middle farther than the next swing carried it *away* again from the middle . . .

It is clear that the formation first of the middle ripples, and the collecting of a mound across the middle of the vessel after prolonged oscillation could only be extensions of the operation just described. This, then, is the way in which the first ridge can form without the aid of any chance excrescence to start it.

When water is kept oscillating over sand, then, the dance of the grains described by Prof. Darwin is not, as he conjectured, a simple swaying to and fro, which would leave each grain where it found it, but for chance inequalities of the surface. *It is*, on the contrary, *a steady periodic advance from places where the horizontal velocity of the water is least to places where it is greatest, each oscillation leaving the grain nearer to their goal than it found them.*

Since water which rises and falls alternately at each end of a vessel, while its level remains nearly constant at the middle, is really oscillating in a stationary wave, of which the middle of the vessel is a loop, as regards horizontal motion, and the ends are nodes, it is interesting to note that the sand, in gathering across the middle of a vessel, is really collecting at a loop of the stationary wave generated by the oscillation of the water.

2. *The Formation of Fresh Ripples beside an Existing One.*— Having established the primary ridge, and found the conditions necessary for its formation from smooth and level sand, the next question that arose was, how are all the other ripples started? Do they depend for their initiation simply on unevennesses of the surface, or are they also subject to some definite law? M. Forel noted that a foreign body in the sand set up, in some way or other, a series of ripples in the sand on either side of it; but how these ripples start – whether all at one instant or each separately – and what is the process of initiation, has not, I think, hitherto been elucidated.

The solution of the problem cost me several weeks of observation and experiment, yet it was absurdly simple when it came. It was that a single ripple, existing alone, in otherwise smooth sand, initiates a ripple on either side of it, that each of these ripples produces another on its farther side – these in their turn originate other ripples on their farther sides, and so on, till the whole sand is ripple-marked. This suggestion having occurred to me, I tried in many ways to make sure that it was correct. For instance, I formed a fairly high ridge at some distance from the middle of the vessel, and watched to see if others followed from that, before the primary ridge in the middle became visible; or, again, I made a ridge of some peculiar shape, such as this in plan, >, and noted whether the succeeding ridges took the same, or nearly the same shape; and they did, the angle in each one being more obtuse than in that formed before it. Thus I felt sure of the *fact*; it only remained to see how it was accomplished.

For this purpose I abandoned my brown sand in favour of silver sand, which I had found to be so mobile that it spun in delicate fairy-like vortices in the lees of some of the ridges. These vortices, which differed widely from those discovered and described by Prof. Darwin,

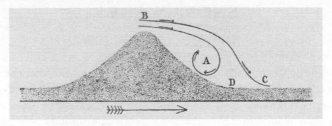

18. A Ripple-forming Vortex with Generating Ridge.

were, as I afterwards found, the true ripple-forming vortices, and I therefore call them ripple vortices.

3. *Structure and Functions of the Ripple Vortex.*—The vortices I saw had horizontal axes and were spiral in shape, and they seemed to scoop sand out from the bases of the ridges, and to push some of it up the ridges while they carried the remainder whirling round with them. In watching them it occurred to me that since each vortex (A) raised sand up the ridge against which it revolved, while the water that flowed over the vortex BC swept sand *away* from that ridge, there must be some neutral line parallel to the ridge at about D, on one side of which sand was being swept in one direction and on the other side in the other: and that, if that were so, a hollow must be formed parallel to the ridge, and reaching to some little distance on either side of the neutral line. Such a hollow must, it seemed to me, form a new furrow, while the wall of it on the side remote from the ridge must ultimately become a new ridge.

Source: Hertha Ayrton, 'The Origin and Growth of Ripple-Mark', *Proceedings of the Royal Society of London, Series A,* 84 (1910), pp. 285–9.

SNOWFLAKES

Although the Chinese scholar Han Ying described the hexagonal nature of snow crystals in 135 BC – 'flowers of plants and trees are generally five-pointed, but those of snow, which are called *ying*, are always six-pointed', he wrote – it was another 1,500 years or more before the structure of snowflakes was rediscovered in the West. One of the first to analyse the crystalline nature of the snowflake was the German astronomer Johannes Kepler (1571–1630), who is best known today for his laws of planetary motion that account for the movements of bodies in orbit. In a playful essay, *On the Six-Cornered Snowflake* (1611), written as a new year's gift for his patron, Kepler revealed a sophisticated understanding of the world of crystalline structures, as well as a realization that a subject as apparently simple as the snowflake offered a showcase for the subtleties of the new mathematics that were beginning to emerge in Europe: 'There must be some definite cause why, whenever snow begins to fall, its initial formations invariably display the shape of a six-cornered starlet', he wrote. 'For if it happens by chance, why do they not fall just as well with five corners or with seven? Why always with six, so long as they are not tumbled and tangled in masses by irregular drifting, but still remain widespread and scattered? When I recently had a conversation with someone about this subject, we first agreed that the cause was not to be looked for in the material, but in an agent. For the stuff of snow is vapour, and when vapour arises from the earth and is wafted up by some inherent heat, it simply hangs together in an almost liquid state, and is not therefore broken up into single starlets of this kind.' Kepler then went on to demonstrate that a range of other natural structures, such as the seed cases of pomegranates, follow the same hexagonal stacking arrangement, exhibiting a numerical constant found in many natural forms that offers an efficient use of space as well as stability (particularly in cold temperatures): could this have been the secret of the snowflake's design?

Three hundred years after Kepler's investigations, a farmer from Vermont, Wilson A. Bentley (1865–1931), established the rule that no two snowflakes are alike. 'Snowflake' Bentley spent years photographing snow crystals through a customized bellows camera, and published dozens of articles in the *Monthly Weather Review* of the American Meteorological Society. This extract is from his last published paper, in which Bentley allows himself some reflection on his life's work, which he describes as 'one of the little romances of science':

The past winter, 1926–27, noted for its mildness and scant snowfall in Vermont, goes on record as a fairly favorable winter for snow-crystal photography. There were 12 favorable snowfalls, and although 10 of them were light, and furnished only about a dozen photographs each, two of them – the storms of January 23 and February 22 – furnished large sets of crystals (over 40 each). December furnished three favorable storms, the 4th, 6th, and 20th; January, six, the 1st, 2d, 5th, 15th, 17th, and 23d; and February, three, the 3d, 9th, and 23d. Of these, four were cold snowfalls, occurring during temperatures ranging from 5° above to 5° F. below zero.

The winter furnished about 200 new photomicrographs, among them more than the usual number of exceptional or 'wonderful' crystals; 40 of them can be classed as such.

The new snow gems were doubly welcome because of the fact that recent winters, since 1920, have been rather unfavorable. The recent winters seemed to have gotten into a habit of periodicity as regards the character of their snowfall . . . I spent most of the winter of 1925 at Morrisburg on the St. Lawrence River, about 80 miles upriver from Montreal, carrying on my snow-crystal work there. I had wished for years to try photographic work a little farther north in Canada, because I thought it might be even more favorable for my work there than at Jericho, Vt.

The winter of 1925, however, happened to be an unfavorable one, so I was left somewhat in doubt as to whether the Canadian location was favorable or not, yet from the data secured I feel sure that the modifying influence of the Great Lakes to the westward is an unfavorable influence. The lakes tend to raise the temperature of the air both at the surface and presumably in cloudland, and thus favor the production of granular snow and granular-covered crystals; so I returned home with the query in my mind, 'Could it be that through some strange freak of accident or providence, that the one man who loves the snowflakes most had been born at the one most favorable spot on earth for the study and photographing of them?' . . .

Resuming once more in 1926 my work at my home location, I have succeeded in photographing many new and wonderful snow crystals the past two winters.

19. In 1885 Vermont farmer Wilson Bentley became the first person to successfully photograph snowflakes, using a bellows camera rigged up to a light microscope. Over the next forty years he built up a collection of many thousands of images, from which he concluded that no two flakes are the same. This image shows a small selection of them.

A brief chronological review and mention of some of the results of my more recent work during the seven-year period 1921–1927 may be of interest. Among the most wonderful crystals of this period is one (3950) that fell November 26 in the early winter of 1922. I have named it the 'good luck,' or horseshoe crystal. It not only has a horseshoe pictured as its nuclear feature, but more wonderful still, it has six surrounding features, each of which resembles a horse's hoof with heel calks.

The beautiful branching one that fell December 9, 1921, No. 3999, is also a masterpiece of crystal architecture.

The winter of 1923 produced among others three very exquisite specimens. One of these (No. 4149), which fell during the very cold snowfall of December 7, 1922, is a very marvellous quasi-trigonal crystal. The thrillingly beautiful one (No. 4215) of January 10, 1920, is also very notable, and also No. 4273 because of the beautiful circular arrangement of its tiny loops and scallops . . . As previously noted, the winter of 1927 was quite favorable. The writer wishes all the readers of the Review could see and enjoy the snowflake masterpieces of this winter . . .

The 1927 snowflake season had an early but wonderfully brilliant closing on February 22. On that date, in the early morning, the clouds for a while showered the earth with starry, fernlike gems such as thrill, amaze, and delight snowflake lovers. Many of these crystals were of extraordinary size, some being one-half inch in diameter. So heavy were they that many of them were broken in alighting upon my blackboard.

The snowflakes in this storm were so substantial that after the snow ceased I took quantities of them indoors and used them for photographing until nearly noon, when sunlight and rising temperature prevented further work. Although many of the crystals were somewhat deformed by unequal evaporation, the set as a whole is of exquisite beauty – a priceless addition to my series of branching crystals. It will be noted that the general effect of the arrangement of the multitudes of secondary and tertiary degree rays around the axial rays is beautifully symmetrical. Yet a closer analysis discloses that no two of the axial and pendant rays are alike, and that the secondary and tertiary degree rays are not always arranged opposite each other in pairs as is often the case. This suggests colloidal crystallization, the use by the growing crystals, in part, of groups of water molecules not completely subject to crystallographic law.

A thorough analysis of this wonderful series of branching crystals leaves one in doubt as to which ones are the most beautiful and interesting. The drooping pattern of No. 4711 recalls some of the drawings of Glaisher. The downward growth of rays of the third degree in No. 4695 forms a lovely, unique pattern. Very interesting also are the branchy rays arranged as peripheral adornments around the solid centerpiece of No. 4716.

Perhaps most interesting of all is No. 4726, because it shows so beautifully the tendency, so often seen in some form, by many hexagonal crystals, to divide into three. In this specimen it will be noted that the main secondary rays of each alternate axial ray have grown farther than those lying between them, thus forming a triangular effect.

This brief account of the newer 'treasures of the snow' will perhaps once more serve to inspire renewed interest in the peerless snow gems and to emphasize the fact that the treasures of the snow are absolutely inexhaustible, almost untouched as yet.

The writer is happy in the thought of having added during recent years so many new snow gems of the 'first water' to his already numerous collection of over 4,700 specimens, of which no two are alike. There is much room also for gratification in the fact that there is an ever increasing interest in snow crystals the world over, as proven by the manner in which they are being featured by the press, magazines, lecturers, museums, textbooks, and moving pictures, as well as the new uses of them as designs in the arts, crafts, and industrial sciences.

As the writer looks back 44 years to the beginning of his seemingly unimportant study of snow crystals, it seems to him remarkable that the work should have produced such undreamed of results. Perhaps it is not too much to say that the results of his studies form one of the 'little romances of science.'

Source: Wilson A. Bentley: 'Some Recent Treasures of the Snow', *Monthly Weather Review* 55:8 (1927), pp. 358–9.

SCIENCE *VERSUS* RELIGION

A great deal has been written and said about the perceived incompatibility of science and religion, and there is certainly much that divides them: science is based on observable evidence, while religious belief is based on faith; the sciences are willing to discard their central tenets, while religions tend to cling fast to theirs; science thrives on doubt and scepticism, while religion deals in the illusions of certainty. But, as the biologist J. B. S. Haldane (1892–1964) pointed out in the following essay, to concentrate on the differences between them is to overlook their long shared history, during which natural science was more or less indistinguishable from theology. Their recent divergence was, he suggests, more aesthetic than ideological, with the nineteenth-century rise of science accompanied by the spread of a distinctively modern cast of mind that preferred testable shared experiences to untestable private convictions. It's an intriguing hypothesis, elegantly argued by one of the wittiest science writers of all time (his celebrated reply to a clergyman's question about what science can infer about the mind of God was 'an inordinate fondness for beetles'):

Religion and science are human activities with both practical and theoretical sides. There is at present a certain degree of conflict between them, and this will undoubtedly continue for some generations. During this conflict the disputants have tended to emphasize the differences between them. But their resemblances are equally interesting, and perhaps throw a good deal of light on the differences.

It is only very recently that they have had a chance of diverging. Readers of the Pentateuch, or of the contemporaneous or earlier religious literature of Egypt or Mesopotamia, will find it very difficult to disentangle the science from the religion. The Pentateuch contains some very good applied science in the sanitary laws of Moses. The palaeontology of Genesis is also correct in many points; particularly in describing a period of the earth's history before the origin of life, followed by the appearance of animals in the seas, and only later on land, man being the last creation. It is of course wrong in putting the origin of plants before that of stars, of birds before that of creeping things, and in several other respects.

Unfortunately, however, since Moses' time science and religion have diverged, not without a certain loss to both. The reason for this divergence can best be seen when we study some typical scientific and religious minds at work. Each starts from a certain experience, and builds up a system of thought to bring it into line with the remainder of experience. The organic chemist says, 'The substance I have just made is a liquid with a characteristic smell, melting at 31º C., boiling at 162º C., and whose compound with phenyl-hydrazine melts at 97º C. I have probably synthesized furfural in a new way.' The saint says, 'I have had an experience very wonderful and rather difficult to describe in detail, but I interpret it to mean that God desires me to devote myself to preaching rather than to shut myself up from the world.'

The scientific man then starts from experiences in themselves emotionally flat, though to him perhaps interesting enough. He may end by producing a theory as exciting as Darwinism, or a practical invention as important as antiseptics or high explosives. The mind of the religious man on the other hand works on a descending scale of emotions. The dogma, prophecy, or good works which he may produce are inevitably less thrilling than his religious experience.

It is more interesting to most minds to read or speculate about the distances of the stars than to measure the positions of their images on a photographic plate. But it is less interesting to read a work on justification by faith or on St. Thomas's theory of transubstantiation than to take part in a well-conducted church service.

Now, the rather dull raw material of scientific thought consists of facts which can be verified with sufficient patience and skill. The theories to which they have given rise are far less certain. They change from generation to generation, even from year to year. And the religious opponents of science tend to scoff at this perpetual change. In scientific thought we adopt the simplest theory which will explain all the facts under consideration and enable us to predict new facts of the same kind. The catch in this criterion lies in the word 'simplest.' It is really an aesthetic canon such as we find implicit in our criticisms of poetry or painting. The layman finds such a law as

$$\frac{\partial x}{\partial t} = k \frac{\partial^2 x}{\partial y^2}$$

much less simple than 'it oozes,' of which it is the mathematical state-ment. The physicist reverses this judgement, and his statement is certainly the more fruitful of the two, so far as prediction is con-cerned. It is, however, a statement about something very unfamiliar to the plain man, namely, the rate of change of a rate of change. Now, scientific aesthetic prefers simple but precise statements about unfa-miliar things to vaguer statements about well-known things. And this preference is justified by practical success. It is more satisfactory scientifically to say: 'The blood-vessels in John Smith's skin are dilated by a soluble toxin produced by a haemolytic streptococcus growing in his pharynx,' than to say that he has scarlet fever. It sug-gests methods of curing and preventing the disease. But only a few people have seen a streptococcus, and no one has seen a toxin in the pure state. In physics, the most developed of the sciences, things have gone so far that many physicists frankly say that they are describing atom models and not atoms. Atoms themselves have the same sort of reality as chairs and tables, because single atoms can be seen if going fast enough. But when we come to their internal structure we can only say that they behave, in some important respects, as if electrons were going round in them with such and such velocities in such and such orbits. If that is the real structure we can calculate the velocities with a great deal more accuracy than that with which a speedometer gives the speed of an automobile; and verifiable predictions based on these calculated speeds come out with very great accuracy. But the speeds are not observable, and physicists are becoming less and less careful as to what hypothetical events they postulate to explain observable phenomena, provided the hypotheses enable them to pre-dict accurately.

Einstein showed that we could explain and predict slightly better if he substituted other conceptions for those of space and time, and his own substitutes will doubtless be replaced in their turn. Now, if Einstein is right, or even partly right, no physicists before his time knew quite what they were talking about when they used the ideas of distance and time, and practically every statement they made which purported to be accurate was false. So presumably is every such statement of a modern physicist . . .

In fact, the experience of the past makes it clear that many of our most cherished scientific theories contain so much falsehood as to deserve the title of myths. Their claims to belief are that they contradict fewer known facts than their predecessors, and that they are of practical use. But there is one very significant feature of the most fully developed scientific theories. They tell us nothing whatever about the inner nature of the units with which they deal. Electrons may be spiritually inert, they may be something like sensations, they may be good spirits or evil spirits. The physicist, however, can only tell us that they repel one another according to a certain law, are attracted by positive charges according to another law, and so on. He can say nothing about their real being, and knows that he cannot.

Marie Stopes

All that may be true on the theoretical level, but in practice the kinship between science and religion can easily break under strain. The career of Marie Carmichael Stopes (1880–1958) was a case in point; her advocacy of family planning in books such as *Married Love* (1918) and *Wise Parenthood* (1918) saw her publicly condemned by Britain's Church leaders, who demanded her prosecution and imprisonment. In her pioneering work, *Contraception* (1923), Stopes attempted to answer her critics along rational, scientific lines, while lamenting the strongly religious atmosphere that seemed to surround the subject of contraception, which was, she argued, in every other respect 'a purely medical and scientific theme'. What made her particularly angry was the fact that her male Christian critics appeared to be outraged by the thought of the alleviation of human suffering, and she likened their attacks on contraception to the clerical objections to chloroform and smallpox vaccination that had been aired in the previous century. 'In our own country inoculation against small-pox was denounced as being "indefensible on religious as well as medical grounds" …"a diabolical operation" …"a discovery sent into the world by the powers of evil." Then clergy preached against vaccination and described it as a "daring and profane violation of our holy religion." Dr. Rowley preached against it, saying — "The law of God prohibits the practice, the law of man and the law of nature loudly exclaim against it." Yet where, to-day, is the cleric who would dare to preach thus to an educated congregation?' Stopes was fond of quoting the words of the Scottish medic James Young Simpson (1811–70), who had pointed out that the clerics who objected to chloroform – which *really* alleviated the pains of childbirth – offered no objections to warm baths, compresses and manual manipulations, which were designed *partly* to do the same. 'By these means they succeeded partially, in times past, in mitigating the sufferings and effects of parturition, and thought they committed no sin', he wrote. 'But a means is discovered by which the sufferings of the mother may be relieved far more effectually and then they immediately denounce this higher amount of relief as a high sin. Gaining your end, according to their religious views, imperfectly was no sin — gaining your end more fully and perfectly, is, they argue, an undiluted and unmitigated piece of iniquity.' 'There is little doubt', observed Stopes, 'that in another twenty years or less those same "arguments" will be used and those same objurgations hurled at some other advance of scientific alleviation of human suffering, and that no priest or cleric will dare to inveigh against birth control then, just as to-day none dares to repeat the sermons of his predecessors against chloroform.'

Sources: J. B. S. Haldane, 'Science and Theology as Art Forms', in *Possible Worlds and Other Essays* (London: Chatto & Windus, 1927), pp. 225–9; Marie Stopes, *Contraception (Birth Control): Its Theory, History and Practice* (London: John Bale, Sons & Danielsson, 1923), pp. 227–40.

PLANCK TIME

The German physicist Max Planck (1858–1947) is often referred to as 'the father of quantum theory', because it was his deceptively simple question, posed in 1900, that set the quantum revolution in train: what if the heat and light emitted by a hot piece of metal (such as a light-bulb filament) did not travel in waves, as the prevailing view suggested, but in discrete packets of energy that he called 'quanta'? If that was true, it would solve a number of mathematical conundrums that physicists had so far been unable to explain, such as the precise mechanism by which the dominant frequency of light in a glowing metal bar increases with a rise in temperature. As will be seen in the next entry, it would be the young Albert Einstein who took up Planck's innocent-seeming query and from it refashioned the whole of twentieth-century physics. Planck, meanwhile, continued with his experiments, as yet unaware that he had laid the foundations for the coming scientific revolution, though his contribution is commemorated in the quintet of fundamental Planck units that include Planck mass (with a value of 2.177×10^{-8} kilograms); Planck length (with a value of 1.616×10^{-35} metres); and Planck time (with a value of 5.391×10^{-44} seconds), each with an unimaginably tiny value, and each defined in precise relation to the five universal physical constants that include gravity and the speed of light.

The following poem, from George Bradley's (b. 1953) first collection, *Terms to be Met* (1986), imagines the coming into being of the physical universe some 13.7 billion years ago, from a tiny speck of pure energy that transformed into an array of subatomic particles of elementary matter in a blink of an eye, in 'no time at all', as Bradley puts it, which is to say, in the first 10^{-43} seconds of the universe's life, the so-called Planck epoch, a period equivalent to 0.0000000000000000000000000000000000000000001 of a second. In this brief moment, which represents the first stage of everything – before the sudden rapid expansion that came one hundred-billionth of a yoctosecond later, inflating the new-born universe from billions of times smaller than a proton to the size of a football pitch in a fraction of a second – all bets are off. For nothing is known of the Planck epoch's physical reality, no laws of physics, not even the universal constants (gravity, the speed of light) can be made to apply, for the four fundamental forces of nature (gravity, the strong and weak nuclear forces, and electromagnetism) had yet to be separated out.

It's almost impossible to imagine, of course, but Planck time really was no time at all:

> Once upon a time, way back in the infinitesimal
> First fraction of a second attending our creation,
> A tiny drop containing all of it, all energy
> And all its guises, burst upon the scene,
> Exploding out of nothing into everything
> Virtually instantaneously, the way our thoughts
> Leap eagerly to occupy the abhorrent void.
> Once, say ten or twenty billion years ago,
> In Planck time, in no time at all, the veil
> Available to our perceptions was flung out
> Over space at such a rate the mere imagination
> Cannot keep up, so rapidly the speed of light
> Lags miraculously behind, producing a series
> Of incongruities that has led our curiosity,
> Like Ariadne's thread, through the dim labyrinth
> Of our conclusions to the place of our beginning.
> In Planck time, everything that is was spread so thin
> That all distance is enormous, between each star,
> Between subatomic particles, so that we are composed
> Almost entirely of emptiness, so that what separates
> This world, bright ball floating in its midnight blue,
> From the irrefutable logic of no world at all
> Has no more substance than the traveller's dream,
> So that nothing can be said for certain except
> That sometime, call it Planck time, it will all just
> Disappear, a parlor trick, a rabbit back in its hat,
> Will all go up in a flash of light, abracadabra,
> An idea that isn't being had anymore.

Source: George Bradley, 'About Planck Time', from *Terms to be Met* (New Haven: Yale University Press, 1986), p. 39.

THE QUANTUM THEORY OF LIGHT

Albert Einstein (1879–1955) did more than most to popularize twentieth-century science, even though none of his publications sold well; 'Isn't it strange that I who have written only unpopular books should be such a popular fellow?' he once asked. He remains, of course, most famous for his theories of relativity, which showed that strong gravitational fields could distort the fabric of space and time, but it was for his work on the quantum nature of light that he won the 1921 Nobel Prize in Physics. Having taken up Max Planck's earlier attempt to explain the 'photoelectric effect', in which electrons are emitted from the surfaces of metals in different ways when heated or lit, Einstein demonstrated that light travelled not in continuous waves, as most nineteenth-century physicists had believed, but as discrete packets of energy, or 'quanta', that would later be known as photons. Because photons exhibit certain wavelike properties, light could therefore be considered as both a wave *and* a particle during different stages of its career. This is a deeply strange idea: no wonder Niels Bohr maintained that anyone who is not shocked by quantum theory has failed to understand it.

The following extract from a history of theoretical physics that Einstein co-wrote at Princeton University with the Polish physicist Leopold Infeld (1898–1968), begins with a striking visual analogy that goes some way to elucidating the quantum/wave conundrum:

Let us consider a wall built along the seashore. The waves from the sea continually impinge on the wall, wash away some of its surface, and retreat, leaving the way clear for the incoming waves. The mass of the wall decreases and we can ask how much is washed away in, say, one year. But now let us picture a different process. We want to diminish the mass of the wall by the same amount as previously but in a different way. We shoot at the wall and split it at the places where the bullets hit. The mass of the wall will be decreased and we can well imagine that the same reduction in mass is achieved in both cases. But from the appearance of the wall we could easily detect whether the continuous sea wave or the discontinuous shower of bullets has been acting. It will be helpful in understanding the phenomena which we are about to describe, to bear in mind the difference between sea waves and a shower of bullets.

We said, previously, that a heated wire emits electrons. Here we shall introduce another way of extracting electrons from metal. Homogeneous light, such as violet light, which is, as we know, light of a definite wave-length, is impinging on a metal surface. The light extracts electrons from the metal. The electrons are torn from the metal and a shower of them speeds along with a certain velocity. From the point of view of the energy principle we can say: the energy of light is partially transformed into the kinetic energy of expelled electrons. Modern experimental technique enables us to register these electron-bullets, to determine their velocity and thus their energy. This extraction of electrons by light falling upon metal is called the *photoelectric effect*.

Our starting-point was the action of a homogeneous light wave, with some definite intensity. As in every experiment, we must now change our arrangements to see whether this will have any influence on the observed effect.

Let us begin by changing the intensity of the homogeneous violet light falling on the metal plate and note to what extent the energy of the emitted electrons depends upon the intensity of the light. Let us try to find the answer by reasoning instead of by experiment. We could argue: in the photoelectric effect a certain definite portion of the energy of radiation is transformed into energy of motion of the electrons. If we again illuminate the metal with light of the same wave-length but from a more powerful source, then the energy of the emitted electrons should be greater, since the radiation is richer in energy. We should, therefore, expect the velocity of the emitted electrons to increase if the intensity of the light increases. But experiment again contradicts our prediction. Once more we see that the laws of nature are not as we should like them to be. We have come upon one of the experiments which, contradicting our predictions, breaks the theory on which they were based. The actual experimental result is, from the point of view of the wave theory, astonishing. The observed electrons all have the same speed, the same energy, which does not change when the intensity of the light is increased.

This experimental result could not be predicted by the wave theory. Here again a new theory arises from the conflict between the old theory and experiment.

Let us be deliberately unjust to the wave theory of light, forgetting its great achievements, its splendid explanation of the bending of light around very small obstacles. With our attention focused on the photoelectric effect, let us demand from the theory an adequate explanation of this effect. Obviously, we cannot deduce from the wave theory the independence of the energy of electrons from the intensity of light by which they have been extracted from the metal plate. We shall, therefore, try another theory. We remember that Newton's corpuscular theory, explaining many of the observed phenomena of light, failed to account for the bending of light, which we are now deliberately disregarding. In Newton's time the concept of energy did not exist. Light corpuscles were, according to him, weightless; each colour preserved its own substance character. Later, when the concept of energy was created and it was recognized that light carries energy, no one thought of applying these concepts to the corpuscular theory of light. Newton's theory was dead and, until our own century, its revival was not taken seriously.

To keep the principal idea of Newton's theory, we must assume that homogeneous light is composed of energy-grains and replace the old light corpuscles by light quanta, which we shall call *photons*, small portions of energy, travelling through empty space with the velocity of light. The revival of Newton's theory in this new form leads to the *quantum theory of light*. Not only matter and electric charge, but also energy of radiation has a granular structure, i.e., is built up of light quanta. In addition to quanta of matter and quanta of electricity there are also quanta of energy.

The idea of energy quanta was first introduced by Planck at the beginning of this century in order to explain some effects much more complicated than the photoelectric effect. But the photo-effect shows most clearly and simply the necessity for changing our old concepts.

It is at once evident that this quantum theory of light explains the photoelectric effect. A shower of photons is falling on a metal plate. The action between radiation and matter consists here of very many single processes in which a photon impinges on the atom and tears out an electron. These single processes are all alike and the extracted electron will have the same energy in every case. We also understand that increasing the intensity of the light means, in our new language, increasing the number of falling photons. In this case, a different number of electrons would be thrown out of the metal plate, but the energy of any single atom would not change. Thus we see that this theory is in perfect agreement with observation.

Source: Albert Einstein and Leopold Infeld, *The Evolution of Physics: The Growth of Ideas from Early Concepts to Relativity and Quanta* (Cambridge: Cambridge University Press, 1938), pp. 272–80.

THE FORCE OF GRAVITY DOES NOT EXIST

As was demonstrated by Arthur Eddington's man in the falling lift (see p. 106), Newtonian mechanics are good for describing everyday interactions between objects, but not so good for accounting for the behaviour of objects in extreme circumstances, such as at speeds approaching the velocity of light. That is where relativity comes in. One of the main differences between the two systems is their competing definitions of gravity. For Newton, gravity was an attractive force that emanated from massive objects, while for Einstein it was not so much a force as a consequence of the bending of space in the neighbourhood of massive objects. In the case of earth's movement round the sun, for example, it is not gravity that sends our planet off on its elliptical orbit, but the fact that we are travelling through a region of curved space. As the American theoretical physicist John A. Wheeler (1911–2008) once said, 'there is nothing in the world except empty curved space. Matter, charge, electromagnetism and other fields are only manifestations of the curvature of space.'

There have been many other attempts to explain Einstein's ideas in accessible language, and one of my favourites is by Marcus Chown, in his eminently readable book *The Quantum Zoo* (2006) (published in the United Kingdom as *Quantum Theory Cannot Hurt You*). In this extract, Chown, a former radio astronomer at the California Institute of Technology, describes the effects of the bending of space-time from the point of view of a colony of ants trying to march across the surface of a trampoline:

The idea that gravity is a fictitious force may sound a little farfetched. However, in other everyday situations, we are perfectly happy to invent forces to make sense of what happens to us. Say you are a passenger in a car that is racing round a sharp corner in the road. You appear to be flung outward and, to explain why, you invent a force — centrifugal force. In reality, however, no such force exists.

All massive bodies, once set in motion, have a tendency to keep travelling at constant speed in a straight line.* Because of this propen-

---

* This is not at all obvious on earth, where frictional forces act to slow a moving body. However, it is apparent in the empty vacuum of space.

sity, known as inertia, unrestrained objects inside the car, including a passenger like you, continue to travel in the same direction the car was travelling before it rounded the bend. The path followed by the car door, however, is a curve. It should be no surprise, then, that you find yourself jammed up against a door. But the car door has merely come to meet you . . . there is no force.

Centrifugal force is known as an inertial force. We invent it to explain our motion because we choose to ignore the truth — that our surroundings are moving relative to us. But, really, our motion is just a result of our inertia, our natural tendency to keep moving in a straight line. It was Einstein's great insight to realise that gravity too is an inertial force. "Can gravitation and inertia be identical?" asked Einstein. "This question leads directly to my theory of gravity."

According to Einstein, we concoct the force of gravity to explain away the motion of apples falling from trees and planets circling the Sun because we ignore the truth — that our surroundings are accelerating relative to us. In reality, things move merely as a result of their inertia. The force of gravity does not exist!

But wait a minute. If the motion we attribute to the force of gravity is actually just the result of inertia, that must mean that bodies like Earth are really just flying through space at constant speed in straight lines. That's patently ridiculous! Earth is circling the Sun and not flying in a straight line, right? Not necessarily. It all depends on how you define a straight line . . .

Gravity bends light because space, in the presence of gravity, is somehow curved. In fact, this is all gravity turns out to be — curved space.

What exactly do we mean by curved space? It is easy to visualise a curved surface like the surface of Earth. But that is because it has only two directions, or dimensions — north-south and east-west. Space is a bit more complicated than that. In addition to three space dimensions — north-south, east-west, and up-down — there is one time dimension — past-future. As Einstein showed, however, space and time are really just aspects of the same thing, so it is more accurate to think of there being four "space-time" dimensions.

Four-dimensional space-time is impossible for us to visualise since we live in a world of three-dimensional objects. This means that the curvature, or warpage, of four-dimensional space-time is doubly impossible to visualise. But that's what gravity is: the warpage of four-dimensional space-time.

Fortunately, we can get some idea of what this means. Imagine a race of ants that spends its entire existence on the two-dimensional surface of a taut trampoline. The ants can only see what happens on the surface and have no concept whatsoever of the space above and below the trampoline — the third dimension. Now imagine that you or I — mischievous beings from the third dimension — put a cannon-ball on the trampoline. The ants discover that when they wander near the cannonball their paths are mysteriously bent towards it. Quite reasonably, they explain their motion by saying that the can-nonball is exerting a force of attraction on them. Perhaps they even call the force gravity.

However, from the God-like vantage point of the third dimen-sion, it is clear the ants are mistaken. There is no force attracting them to the cannonball. Instead, the cannonball has made a valley-like depression in the trampoline, and this is the reason the paths of the ants are bent towards it.

Einstein's genius was to realise that we are in a remarkably simi-lar position to the ants on the trampoline. The path of Earth as it travels through space is constantly bent towards the Sun, so much so that the planet traces out a near-circular orbit. Quite reasonably, we explain away this motion by saying that the Sun exerts a force of attraction on Earth — the force of gravity. However, we are mis-taken. If we could see things from the God-like perspective of the fourth dimension — something that is as impossible for us to do as it is for the ants to see things from the third dimension — we would see there is no such force. Instead, the Sun has created a valleylike depression in the four-dimensional space-time in its vicinity, and the reason Earth follows a near-circular path around it is because this is the shortest possible path through the warped space.

There is no force of gravity. Earth is merely following the straight-est possible line through space-time. It is because space-time near the Sun is warped that that line happens to be a near-circular orbit. According to physicists Raymond Chiao and Achilles Speliotopou-los: "In general relativity, no 'gravitational force' exists. What we normally associate with the force of gravity on a particle is not a force at all: The particle is simply travelling along the 'straightest' possible path in curved space-time."

A body travelling along the "straightest" possible path through space-time is in free fall. And, since it is in free fall, it experiences no gravity. Earth is in free fall around the Sun. Consequently, we do not feel the Sun's gravity on Earth. The astronauts on the International Space Station are in free fall around Earth. Consequently, they do not feel Earth's gravity.[*]

Source: Marcus Chown, *The Quantum Zoo: A Tourist's Guide to the Neverending Universe* (Washington, D.C.: Joseph Henry Press, 2006), pp. 122–7.

[*] Most people assume that astronauts orbiting Earth are weightless because there is no gravity in space. However, at the 500-kilometre-or-so height of the International Space Station, gravity is only about 15 percent weaker than on Earth's surface. The real reason astronauts are weightless is that they and their spacecraft are in free fall just as surely as someone in an elevator when the cable breaks. The difference is that they never hit the ground. Why? Because Earth is round and, as fast as they fall toward the surface, the surface curves away from them. They, therefore, fall forever in a circle.

BOMBARDING THE ATOM

The transmutation of one element into another had been the goal of the ancient alchemists, but it was not until the early twentieth century that it became anything like a reality. In 1919, Ernest Rutherford (1871–1937) – then Professor of Physics at the University of Manchester – conducted one of the most famous experiments in the history of atomic physics, when he bombarded nitrogen gas with alpha-particles from a piece of radioactive polonium. When an α-particle struck a nitrogen nucleus, a proton was ejected, and an oxygen nucleus formed in its wake. Rutherford knew at once that he had achieved the first artificial nuclear transmutation in history – no wonder he called his last book *The Newer Alchemy*.

New Zealand-born Rutherford was an extraordinary scientist, 'a force of nature', as one of his colleagues described him, whose career coincided with the end of amateurism and the emergence of state-funded 'big science'. As Director of the Cavendish Laboratory at Cambridge, he presided over a period of unprecedented achievement in laboratory science, including the celebrated 'splitting' of the atom in April 1932, a few weeks after the discovery of the neutron – also in Cambridge – and shortly before the discovery of the positron (the positive electron) at the California Institute of Technology. Rutherford refers to all these breakthroughs in the following excerpt from a BBC national radio lecture, 'The Transmutation of the Atom', that he delivered in October 1933. As he pointed out at the end of the talk, this journey of discovery into the sub-atomic realm constituted one of the most enthralling adventures of the human imagination:

The broad features of the constitution of all atoms are now well established. As a result of the splendid work of Bohr and those who have followed him, we are able to understand the arrangement and motions of the planetary electrons and the way in which light or X-rays are emitted when the atom is disturbed. Unfortunately we have much less information about the constitution of the minute central nucleus. We know the value of the nuclear charge and the mass of each atom, but we have no precise information on the nature and arrangement of the particles composing it. Until a year or so ago, it was generally supposed that the nucleus of an atom was ultimately composed of two electrical units, the negative electrons of small mass and the positively charged protons of mass 1. At the same time it became clear that secondary units were also present and that the helium nucleus of mass 4 – the α-particle – played a prominent part. Recently, however, we have had to extend our views, for undoubted evidence has been obtained of the existence of a new type of particle called the neutron which has a mass of 1 but no electrical charge. At the same time, the discovery this year of what is believed to be the positive electron of light mass – the counterpart of the negative electron – has complicated the problem. However, we may, I think, assume with some confidence that the nucleus of a heavy atom is in general composed of a large number of particles, some charged like the α-particle and proton, and others like the neutron electrically neutral. These are held together by powerful forces in an extraordinarily minute volume and form a very stable structure. We have, however, little to guide us in seeking a more detailed knowledge of the number, arrangements and motions of these constituent particles.

In order to transmute one atom into another, it appears essential to alter the charge on the nucleus. This can be done in imagination by adding another charged particle, say a proton or α-particle, to the nucleus, or removing a charged particle from it. We must, however, bear in mind that the nucleus is a strongly guarded structure held firmly together by strong attractive forces. In order to disrupt a nucleus, it thus seemed likely that very intense forces must be brought to bear directly upon it. One method of accomplishing this is to bombard the nucleus with very swift particles. Now the α-particle, which is spontaneously ejected from radium, is one of the most energetic particles known to science. It was recognised that if a stream of swift α-particles fell on matter, there was a small chance that one out of a great number might have an almost head-on collision with a nucleus. Under these conditions it must approach very near to it before it was turned back by the strong repulsive forces due to the electric charges on the two particles. It must be emphasised that the close collision of an α-particle with a nucleus involves the setting up of gigantic forces between the two nuclei concerned. In the case of light atoms where the nuclear charge is small, calculation indicated that the colliding α-particle, if it did not enter the struck nucleus, must at least approach sufficiently near to distort greatly its electrical structure. Under such disturbing forces, the nucleus might be expected to become unstable and then break up into other nuclei.

Actuated by these general ideas, I made in 1919 some experiments to test whether any evidence of transformation could be obtained when α-particles were used to bombard matter. The experiments were of a simple type; a preparation of radium served as a source of α-rays and the scintillation method was used to detect the presence of any new types of particles. It is well known that each α-particle falling on a preparation of zinc sulphide gives a flash of light, a scintillation, which is easily seen in a darkened room, and it was to be expected that any fast charged particle liberated from the bombarded matter would indicate its presence by a scintillation. When α-particles were used to bombard the gas oxygen, no new effect was observed. When, however, nitrogen gas was substituted, a number of scintillations were observed far beyond the distance of travel of the α-particles. Special experiments showed that these scintillations were produced by charged hydrogen atoms which we now call protons. The appearance of fast protons in these experiments could only be explained by supposing that they arose from the transformation of some of the nitrogen nuclei as a result of α-particle bombardment. This was the first time that definite evidence was obtained that an atom could be transformed by artificial methods. In the light of later experiments by Blackett, the general mechanism of this transformation became clear. It was found that the α-particle must actually penetrate into the nitrogen nucleus and be captured by it. As a consequence of this profound disturbance, a proton was ejected with high speed from the new nucleus.

Let us for a moment consider the simple arithmetic of this process. The mass of the nitrogen nucleus is 14 and its nuclear charge 7 units. The capture of an α-particle of mass 4 and nuclear charge 2 raises the mass to 18 and the charge to 9, while the loss of a proton of mass 1 and charge 1 results in the formation of an atom of mass 17 and charge 8. Now the oxygen nucleus has a charge 8, so that as a result of the interaction with an α-particle, the nitrogen nucleus is changed into the nucleus of oxygen . . .

My listeners may quite naturally ask why these experiments on transmutation should excite such interest in the scientific world. It is not that the experimenter is searching for a new source of power or the production of rare and costly elements by new methods. The real reason lies deeper, and is bound up with the urge and fascination of a search into one of the deepest secrets of Nature. Until a few years ago, we had to be content with the knowledge that the whole of matter in the universe, including our own bodies, was made up of ninety or more distinct chemical elements, but we had little definite knowledge of the inner structure of their atoms or of the processes by which one element could be converted into another. Now, for the first time, we are able to investigate these problems by direct experiments in the laboratory, and we are hopeful we shall soon add widely to our knowledge. The information so gained cannot but widen our outlook on the nature of matter, but must also have a direct bearing on many problems of cosmical physics. For example, in the furnace of the sun and other hot stars, the electrons, protons, neutrons and atoms present must be endowed with high average velocities owing to thermal agitation . . . It is well known that the abundance of the elements in our earth's crust varies very widely. Some elements like iron, nickel and oxygen are abundant, whilst others like lithium, platinum and gold are relatively rare. The information to be gained in our laboratories on the efficiency of various types of agencies in transforming atoms may help us to throw light on the reason for the relative abundance of different elements in our earth, and thus in the sun from which our earth is believed to be derived.

As one whose scientific life has been largely devoted to investigations on the structure and transformation of the atom, I watch with much interest and enthusiasm the development of these beautiful experiments to add to our knowledge of the constitution of nuclei. No one can be certain what strange particles or unexpected phenomena may not appear. I know of no more enthralling adventure of the human mind than this voyage of discovery into the almost unexplored world of the atomic nucleus.

Source: Ernest Rutherford, *The Transmutation of the Atom: The Thirteenth of the Broadcast National Lectures delivered on 11 October 1933* (London: The British Broadcasting Corporation, 1933), pp. 12–16; 25–8.

NUCLEAR FISSION

In the closing pages of his last book, *The Newer Alchemy* (1937), Ernest Rutherford had cautioned his readers that 'the outlook for gaining useful energy from atoms by artificial processes of transformation does not look very promising.' One year later, the discovery of nuclear fission kick-started the escalation in atomic research that led to the development of the Bomb. In this remarkable passage from his memoirs, the Austrian-born physicist Otto Frisch (1904–79) looks back over the role he played in the story of fission. At the heart of nuclear fission is the concept of the 'chain reaction', in which the release of high-energy neutrons from an initial rapid nuclear decay in turn triggers another decay, thus releasing further high-energy neutrons, until all the contained material suddenly splits instantaneously with an enormous burst of energy. As Frisch points out below, all this proved to be in full agreement with Einstein's famous equation ($E = mc^2$), but at the time it was still unknown territory for the many scientists involved, including Frisch's illustrious aunt, the physicist Lise Meitner (1878–1968).

Meitner had been a member of the German team that had made the unsettling discovery that bombarding uranium atoms produced two new neutrons for every neutron absorbed, releasing unexpectedly large amounts of energy in the process. The full implications of the discovery remained unrealized in 1938, but by then Meitner had fled Nazi Germany for Sweden, from where she continued to correspond with her scientific colleagues. As Frisch recalled, it was a letter from the chemist Otto Hahn in Berlin that first suggested to Meitner the likely mechanics of nuclear fission; it also gave rise to the disturbing thought that the German military would be keen to exploit its potential. At the end of his stay with his aunt, Frisch made his way back to Copenhagen where he passed on Meitner's calculations to his mentor, Niels Bohr, who took them with him to the United States, where, under the assumed name of Nicholas Baker, he took up his appointment as a senior consultant at the Los Alamos nuclear weapons laboratory in the New Mexico desert:

Until 1938 nobody dreamt that there was yet another way for a heavy nucleus to react to the mutual repulsion of its many protons, namely by dividing itself into two roughly equal halves. It was mere chance that I became involved in the discovery of that 'nuclear fission', which for the first time showed a way to make huge numbers of nuclei give up their hidden energy; the way to the atom bomb and to atomic power.

The occupation of Austria in March 1938 changed my aunt, the physicist Lise Meitner – technically – from an Austrian into a German. She had acquired fame by many years' work in Germany, but now had to fear dismissal as a descendant of a Jewish family. Moreover, there was a rumour that scientists might not be allowed to leave Germany; so she was persuaded – or perhaps stampeded – into leaving at very short notice, assisted by friends in Holland, and in the autumn she accepted an invitation to work in Stockholm, at the Nobel Institute led by Manne Siegbahn. I had always kept the habit of celebrating Christmas with her in Berlin; this time she was invited to spend Christmas with Swedish friends in the small town of Kungälv (near Gothenburg), and she asked me to join her there. That was the most momentous visit of my whole life.

Let me first explain that Lise Meitner had been working in Berlin with the chemist Otto Hahn for about thirty years, and during the last three years they had been bombarding uranium with neutrons and studying the radioactive substances that were formed. Fermi, who had first done that, thought he had made 'transuranic' elements – that is, elements beyond uranium (the heaviest element then known to chemists), and Hahn the chemist was delighted to have a lot of new elements to study. But Lise Meitner saw how difficult it was to account for the large number of different substances formed, and things got even more complicated when some were found (in Paris) that were apparently lighter than uranium. Just before Lise Meitner left Germany, Hahn had confirmed that this was so, and that three of those substances behaved chemically like radium. It was hard to see how radium – four places below uranium – could be formed by the impact of a neutron, and Lise Meitner wrote to Hahn, imploring him not to publish that incomprehensible result until he was completely sure of it. Accordingly Hahn, together with his collaborator, the chemist Fritz Strassmann, decided to carry out thorough tests in order to make quite sure that those substances were indeed of the same chemical nature as radium.

When I came out of my hotel room after my first night in Kungälv I found Lise Meitner studying a letter from Hahn and obviously worried by it. I wanted to tell her of a new experiment I was planning, but she wouldn't listen; I had to read that letter. Its content was indeed so startling that I was at first inclined to be sceptical. Hahn and Strassmann had found that those three substances were not radium, chemically speaking; indeed they had found it impossible to separate them from the barium which, routinely, they had added in order to facilitate the chemical separations. They had come to the conclusion, reluctantly and with hesitation, that they were isotopes of barium.

Was it just a mistake? No, said Lise Meitner; Hahn was too good a chemist for that. But how could barium be formed from uranium? No larger fragments than protons or helium nuclei (alpha particles) had ever been chipped away from nuclei, and to chip off a large number not nearly enough energy was available. Nor was it possible that the uranium nucleus could have been cleaved right across. A nucleus was not like a brittle solid that can be cleaved or broken; George Gamow had suggested early on, and Bohr had given good arguments that a nucleus was much more like a liquid drop. Perhaps a drop could divide itself into two smaller drops in a more gradual manner, by first becoming elongated, then constricted, and finally being torn rather than broken in two? We knew that there were strong forces that would resist such a process, just as the surface tension of an ordinary liquid drop tends to resist its division into two smaller ones. But the nuclei differed from ordinary drops in one important way: they were electrically charged, and that was known to counteract the surface tension.

At that point we both sat down on a tree trunk (all that discussion had taken place while we walked through the wood in the snow, I with my skis on, Lise Meitner making good her claim that she could walk just as fast without), and started to calculate on scraps of paper. The charge of a uranium nucleus, we found, was indeed large enough to overcome the effect of the surface tension almost completely; so the uranium nucleus might indeed resemble a very wobbly, unstable drop, ready to divide itself at the slightest provocation, such as the impact of a single neutron.

But then there was another problem. After separation, the two drops would be driven apart by their mutual electric repulsion and would acquire high speed and hence a very large energy, about 200 MeV [200 million electron-volts] in all; where could that energy come from? Fortunately Lise Meitner remembered the empirical formula for computing the masses of nuclei and worked out that the two nuclei formed by the division of a uranium nucleus together would be lighter than the original uranium nucleus by about one-fifth the mass of a proton. Now whenever mass disappears energy is created, according to Einstein's formula $E = mc^2$, and one-fifth of a proton mass was just equivalent to 200 MeV. So here was the source for that energy; it all fitted!

A couple of days later I travelled back to Copenhagen in considerable excitement. I was keen to submit our speculations – it wasn't really more at the time – to Bohr, who was just about to leave for the U.S.A. He had only a few minutes for me; but I had hardly begun to tell him when he smote his forehead with his hand and exclaimed: 'Oh what idiots we all have been! Oh but this is wonderful! This is just as it must be! Have you and Lise Meitner written a paper about it?' Not yet, I said, but we would at once; and Bohr promised not to talk about it before the paper was out. Then he went off to catch his boat.

The paper was composed by several long-distance telephone calls, Lise Meitner having returned to Stockholm in the meantime. I asked an American biologist who was working with Hevesy what they call the process by which single cells divide in two; 'fission', he said, so I used the term 'nuclear fission' in that paper . . .

Lise Meitner felt that probably most of the radioactive substances which had been thought to lie beyond uranium – those 'transuranic' substances which Hahn thought they had discovered – were also fission products; a month or two later she came to Copenhagen and we proved that point by using a technique of 'radioactive recoil' which she had been the first to use, about thirty years previously. Yet transuranic elements were also formed; that was proved in California by Ed McMillan, with techniques much more sensitive than those available to Hahn and Meitner.

In all this excitement we had missed the most important point: the chain reaction. It was Christian Møller, a Danish colleague, who first suggested to me that the fission fragments (the two freshly formed nuclei) might contain enough surplus energy each to eject a neutron or two; each of these might cause another fission and generate more neutrons. By such a 'chain reaction' the neutrons would multiply in uranium like rabbits in a meadow! My immediate answer was that in that case no uranium ore deposits could exist: they would have blown up long ago by the explosive multiplication of neutrons in them. But I quickly saw that my argument was too naïve; ores contained lots of other elements which might swallow up the neutrons; and the seams were perhaps thin, and then most of the neutrons would escape. So, from Møller's remark the exciting vision arose that by assembling enough pure uranium (with appropriate care!) one might start a controlled chain reaction and liberate nuclear energy on a scale that really mattered. Many others independently had the same thought, as I soon found out. Of course the spectre of a bomb – an uncontrolled chain reaction – was there as well; but for a while anyhow, it looked as though it need not frighten us. That complacency was based on an argument by Bohr, which was subtle but appeared quite sound.

In a paper on the theory of fission that he wrote in the U.S.A. with John Wheeler, Bohr concluded that most of the neutrons emitted by the fission fragments would be too slow to cause fission of the chief isotope, uranium-238. Yet slow neutrons did cause fission; this he attributed to the rare isotope uranium-235. If he was right the only chance of getting a chain reaction with natural uranium was to arrange for the neutrons to be slowed down, whereby their effect on uranium-235 is increased. But in that manner one could not get a violent explosion; slow neutrons take their time, and even if the conditions for rapid neutron multiplication were created this would at best (or at worst!) cause the assembly to heat up and disperse itself, with only a minute fraction of its nuclear energy liberated.

All this was quite correct, and the development of nuclear reactors followed on the whole the lines which Bohr foresaw. What he did not foresee was the fanatical ingenuity of the allied physicists and engineers, driven by the fear that Hitler might develop the decisive weapon before they did. I was in England when the war broke out, and in Los Alamos when I saw Bohr again. By that time it was clear that there were even two ways for getting an effective nuclear explosion: either through the separation of the highly fissile isotope uranium-235 or by using the new element plutonium, formed in a nuclear reactor. But I am again getting ahead of my story.

Frisch was indeed getting ahead of his story, and there was a great deal more highly sensitive research to be undertaken before a containable reaction could be parcelled up into a usable atomic bomb. Those five or six years' work that followed the discovery that Frisch has just described constitute one of the most intense and productive phases in the history of physics, involving many thousands of scientists across the world: the Manhattan Project alone employed more than 100,000 people across thirty separate sites in the United States, Canada and Britain. Yet the moral implications of this collaborative work would be lasting and profound, and the following section explores some of the misgivings of scientists both during and after the Second World War.

Source: Otto R. Frisch, *What little I remember* (Cambridge: Cambridge University Press, 1979), pp. 113–19.

THE IMPACT OF THE WAR ON
SCIENTISTS

'Now I am become Death, the destroyer of worlds', as J. Robert Oppenheimer (1904–67) claimed to have whispered as he watched the detonation
of the first atomic bomb at the Trinity test site in New Mexico on 16 July
1945; 'Now we are all sons of bitches', as the test director Kenneth Bainbridge is reputed to have said in response. The two comments shed some
light on the mixed feelings of the thousands of military scientists for whom
the atomic bomb was the successful culmination of years of stressful work.
Never before had such theoretical know-how been put towards military
ends, and, once the bomb had actually been used against Japan, the implications for post-war science were complex and profound, especially for
physicists, who felt a particular responsibility for their work on atomic
weapons. 'In some sort of crude sense which no vulgarity, no humor, no
overstatement can quite extinguish', wrote Oppenheimer, 'the physicists have
known sin; and this is a knowledge which they cannot lose.' Yet, as Jacob
Bronowski famously observed in the course of The Ascent of Man (1973), the
bomb was 'not the tragedy of scientists: it is the tragedy of mankind.'

For this section, I have chosen the words of three mid-twentieth-century
scientists for whom the Second World War exercised a shaping influence on
their subsequent work and ideas.

Erwin Chargaff

The Austrian-born American biochemist Erwin Chargaff (1905–2002) was a
famously outspoken commentator on scientific matters. ('If at one time or
another I have brushed a few colleagues up the wrong way, I must apologize',
he wrote; 'I had not realized that they were covered with fur.') In this extract
from his memoirs, published in 1978, he describes his reaction to what he
called 'the Devil's carnival' of post-war nuclear research:

It is difficult to describe the effect that the triumph of nuclear physics
had on me. (I have recently seen a film made by the Japanese at that
time, and all the horror was revived, if "revive" is the correct word

in front of mega-death.) It was an early evening in August, 1945 – was it the sixth? – my wife, my son, and I were spending the summer in Maine, in South Brooksville, and we had gone on an after-dinner walk where Penobscot Bay could be seen in all its sunset loveliness. We met a man who told us that he had heard something on the radio about a new kind of bomb which had been dropped on Japan. Next day, the *New York Times* had all the details. But the details have never stopped coming in since that day.

The double horror of two Japanese city names grew for me into another kind of double horror: an estranging awareness of what the United States was capable of, the country that five years before had given me its citizenship; a nauseating terror at the direction the natural sciences were going. Never far from an apocalyptic vision of the world, I saw the end of the essence of mankind; an end brought nearer, or even made possible, by the profession to which I belonged. In my view, all natural sciences were as one; and if one science could no longer plead innocence, none could. The time had long gone when you could say that you had become a scientist because you wanted to learn more about nature. You would immediately be asked: "Why do you want to know more about nature? Do we not know enough?" – and you would be lured into the expected answer: "No, we don't know enough; but when we do, we shall improve; we shall exploit nature. We shall be the masters of the universe." And even if you did not give this silly answer, you felt inwardly that the evil do-gooders might get away with such talk, were it not for death, the great eraser of stupidities. For had not Bacon assured me that knowledge was power, and Nietzsche – or rather his misinterpreters, his sister and the other exploiters of the silenced great man – that this was what I had wanted all my life? Of course, they were completely wrong, as far as I am concerned; and there is more wisdom in one of Tolstoy's folk tales than in the entire *Novum Organum* (with *Zarathustra* added without regret).

In 1945, therefore, I proved a sentimental fool; and Mr. Truman could safely have classified me among the whimpering idiots he did not wish admitted to the presidential office. For I felt that no man has the right to decree so much suffering, and that science, in providing and sharpening the knife and in upholding the ram, had incurred a guilt of which it will never get rid. It was at that time that the nexus between science and murder became clear to me. For several years after the somber event, between 1947 and 1952, I tried desperately to find a position in what then appeared to me as bucolic Switzerland, – but I had no success . . .

The impact that the discovery, the bloodstained discovery, of nuclear energy had on me I have tried to describe in the first pages of this account. From that time the Devil's carnival was on, for me at any rate. As the dances became more frenetic, the air turned thinner and harder to breathe. That science, the profession to which I had devoted my life – and a life is the heaviest investment a man can make – that science should engage in such misdeeds was more than I could bear. I had to speak out, for I was bound to ask myself: is this still the same kind of science that I thought of getting into more than fifty years ago? And I had to reply: it is not.

Kathleen Lonsdale

On 22 March 1945 the crystallographer Kathleen Lonsdale (1903–71) became the first woman scientist to be elected to the fellowship of the Royal Society of London (the biochemist Marjory Stephenson (1885–1948) was also elected that day, but Lonsdale was first, by virtue of her position in the alphabet). As a Quaker, Lonsdale was a convinced pacifist, and spent a month in prison in 1943 for refusing to pay a fine incurred by her failure to register for war duties. The following extract is from her pacifist manifesto, *Is Peace Possible?* (1957), which she wrote in response to a worldwide increase in nuclear weapons testing. In it, she attempts to answer a question that was put to her in 1945, in the wake of the bombing of Japan: 'Do you see what you scientists have done now?':

When I became a research student, training under Sir William Bragg in the very place where Sir Humphry Davy, Michael Faraday, John Tyndall, Sir James Dewar, and other world-famous scientists had carried out their researches, the [1914–18] war was over and, as we thought, won. We genuinely hoped for a peace settlement that would end all war. Terrible things had happened, but we believed that there were plenty of good Germans, and that they would now have a chance to come out on top. Terrible things had happened and were perhaps still happening in Russia, but other countries, America and France, for instance, had had pretty ghastly revolutions too, and then settled down. It might take time. Meanwhile my work was fun. I often ran the last few yards to the laboratory. Later on I took my mathematical calculations with me to the nursing-homes where my babies were born: it was exciting to find out new facts.

Now science seems to have become something of a Frankenstein. Chunks of it have become secret; slightly indecent, as it were. For a time, indeed, during the war and for a few years after, secrecy became a disease. If a discovery had any practical value at all, it must be kept secret. If good, it must not be shared with our enemies or competitors. If bad, they must not be allowed to copy it or to discover the antidote. What does this enmity and competition involve?

Scientific discoveries of any kind are certainly a power and a responsibility. The world's resources are very unevenly distributed. If a new use is found for some raw material that is the monopoly of one or a few nations, those nations may become wealthy overnight, or they may become a prey to more powerful neighbours. That was brought home to me very forcibly after World War II. I had gone to give lectures in Paris. My husband went to a scientific congress in Brussels. The shortages of food and of almost all other commodities were still acute in France. Not so, apparently, in Belgium. Why? Both had suffered during the war. But Belgium now had uranium to sell, from rich mines in the Belgian Congo. France was obliged to export her dairy produce. The uranium from Belgium was going to the USA for dollars, and some of it was sold to Britain . . .

The bomb that Britain proposes to make and test will be, or so it is reported, a fission-fusion-fission bomb; what is sometimes called a 'rigged' bomb: the most dangerous of all kinds yet produced.

This has an atomic bomb as detonator, or trigger, and a shell outside of ordinary uranium of atomic weight 238. The main body of explosive material is hydrogen: not ordinary hydrogen but, in one type of bomb anyhow, a mixture of heavy isotopes of hydrogen under high pressure, generated by a solid compound, lithium-6-deuteride.

The process in this explosive mixture is not fission but fusion, the synthesis of light atoms to form a heavier one: the process that is believed to be the source of the heat, light, and other radiation from the sun. This fusion process can only take place at a very high temperature and hence the use of the fission (atomic) bomb to trigger it off. At the same time it produces such high-energy neutrons that they can cause the complete fission of the uranium 238 outer shell of the bomb.

The atomic bomb itself cannot exceed a maximum size, but the hydrogen bomb, whether rigged or not, can be made as big as there are means to deliver it. Since the United States Air Force has confirmed that they have already successfully flown an atomic reactor in a B.36 bomber (not as a means of propulsion but for experimental purposes), this means that the maximum size of such a bomb could be very big indeed. A 20-megaton fission-fusion-fission bomb, equal in explosive power to 20 million-ton block-busters, has already been tested. Its performance apparently exceeded the expectations of the scientists who designed it.

Edward Bullard

Sir Edward Crisp Bullard (1907–80) was a marine geophysicist who worked on ocean-floor sediments during the late 1930s. At the outbreak of war, Bullard joined the Royal Naval Mine and Torpedo School at Portsmouth, where he worked, with great success, on technology designed to protect ships from magnetic and acoustic mines. He always regarded his wartime work, for which he was knighted in 1953, as the most important of his life; the following extract is from the transcript of a seminar on the effects of the war on scientific research that was held at the Royal Society in March 1974:

Most of the scientists in their twenties and thirties who went in 1939 to work on wartime problems were profoundly affected by their experience. The sudden shift from work which only you and a few of your friends understood or cared about to being a magician whom admirals must consult and who could have practically any facility or assistance that he asked for was an exhilarating and maturing experience. For myself, I found that having worked for two years in the Cavendish Laboratory made all the difference; not for anything I learnt there, but for the confidence it gave. The belief that Rutherford's boys were the best boys, that we could do anything that was do-able and could master any subject in a few days was of enormous value. We were, I suppose, arrogant and insufferable but, for myself, I always had a feeling that I was really the Wizard of Oz, the best wizard there was around and able to put up a tremendous front, but behind it all a bit of a fraud.

Scientific work is perhaps the most competitive activity there is, in its nature it is elitist in the best sense; this competitiveness for jobs, facilities, good students and money is, to most people, a strain. The sudden transition to a different society and to cooperative aims of universally acknowledged desirability was, in some ways, a relief; I found the Navy much less competitive and more cooperative than the laboratories I had known. When you got to know people and they trusted you, you could do almost anything (I once took over a French battleship with no authority at all and with no murmur of disapproval).

The important lessons for post-war science were how to use the Government machine, how to get one's way with committees, how to persuade people with arguments suitable to their backgrounds and prejudices and how realistically to assess the means needed for a given end. I often noticed after the war the contrast between the effectiveness of those who had had this training and those who had not. I remember, for example, an early meeting of the D.S.I.R. Geology and Geophysics grants committee where Cambridge came out with two-thirds of the total money and my Department with half of it. I felt that I had been robbing a piggy-bank. The nuclear physicists in particular learnt the importance of supporting each other's projects and not, as often happened in other subjects, criticizing the other man in order to make your own project seem better.

After the entry of Russia and the United States into the war victory was certain and there was, for many people, a growing feeling that the urgency of 1940 had gone. We began to talk and think more and more about the post-war world, to discuss priorities for research and how we would now be able to use our new understanding of the workings of Government to get funds and to do what we wanted on a scale that would be worthwhile.

In most subjects this wish for a change of scale and tempo was opportune. Certainly in physics what could be got with the techniques of J. J. Thomson and Rutherford had been got and the line of progress lay, in the main, through operations on a larger scale. The difficulty, of course, lay in combining the large operations with the spontaneity and creativeness of the bright young men. There are many salutary examples of the dangers of 'big science' to creativity.

I suspect that the most important effect of World War II on physical science lay in the change of attitude of people to science. The politicians and the public were convinced that science was useful and were in no position to argue about the details. A professor of physics might be more sinister than he was in the 1930s, but he was no longer an old fool with a beard in a comic-strip. The scientists, or at any rate the physicists, had changed their attitude. They not only believed in the interest of science for themselves, they had acquired also a belief that the tax-payer should and would pay for it and would, in some unspecified length of run, benefit by it.

Today the impetus is largely spent, the expansion has levelled off and the weathercock of public opinion has swung round. There is now a real danger that, because certain problems have been neglected, the public will believe they are insoluble by scientific means and that we must settle down to cultivate our garden and leave most of the world to starve, freeze or blow itself up.

Sources: Erwin Chargaff, *Heraclitean Fire: Sketches from a Life before Nature* (New York: Rockefeller University Press, 1978), pp. 3–4; 183; Kathleen Lonsdale, *Is Peace Possible?* (Harmondsworth: Penguin, 1957), pp. 13–47; Sir Edward C. Bullard, 'The Effect of World War II on the Development of Knowledge in the Physical Sciences', *Proceedings of the Royal Society of London*, Series A, 342 (1975), pp. 531–3.

ANTI-SCIENCE

Hostility towards science takes many forms and emanates from a variety of outlooks, whether religious, political, social or aesthetic ('science is the religion of the suburbs', as W. B. Yeats once observed). Some view science as 'a dangerous weapon', as Jean-Jacques Rousseau described it in the 1760s, one that ought to be taken out of our hands before it causes any further harm; while others view it as a materialist assault on the spiritual and religious values that make us uniquely human. Some view scientific rationality as incompatible with human creativity, while others deny that it even exists, that 'objectivity' is a philosophical sham dreamed up to legitimize the technocratic arrogance of the West.

The following extracts have been chosen to illustrate some of this range of anti-scientific sentiment. (I had intended to end this section with a short discussion of science in the media from Ben Goldacre's insightful 'Bad Science' column in the *Guardian* newspaper; unfortunately the fee demanded for reproducing the extract was so high that it had to be omitted):

Heinrich Cornelius Agrippa

Heinrich Cornelius Agrippa (1486–1535) was a German magician and theologian, who studied the occult arts and sciences from a young age. Tellingly, he was much admired by Mary Shelley's Victor Frankenstein; having chanced upon a volume of his alchemical writings, Frankenstein recalled how 'a new light seemed to dawn upon my mind; and, bounding with joy, I communicated my discovery to my father. My father looked carelessly at the title page of my book, and said, "Ah! Cornelius Agrippa! My dear Victor, do not waste your time upon this; it is sad trash."' Agrippa's best-known work was a sceptical assault on science, *On the Uncertainty and Vanity of the Sciences and the Arts* (1527), the purpose of which, as he announced in the preface, was 'to wage War against the Giant-like Opposition of all the *Arts* and *Sciences*; and thus to challenge the stoutest Hunters of Nature.' The following excerpt is from an English translation published in 1694:

It is an old Opinion, and the concurring and unanimous judgment almost of all Philosophers, whereby they uphold, that every Science addeth so much of a sublime Nature to Man himself, according to the Capacity and Worth of every Person, as many times enables them to Translate themselves beyond the Limits of Humanity, even to the Celestial Seats of the Blessed. From hence have proceeded those various and innumerable Encomiums of the Sciences, whereby every one hath endeavour'd, in accurate, as well as long Orations, to prefer, and as it were to extol beyond the Heavens themselves, those Arts and Mysteries, wherein, with continual labour, he hath exercised the strength and vigour of his Ingenuity or Invention. But I, perswaded by reasons of another nature, do verily believe, that there is nothing more pernicious, nothing more destructive to the well-being of Men, or to the Salvation of our Souls, than the Arts and Sciences themselves. And therefore quite contrary to what has been hitherto practised, my Opinion is, That these Arts and Sciences are so far from being to be extolled with such high applauses and Panegyricks, that they are rather for the most part to be disprais'd and vilifi'd: And that indeed there is none which does not merit just cause of Reproof and Censure; nor any one which of it self deserves any praise or commendation, unless what it may borrow from the Ingenuity and Virtue of the first possessor . . .

And it behoves us to shew how intolerable the blindness of Men is, to wander from the Truth, misguided by so many Sciences and Arts, and by so many Authors and Doctors thereof. For how great a boldness is it, what an arrogant presumption, to prefer the Schools of Philosophers before the Church of Christ? and to extol or equal the Opinions of Men, to the Word of God? Lastly, how impious a piece of Tyranny it is, to captivate the Wits of Students to prefixed Authors, and to deprive their Disciples of the liberty of searching after and following the Truth? All which things being so manifest, they cannot be denied, I may be the more easily pardoned, if I seem to have more freely and bitterly enveighed against some sorts of *Sciences* and their Professors.

Thomas Love Peacock

This classic expression of the 'science as a dangerous weapon' argument comes from Peacock's last novel, *Gryll Grange* (1860). The novel tracks the many conversational twists and turns during a long weekend at a country house; here, two of the guests, the scientifically minded Lord Curryfin and the constitutionally cautious Revd Opimian, disagree over the benefits of science:

Lord Curryfin: . . . We ought to have more wisdom, as we clearly have more science. –
The Rev. Dr. Opimian: Science is one thing and wisdom is another. Science is an edged tool with which men play like children and cut their own fingers. If you look at the results which science has brought in its train, you will find them to consist almost wholly in elements of mischief. See how much belongs to the word Explosion alone, of which the ancients knew nothing. Explosions of powder-mills and powder magazines; of coal gas in mines and in houses; of high-pressure engines in ships and boats and factories. See the complications and refinements of modes of destruction, in revolvers and rifles and shells and rockets and cannon. See collisions and wrecks and every mode of disaster by land and by sea, resulting chiefly from the insanity for speed, in those who for the most part have nothing to do at the end of the race, which they run as if they were so many Mercuries speeding with messages from Jupiter. Look at our scientific drainage, which turns refuse into poison. Look at the subsoil of London, whenever it is turned up to the air, converted by gas leakage into one mass of pestilent blackness, in which no vegetation can flourish, and above which, with the rapid growth of the ever-growing nuisance, no living thing will breathe with impunity. Look at our scientific machinery, which has destroyed domestic manufacture, which has substituted rottenness for strength in the thing made, and physical degradation in crowded towns for healthy and comfortable country life in the makers. The day would fail, if I should attempt to enumerate the evils which science has inflicted on mankind. I almost think it is the ultimate destiny of science to exterminate the human race.

Maria Mitchell

The American astronomer Maria Mitchell (1818–89) was appointed Professor of Astronomy at Vassar College in 1848, a year after discovering the comet which bears her name ('Miss Mitchell's Comet', or C/1847 T1). A lifelong advocate of women's education, she was also a witty and acerbic observer of scientific ignorance, as can be seen in the following extract from her posthumously published journals:

When crossing the Atlantic, an Irish woman came to me and asked me if I told fortunes; and when I replied in the negative, she asked me if I were not an astronomer. I admitted that I made efforts in that direction. She then asked me what I could tell, if not fortunes. I told her that I could tell when the moon would rise, when the sun would rise, etc. She said, 'Oh,' in a tone which plainly said, 'Is *that* all?' . . .

One of the unfavorable results of the attempt to popularize science is this: the reader of popular scientific books is very likely to think that he understands the science itself, when he merely understands what some writer says about science.

Take, for example, the method of determining the distance of the moon from the earth—one of the easiest problems in physical astronomy. The method can be told in a few sentences; yet it took a hundred years to determine it with any degree of accuracy—and a hundred years, not of the average work of mankind in science, but a hundred years during which able minds were bent to the problem.

Still, with all the school-masters, and all the teaching, and all the books, the ignorance of the unscientific world is enormous; they are ignorant both ways—they underrate the scientific people and they overrate them. There is, on the one hand, the Irish woman who is disappointed because you cannot tell fortunes, and, on the other hand, the cultivated woman who supposes that you must know *all* science.

I have a friend who wonders that I do not take my astronomical clock to pieces. She supposes that because I am an astronomer, I must be able to be a clock-maker, while I do not handle a tool if I can help it! She did not expect to take her piano to pieces because she was musical! She was as careful not to tinker it as I was not to tinker the clock, which only an expert in clock-making was prepared to handle . . .

Then, too, the uneducated assume the unvarying exactness of mathematical results; while, in reality, mathematical results are often only approximations. We say the sun is 91,000,000 miles from the earth, plus or minus a probable error; that is, we are right, probably, within, say, 100,000 miles; or, the sun is 91,000,000 minus 100,000 miles, or it is 91,000,000 plus 100,000 miles off; and this probable error is only a probability.

If we make one more observation it cannot agree with any one of our determinations, and it changes our probable error.

This ignorance of the masses leads to a misconception in two ways; the little that a scientist can do, they do not understand, — they suppose him to be godlike in his capacity, and they do not see results; they overrate him and they underrate him — they underrate his work.

There is no observatory in this land, nor in any land, probably, of which the question is not asked, 'Are they doing anything? Why don't we hear from them? They should make discoveries, they should publish.'

D. H. Lawrence

The following is a passionately argued denunciation of the spiritual emptiness of the scientific approach to nature. Overly rational, materialist and dehumanizing, science robs both nature and humanity of their sublime mystery and magic: at least that's what D. H. Lawrence (1885–1930) claims in this extract from an essay published in the last year of his life. Funnily enough, I used to agree with this kind of thing wholeheartedly when I was a teenager; now I just find it annoying:

The Universe is dead for us, and how is it to come alive again? "Knowledge" has killed the sun, making it a ball of gas, with spots; "knowledge" has killed the moon, it is a dead little earth fretted with extinct craters as with smallpox; the machine has killed the earth for us, making it a surface, more or less bumpy, that you travel over. How, out of all this, are we to get back the grand orbs of the soul's heavens, that fill us with unspeakable joy? How are we to get back Apollo, and Attis, Demeter, Persephone, and the halls of Dis? How even see the stars Hesperus, or Betelguese?

We've got to get them back, for they are the world our soul, our greater consciousness, lives in. The world of reason and science, the moon, a dead lump of earth, the sun, so much gas with spots: this is the dry and sterile little world the abstracted mind inhabits. The world of our little consciousness, which we know in our pettifogging *apartness*. This is how we know the world when we know it apart from ourselves, in the mean separateness of everything. When we know the world in togetherness with ourselves, we know the earth hyacinthine or Plutonic . . . There are many ways of knowing, there are many sorts of knowledge. But the two ways of knowing, for man, are knowing in terms of apartness, which is mental, rational, scientific; and knowing in terms of togetherness, which is religious and poetic.

So far, so high-minded. The chemist Anthony Standen (1907–93), by contrast, seemed to regard the majority of his scientific colleagues as mere idiotic tinkerers, and in his book *Science is a Sacred Cow* (1950), he launched an angry, sceptical assault on the elevated position of public science in the wake of the Second World War. One of Standen's stories, about a fisheries biologist who worked in the world-renowned aquarium in Naples, described how the scientist recorded the tail-movements of fish by wiring them up to an electrical device: 'Some of them were entire fish, others had had portions of their brains removed. The brainless fish made smooth regular curves on the recording instrument, but the tails of the untreated fish made distressingly irregular movements. The scientist was delighted with this discovery. "See," he said, "it is the higher brain centers that *disturb* the basic movement pattern of the muscles. The brainless fish make beautifully regular movements." This scientist was unable to see any sort of connection between what he was doing in the aquarium and what was going on around him in Italy, where Mussolini made the trains run regularly, but did not allow anyone to think a single thought that was contrary to the ideas of his regime. This scientist was an ichthyologist, and therefore his experiments were harmless (except to the fish).' It just goes to show, thought Standen, what fools most scientists are.

Sources: Henry Cornelius Agrippa, *The Vanity of Arts and Sciences* (London: R. Bentley, 1694), pp. 1–3; Thomas Love Peacock, *Gryll Grange* (London: Parker, Son and Bourn, 1861), pp. 160–61; Phebe Mitchell Kendall (ed.), *Maria Mitchell: Life, Letters, and Journals* (Boston: Lee and Shepard, 1896), pp. 220–23; D. H. Lawrence, *A Propos of Lady Chatterley's Lover* (London: Mandrake Press, 1930), pp. 54–5.

THE TWO CULTURES

The Two Cultures was one of the most influential essays of the twentieth century, its title having entered the language as a description of the growing cultural chasm between the sciences and the humanities. Its author, Charles Percy Snow (1905–80), was a physicist, novelist and government adviser, who in 1959 was invited to deliver the prestigious Rede Lecture at the University of Cambridge. His lecture – 'The Two Cultures' (later published in book form as *The Two Cultures and the Scientific Revolution*) – argued, somewhat pessimistically, that the gulf between scientists and 'literary intellectuals' was more than simply a failure of communication across the disciplinary divide, it was a cultural and political catastrophe that was hindering human progress.

Like all powerful new ideas, traces of it had been lurking in the culture for some time, and a version of it had already appeared in the astronomer Fred Hoyle's science-fiction novel *The Black Cloud* (1957), in which the pipe-smoking protagonist, Dr Marlowe, laments that government-funded scientists were fated to work for 'an archaic crowd of nitwits': 'it isn't just a case of scientists versus the rest. The matter goes deeper. It's a clash between two totally different modes of thinking. Society today is based in its technology on thinking in terms of numbers. In its social organization, on the other hand, it is based on thinking in terms of words. It's here that the real clash lies, between the literary mind and the mathematical mind. You ought to meet the Home Secretary. You'd see straight away what I mean.'

Snow's lecture stated the case even more persuasively than Fred Hoyle's 'frolic' (as he wryly described his novel), and more than fifty years later it remains widely read and discussed. Its best-known line (it appears below), in which he challenges his literary friends to define the Second Law of Thermodynamics, became a recurring dinner-party talking point:

There have been plenty of days when I have spent the working hours with scientists and then gone off at night with some literary colleagues. I mean that literally. I have had, of course, intimate friends among both scientists and writers. It was through living among these groups and much more, I think, through moving regularly from one to the other and back again that I got occupied with the problem of what, long before I put it on paper, I christened to myself as the 'two cultures'. For constantly I felt I was moving among two groups —

comparable in intelligence, identical in race, not grossly different in social origin, earning about the same incomes, who had almost ceased to communicate at all, who in intellectual, moral and psychological climate had so little in common that instead of going from Burlington House or south Kensington to Chelsea, one might have crossed an ocean.

In fact, one had travelled much further than across an ocean – because after a few thousand Atlantic miles, one found Greenwich Village talking precisely the same language as Chelsea, and both having about as much communication with M.I.T. as though the scientists spoke nothing but Tibetan. For this is not just our problem; owing to some of our educational and social idiosyncrasies, it is slightly exaggerated here, owing to another English social peculiarity it is slightly minimised; by and large this is a problem of the entire West.

By this I intend something serious. I am not thinking of the pleasant story of how one of the more convivial Oxford greats dons – I have heard the story attributed to A. L. Smith – came over to Cambridge to dine. The date is perhaps the 1890's. I think it must have been at St. John's, or possibly Trinity. Anyway, Smith was sitting at the right hand of the President – or Vice-Master – and he was a man who liked to include all round him in the conversation, although he was not immediately encouraged by the expressions of his neighbours. He addressed some cheerful Oxonian chit-chat at the one opposite to him, and got a grunt. He then tried the man on his own right hand and got another grunt. Then, rather to his surprise, one looked at the other and said, 'Do you know what he's talking about?' 'I haven't the least idea.' At this, even Smith was getting out of his depth. But the President, acting as a social emollient, put him at his ease by saying, 'Oh, those are mathematicians! We never talk to *them.*'

No, I intend something serious. I believe the intellectual life of the whole of western society is increasingly being split into two polar groups . . . at one pole we have the literary intellectuals, who incidentally while no one was looking took to referring to themselves as 'intellectuals' as though there were no others. I remember G. H. Hardy once remarking to me in mild puzzlement, some time in the 1930's: 'Have you noticed how the word "intellectual" is used nowadays? There seems to be a new definition which certainly doesn't include Rutherford or Eddington or Dirac or Adrian or me. It does seem rather odd, don't y' know.'

Literary intellectuals at one pole – at the other scientists, and as the most representative, the physical scientists. Between the two a gulf of mutual incomprehension – sometimes (particularly among the young) hostility and dislike, but most of all lack of understanding . . .

The degree of incomprehension on both sides is the kind of joke which has gone sour. There are about fifty thousand working scientists in the country and eighty thousand professional engineers or applied scientists. During the war and in the years since, my colleagues and I have had to interview somewhere between thirty to forty thousand of these – that is, around 25 per cent. The number is large enough to give us a fair sample, though of the men we talked to most would still be under forty. We were able to find out a certain amount of what they read and thought about. I confess that even I, who am fond of them and respect them, was a bit shaken. We hadn't quite expected that the links with the traditional culture should be so tenuous, nothing more than a formal touch of the cap.

As one would expect, some of the very best scientists had and have plenty of energy and interest to spare, and we came across several who had read everything that literary people talk about. But that's very rare. Most of the rest, when one tried to probe for what books they had read, would modestly confess, 'Well, I've *tried* a bit of Dickens', rather as though Dickens were an extraordinarily esoteric, tangled and dubiously rewarding writer, something like Rainer Maria Rilke. In fact that is exactly how they do regard him: we thought that discovery, that Dickens had been transformed into the type-specimen of literary incomprehensibility, was one of the oddest results of the whole exercise . . .

But what about the other side? They are impoverished too – perhaps more seriously, because they are vainer about it. They still like to pretend that the traditional culture is the whole of 'culture', as though the natural order didn't exist. As though the exploration of the natural order was of no interest either in its own value or its consequences. As though the scientific edifice of the physical world was not, in its intellectual depth, complexity and articulation, the most beautiful and wonderful collective work of the mind of man. Yet most non-scientists have no conception of that edifice at all. Even if they want to have it, they can't. It is rather as though, over an immense range of intellectual experience, a whole group was tone-deaf. Except that this tone-deafness doesn't come by nature, but by training, or rather the absence of training.

As with the tone-deaf, they don't know what they miss. They give a pitying chuckle at the news of scientists who have never read a major work of English literature. They dismiss them as ignorant specialists. Yet their own ignorance and their own specialisation is just as startling. A good many times I have been present at gatherings of people who, by the standards of the traditional culture, are thought highly educated and who have with considerable gusto been expressing their incredulity at the illiteracy of scientists. Once or twice I have been provoked and have asked the company how many of them could describe the Second Law of Thermodynamics. The response was cold: it was also negative. Yet I was asking something which is about the scientific equivalent of: *Have you read a work of Shakespeare's?*

I now believe that if I had asked an even simpler question – such as, What do you mean by mass, or acceleration, which is the scientific equivalent of saying, *Can you read?* – not more than one in ten of the highly educated would have felt that I was speaking the same language. So the great edifice of modern physics goes up, and the majority of the cleverest people in the western world have about as much insight into it as their neolithic ancestors would have had.

Source: C. P. Snow, *The Two Cultures and a Second Look: An Expanded Version of the Two Cultures and the Scientific Revolution* (Cambridge: Cambridge University Press, 1964), pp. 2–16.

THE DOUBLE HELIX

A plaque on the wall of the Eagle pub on Bene't Street, Cambridge, just around the corner from the old Cavendish Laboratory, records the moment, on 28 February 1953, when Francis Crick and James Watson burst into the saloon bar and announced to the bemused lunchtime drinkers that they had discovered the secret of life. If the DNA story is one of the best known in the history of science, it is partly because of the nature of the people involved ('two loudmouthed young men who devoted more time to talking and drinking than to experiment', as one senior scientist described Crick and Watson), and partly because the rivalry between the Cambridge and London teams added a frisson of skulduggery to proceedings. There was no love lost between the four main protagonists, as is clear from this scene-setting extract from Watson's notorious memoir, *The Double Helix* (1968); here are Crick, Wilkins, Franklin, and Watson himself, whose unmistakeable voice and temperament dominate the telling of the story:

Before my arrival in Cambridge, Francis only occasionally thought about deoxyribonucleic acid (DNA) and its role in heredity. This was not because he thought it uninteresting. Quite the contrary. A major factor in his leaving physics and developing an interest in biology had been the reading in 1946 of *What Is Life?* by the noted theoretical physicist Erwin Schrödinger. This book very elegantly propounded the belief that genes were the key components of living cells and that, to understand what life is, we must know how genes act. When Schrödinger wrote his book (1944), there was general acceptance that genes were special types of protein molecules. But almost at this same time the bacteriologist O. T. Avery was carrying out experiments at the Rockefeller Institute in New York which showed that hereditary traits could be transmitted from one bacterial cell to another by purified DNA molecules.

Given the fact that DNA was known to occur in the chromosomes of all cells, Avery's experiments strongly suggested that future experiments would show that all genes were composed of DNA. If true, this meant to Francis that proteins would not be the Rosetta Stone for unravelling the true secret of life. Instead, DNA would have to provide the key to enable us to find out how the genes determined, among other characteristics, the colour of our hair, our eyes, most likely our comparative intelligence, and maybe even our potential to amuse others.

Of course there were scientists who thought the evidence favouring DNA was inconclusive and preferred to believe that genes were protein molecules. Francis, however, did not worry about these sceptics. Many were cantankerous fools who unfailingly backed the wrong horses. One could not be a successful scientist without realizing that, in contrast to the popular conception supported by newspapers and mothers of scientists, a goodly number of scientists are not only narrow-minded and dull, but also just stupid.

Francis, none the less, was not then prepared to jump into the DNA world. Its basic importance did not seem sufficient cause by itself to lead him out of the protein field which he had worked in only two years and was just beginning to master intellectually. In addition, his colleagues at the Cavendish were only marginally interested in the nucleic acids, and even in the best of financial circumstances it would take two or three years to set up a new research group primarily devoted to using X-rays to look at the DNA structure.

Moreover, such a decision would create an awkward personal situation. At this time molecular work on DNA in England was, for all practical purposes, the personal property of Maurice Wilkins, a bachelor who worked in London at King's College. Like Francis, Maurice had been a physicist and also used X-ray diffraction as his principal tool of research. It would have looked very bad if Francis had jumped in on a problem that Maurice had worked over for several years. The matter was even worse because the two, almost equal in age, knew each other and, before Francis remarried, had frequently met for lunch or dinner to talk about science.

It would have been much easier if they had been living in different countries. The combination of England's cosiness – all the important people, if not related by marriage, seemed to know one another – plus the English sense of fair play would not allow Francis to move in on Maurice's problem. In France, where fair play obviously did not exist, these problems would not have arisen. The States also would not have permitted such a situation to develop. One would not expect someone at Berkeley to ignore a first-rate problem merely because someone at Cal Tech had started first. In England, however, it simply would not look right.

Even worse, Maurice continually frustrated Francis by never seeming enthusiastic enough about DNA. He appeared to enjoy slowly understating important arguments. It was not a question of intelligence or common sense. Maurice clearly had both; witness his seizing DNA before almost everyone else. It was that Francis felt he could never get the message over to Maurice that you did not move cautiously when you were holding dynamite like DNA. Moreover, it was increasingly difficult to take Maurice's mind off his assistant, Rosalind Franklin.

Not that he was at all in love with Rosy, as we called her from a distance. Just the opposite – almost from the moment she arrived in Maurice's lab, they began to upset each other. Maurice, a beginner in X-ray diffraction work, wanted some professional help and hoped that Rosy, a trained crystallographer, could speed up his research. Rosy, however, did not see the situation this way. She claimed that she had been given DNA for her own problem and would not think of herself as Maurice's assistant.

I suspect that in the beginning Maurice hoped that Rosy would calm down. Yet mere inspection suggested that she would not easily bend. By choice she did not emphasize her feminine qualities. Though her features were strong, she was not unattractive and might have been quite stunning had she taken even a mild interest in clothes. This she did not. There was never lipstick to contrast with her straight black hair, while at the age of thirty-one her dresses showed all the imagination of English blue-stocking adolescents . . .

Clearly Rosy had to go or be put in her place. The former was obviously preferable because, given her belligerent moods, it would be very difficult for Maurice to maintain a dominant position that would allow him to think unhindered about DNA. Not that at times he didn't see some reason for her complaints – King's had two combination rooms, one for men, the other for women, certainly a thing of the past. But he was not responsible, and it was no pleasure to bear the cross for the added barb that the women's combination room remained dingily poky whereas money had been spent to make life more agreeable for him and his friends when they had their morning coffee.

Source: James D. Watson, *The Double Helix: A Personal Account of the Discovery of the Structure of DNA* (London: Weidenfeld and Nicolson, 1968), pp. 23–7.

CONTINENTAL DRIFT

The advent of plate tectonic theory in the 1960s had its origins in the theory of continental drift that was advanced by the German geophysicist Alfred Wegener (1880–1930). Wegener, who started his scientific career as an astronomer in Berlin, began to take an interest in geophysical questions while on a two-year posting to Greenland with a team of Danish scientists. Leafing through an atlas one day, he was struck by the fact that the shape of the coastlines on either side of the Atlantic made a near-perfect fit. The thought began to germinate, and after returning wounded from the First World War, he developed his theory that the drifting continents had once been joined together as a single landmass that he named Pangea (from the Greek for 'all lands'). Translations of the first edition of his book, *The Origin of Continents and Oceans* (1915), attracted near-universal hostility – he was jeered at and insulted by his fellow delegates at an international geology symposium in New York – but he persisted in his efforts to collect new evidence in support of his hypothesis, issuing heavily revised editions of the book every few years. The following extract is from a translation of the fourth edition of 1929.

Sadly, Wegener did not live to see his theory vindicated. In 1930, he was back in Greenland, leading a large scientific expedition on the ice. On 1 November, after celebrating his fiftieth birthday at a campsite in the frozen interior, Wegener and a companion headed back towards the expedition's headquarters as a storm began to build. It was the last time either of them was seen alive; six months later, Wegener's body was discovered, carefully buried, marked by a pair of skis. The body of his companion was never found.

The first concept of continental drift first came to me as far back as 1910, when considering the map of the world, under the direct impression produced by the congruence of the coastlines on either side of the Atlantic. At first I did not pay any attention to the idea because I regarded it as improbable. In the fall of 1911, I came quite accidentally upon a synoptic report in which I learned for the first time of palæontological evidence for a former land bridge between Brazil and Africa. As a result I undertook a cursory examination of relevant research in the fields of geology and palæontology, and this provided immediately such weighty corroboration that a conviction of the fundamental soundness of the idea took root in my mind . . .

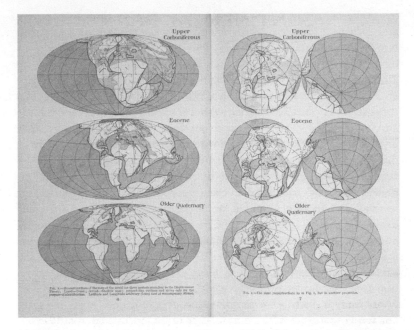

20. The famous map of the break-up of Pangea ('all Earth'), the ancient supercontinent first hypothesized by Alfred Wegener in the 1920s.

This is the starting point of displacement or drift theory. The basic 'obvious' supposition common to both land-bridge and permanence theory — that the relative position of the continents, disregarding their variable shallow-water cover, has never altered — must be wrong. The continents must have shifted. South America must have lain alongside Africa and formed a unified block which was split in two in the Cretaceous; the two parts must then have become increasingly separated over a period of millions of years like pieces of a cracked ice floe in water. The edges of these two blocks are even today strikingly congruent. Not only does the large rectangular bend formed by the Brazilian coast at Cape São Roque mate exactly with the bend in the African coast at the Cameroons, but also south of these two corresponding points every projection on the Brazilian side matches a congruent bay on the African, and conversely. A pair of compasses and a globe will show that both sides are precisely commensurate.

In the same way, North America at one time lay alongside Europe and formed a coherent block with it and Greenland, at least from Newfoundland and Ireland northwards. This block was first broken up in the later Tertiary, and in the north as late as the Quaternary, by a forked rift at Greenland, the sub-blocks then drifting away from each other. Antarctica, Australia and India up to the beginning of the Jurassic lay alongside southern Africa and formed together with it and South America a single large continent, partly covered by shallow water. This block split off into separate blocks in the course of the Jurassic, Cretaceous and Tertiary, and the sub-blocks drifted away in all directions. Our three world maps for the Upper Carboniferous, Eocene and Lower Quaternary show this evolutionary process. In the case of India the process was somewhat different: originally it was joined to Asia by a long stretch of land, mostly under shallow water. After the separation of India from Australia on the one hand (in the early Jurassic) and from Madagascar on the other (at the transition from Tertiary to Cretaceous), this long junction zone became increasingly folded by the continuing approach of present-day India to Asia; it is now the largest folded range on earth, i.e., the Himalaya and the many other folded ranges of upland Asia.

There are also other areas where the continental drift is linked causally with orogenesis (mountain building). In the westward drift of both Americas, their leading edges were compressed and folded by the frontal resistance of the ancient Pacific floor, which was deeply chilled and hence a source of viscous drag. The result was the vast Andean range which extends from Alaska to Antarctica. Consider also the case of the Australian block, including New Guinea, which is separated only by a shelf sea: on the leading side, relative to the direction of displacement, one finds the high-altitude New Guinea range, a recent formation. Before this block split away from Antarctica, its direction was a different one, as our maps show. The present-day east coastline was then the leading side. At that time New Zealand, which was directly in front of this coast, had its mountains formed by folding. Later, as a result of the change in direction of displacement, the mountains were cut off and left behind as island chains. The present-day cordilleran system of eastern Australia was formed in still earlier times; it arose at the same time as the earlier folds in South and North America, which formed the basis of the Andes (pre-cordilleras), at the leading edge of the continental blocks, then drifting as a whole before dividing . . .

We may, however, assume one thing as certain: *The forces which displace continents are the same as those which produce great fold-mountain ranges.* Continental drift, faults and compressions, earthquakes, volcanicity, transgression cycles and polar wandering are undoubtedly connected causally on a grand scale. Their common intensification in certain periods of the earth's history shows this to be true. However, what is cause and what effect, only the future will unveil.

Source: Alfred Wegener, *The Origin of Continents and Oceans* 4th edn (1929), trans. John Biram (New York: Dover Publications, 1966), pp. 1; 17; 179.

FOSSIL COLLECTING

'Collecting is the first and most obvious activity for anyone interested in rocks and minerals', wrote Herbert S. Zim in the opening pages of *Rocks and Minerals* (1957), a popular and much reprinted field guide intended for school-age geology enthusiasts; 'it's fun to find, buy, and swap specimens. Collecting takes you out-of-doors; it also paves the way for more serious studies in science or engineering.' How many scientific careers have been sparked by such seemingly mundane advice?

A Child's View of Science

Joe Lake, the troubled twelve-year-old narrator of Todd McEwen's novel *Arithmetic* (1998), is devoted to Science with a capital 's', and dreams of one day working in a laboratory among 'really really Scientific glass and chemicals' where 'I would stir colored water and invent things with dials on them.' His hero and mentor is Herbert S. Zim (1909–94), real-life editor of the Golden Field Guides series, and author of *Rocks and Minerals*, under the influence of which Joe begins to collect geological specimens. In the following scene from the novel (which is set in the 1950s), Joe and his father drive out into the California desert in search of new additions to the collection:

I've been out in the natural world, I thought, because of Science, because of Dad, because of HERBERT S. ZIM. If the sun shone warm and early on the soapy Natal plums outside the kitchen window, you knew you would end up in Trabuco Cañon, the big smell of sage up your nose with a lot of dust. I decided to collect rocks when I dedicated myself to Science. All through the natural world, nestled in the groins of riverside mountains, under oaks, TREES, in wide places in the road, were ROCK SHOPS. Inside, row upon row of snow-white boxes with cotton in them, each displaying a polished SPECIMEN unlike anything Herbert S. Zim could have known. Some of them weren't rocks at all but fossils they made out back, or colored glass. Usually the guy tried to make himself look like a prospector, dungarees, beard, hat, even if he got his rocks from the EDMUND SCIENTIFIC COMPANY OF BARRINGTON, NEW JERSEY. All along the

geologist's hammer

old newspapers

cold chisel

field bag

magnifying glass

compass

notebook and pencil

heavy gloves

21. 'Herbert S. Zim showed in a watercolor how you collect SPECIMENS.' Zim's *Rocks and Minerals* (1957) was a popular field guide aimed at young Americans like Joe Lake, the twelve-year-old narrator of Todd McEwen's novel *Arithmetic*, for whom Herbert S. Zim was a byword for scientific knowledge and adventure.

highways I watched for rock shops and Dad knew I was watching. Right in the middle of nowhere you saw the sign, ROCKS GEMS MINERALS CURIOS POP and he knew he would have to stop.

Dad bought me a small cabinet with drawers in it for SPECIMENS, a word I learned from Herbert S. Zim, and a mineral hammer. Out of an old Desert waterbag Dad made me a holster for it. When I put on my holster I felt mighty and Scientific . . .

Beyond the orange groves we came to a carrot farm, they processed the carrots right there by cooking and cooking them until you couldn't think straight. Then the hills rose up off the dry flat floor of our county. Humps with dots of scrub. Dad drove with confidence through the tiny cañons, the walls crumbly with shale, which didn't interest me. A guy who really wanted to hammer away, but there was no granite or limestone for him to hit. Probably a lot of geologists would be surprised to hear this, that the hills of California were made of DIRT. I would tell them.

Herbert S. Zim showed in a watercolor how you collect SPECIMENS. You have a hat and glasses and a white shirt with the sleeves rolled up. It helped to have a hat and glasses. I didn't have a hat or glasses but I hoped I would need them one day, for my career, like Anita's father's career, only not as MEAN. You have a mineral hammer and you get out there, in the field, and you chisel something off something else. You put it in your canvas satchel. You eat your

Monterey Jack cheese sandwich out of brown paper. You wrap up the specimen in the sandwich paper and take it home and determine its place on MOH'S SCALE. You look it up in *Rocks and Minerals* by Herbert S. Zim. Then you paint a small yellow disk on one of its lesser planes and when that is dry you label it in India ink with the specimen number, and enter that in your serious-looking canvas logbook. I had a small binder I bought at the variety store, exciting and blank. I was shy about asking for the paint, Dad was afraid I would PERMANENTLY HARM THE GARAGE. He alternately shut and opened the windows of the brown car . . .

My hammer got hot in the sun and it banged against my bare knee, singeing me, as I tried to keep up with Dad. I kept looking for tough rock to hammer, but only the trembling shales surrounded us, you could have used a feather duster, not a hammer. This was the only rock you could collect in our county, outside the white boxes lined with cotton. The heat and the heady smells and the rhythm of the hot hammer against my leg, no water, the colors . . . Whenever I was out with Dad I began to worry that maybe I was afraid of Science, that even if I were eventually awarded a white shirt and a hat I could not be Herbert S. Zim. Driving down Mount Palomar from the mighty telescope, Blugggh! In the depths of the Sea Lion Cave, Oregon, Blugggh! In an elevator at Pacific Ocean Park designed to make you think you're going beneath the sea, Blugggh!

Time for a break, I'd say, son.

How to Collect Fossils . . .

The palaeontologist Ellis L. Yochelson (1928–2006), who worked at the Smithsonian Institution's National Museum of Natural History for most of his career, was a wry and witty observer of the foibles of his chosen discipline, as illustrated in the following two excerpts from the transcript of a paper he presented at the 'Natural History Collections, Past – Present – Future' symposium, held at the Biological Society of Washington in October 1968:

When one asks a palaeontologist how he collects fossils, the answer is generally a curt reply such as 'meticulously.' There are a variety of techniques, governed mainly by the kind of fossils and the kind of sediment which encloses them. Some people swear by a 1-pound hammer with a chisel end and a 14-inch handle; others swear at it. So many common-sense features are involved in collecting that a brief general summary on the subject was reviewed as being 'downright inane.'

In spite of this opinion, I believe that much remains to be discussed and written on the subject of fossil collecting. Although professional collectors have been employed permanently, this luxury is largely a thing of the past in the United States. Today, people continue to provide inadequate locality information when they submit collections of fossils for examination and do such silly things as write labels in water-soluble ink. If anything, the ability to obtain useful fossils, ship them, and have the collection arrive in reasonable shape and containing the proper information has lessened as interest in fossils has declined among nonpalaeontologists.

Collecting may be reduced to two fundamentals. First, find a specimen, and second, retain it at least for a significant time interval. Expressions commonly heard are that collections were made, but after several years of having them take up space, the fossils were discarded. Alternatively, one hears of the proverbial mountain slope littered with fossils, but they were not collected because the age of the formation was known. These are the hallmarks that distinguish the mere seeker of geologic data from the true palaeontologist.

The real trick in the field is finding the first fossil in a sedimentary rock. Once this has been collected at the outcrop, the others come far more readily. Even knowing that years ago fossils were collected in the general area is a help. If the rock is a shale that breaks down to a mud and washes away, this first key fossil may be left as a lag deposit. Crawling on hands and knees, with nose at ground level, is the time-honored way of locating it. If the rock is harder, the hammer comes into operation. It may be more poetic to 'bring the hammer into play,' but even when the day is cool and the rock fairly friable, pounding on an outcrop for extended time periods is hard work.

The point here is that both of these operations have built-in limitations on the number of specimens that may be collected in a short time. The available collecting time at the outcrop, and the weight one person can easily carry for a short distance, have been the factors governing the amount of material that leaves the outcrop. To collect more than three or four bags of fossils at any one outcrop is unusual. If bulk samples can be collected rapidly, they are commonly of the type that requires extensive preparation prior to detailed studies of fossil content. Thus, this kind of upper limit also holds for those who study microfossils. The time involved in taking a channel sample or digging a trench to collect fresh material may become significant. In comparison, for example, with a marine zoologist accustomed to collecting on a shallow-water reef, the palaeontologist is a modest collector.

Extrinsic factors important to fossil collecting are not well understood. It is a general rule that one side of a roadcut will yield more specimens than another. Whether this is a regional feature or whether the phenomenon is related to local factors such as vegetation, runoff, or microclimate have never been investigated. Why some fossils in some rocks may be replaced by other minerals is a major mystery. The conditions that dissolve shells but leave their impressions are poorly understood in detail. The list could be continued.

Intrinsic factors also enter into collecting. There is no substitute for experience; some rocks just look right for a particular kind of fossil. In many respects, this is the same as a biologist knowing the life habits of a desired living animal or plant specimen. This type of information can seldom be imparted except by word of mouth on the outcrop. At least one attempt was made to gather these esoteric tidbits as part of general work on techniques, but the results were far from satisfactory. The principal point distilled is that collecting is a full-time activity. It is possible and often necessary for the palaeontologist to carry on more purely geologic work, such as mapping the area or measuring the thickness of a rock layer, but these have to be done before or after the collecting.

There is an interesting minor support of this hypothesis. In fieldwork in the western United States, a palaeontologist visiting a field may find more arrowheads in a few days than the field geologist finds in a season. The geologist strides across the landscape to get the big picture, but the palaeontologist stays at one spot or shuffles along looking at the ground for his pet objects. Slow motion is also a fine way to avoid most rattlesnakes.

...and how to store them

In the how and why of collecting, the why is the easier to answer, or to at least open the floodgates of rhetoric. Once the fossils are safely inside a building the how to store them is far easier than the why of retention. Compared with other natural-history objects, fossils are paradise for a curator.

Naturally, catastrophic events may cause serious losses. Type specimens lost during the great Chicago fire and the flood at Dayton, Ohio, still cause problems to a few specialists, but, hopefully, natural-history specimens today are as safe from such events as might be expected. Good collections are still in temporary repositories, and undoubtedly a quantity of important material will be discarded as some universities remove palaeontology from the curriculum, but increasingly the odds against accidental loss are being lowered.

For convenience, collected natural history subjects may be divided into three categories. First, living organisms, which are stored with great difficulty in zoos and arboreta. Second, recently dead objects, which must be pressed, vermin-protected, or bottled. Finally, dead things, which do not require watering and which do not deteriorate. About the only difficulty in prospect for a museum fossil is a coating of the ever pervading dust. The present-day air-conditioning expert would try to seduce us into believing that this problem has been solved; it is better to put one's faith and one's specimens in closed cases.

With fossils, one is not troubled by evaporation among alco-holics, which to the museum-oriented person does not mean unexplained staff absenteeism. One is not concerned with material drying to powder. Except for rare specimens replaced by pyrite, fossils do not pick up moisture from the air. Fossils are not edible, and though occasional labels and locality numbers may be lost to particularly desperate cockroaches or rats, such events have been fairly rare in the past and are essentially a thing of the past. Fossils do not change color after years of storage, nor do they smell.

About the only obvious and painful drawback to fossil storage is weight. The average collection of fossils, microfossils excepted, is heavier than the average collection of almost anything else in a museum. One drawer, 28 inches by 22 inches, full of particularly stony fossils, like colonial corals, requires complete attention during a moving operation. Drawers of fossils can be stored to a height of 9 feet, but an administrator, before making a decision for high-level storage, should be required to carry at least one drawer to the floor. There is a general rule of nature (Gumperson's Law) that the heaviest drawers are always at the top; for any case over 5 feet high this may become hazardous. It is also well known that museums that stack drawers rather than place them in cases, keep the needed specimens in the bottom drawer of a stack (Saunders' Corollary).

It is a wise idea to remember always that even though fossils are thoroughly dead, they still retain the ability to move. When specimens hop from one tray to another, the net result may be that two otherwise useful collections will have to be discarded. Trays with deep sides are not a luxury item. Because it is simply no longer feasible to put locality numbers on every specimen, stuffing the smaller specimens in glass bottles has been a technical breakthrough. Clear plastic boxes may well be worth however much more they cost; if they do come into general use in the near future, it will be about five decades since palaeontologists stopped putting their prize fossils into cardboard pillboxes. Folded stand-up labels, in contrast to those that lie flat, are such a menace to retaining fossils where they belong and so antediluvian that examples should be put on special exhibit in the chamber of horrors.

There has been a tendency in unsympathetic administrative environments to equate storage of dead items with dead storage. If fossils cannot be seen easily, they will not be studied. Some of the greatest advances that have been made in palaeontology stem from some things no more complex than making aisles wide enough so that drawers may be moved in and out of cases easily. Lighting adequate to permit specimen examination in a storage area has done more for overall clarification of species problems than the most sophisticated hardware of biometry.

Sources: Todd McEwen, *Arithmetic* (London: Jonathan Cape, 1998), pp. 131–6; Herbert S. Zim and Paul R. Shaffer, *Rocks and Minerals: A guide to familiar minerals, gems, ores and rocks*, illustrated by Raymond Perlman (New York: Golden Press, 1957), pp. 10–11; Ellis L. Yochelson, 'Fossils – the How and Why of Collecting and Storing', *Proceedings of the Biological Society of Washington* 82 (1969), pp. 588–93.

COMPASS-BUILDING BACTERIA

Magnetotactic bacteria are microscopic aquatic organisms that build tiny internal compasses from chains of iron minerals which they use to navigate along earth's magnetic field lines. They were first described in the 1970s by the American microbiologist Richard Blakemore, who discovered some of them lurking in sediments collected near Wood's Hole, Massachusetts. Once Blakemore realized that the bacteria's unusual behaviour was somehow connected to biomagnetism, he teamed up with the physicist Richard B. Frankel of the National Magnet Laboratory at MIT, and together they worked out the mechanism behind this amazing compass-building creature.

It was Frankel who supplied the details of the story to the palaeontologist and popular-science writer Stephen Jay Gould (1941–2002), who retold it in one of his celebrated monthly essays in *Natural History* magazine; Gould's essays were collected over the years in a series of award-winning books, including *The Panda's Thumb* (1980), from which this extract has been taken:

Frankel and his colleagues found that each bacterium builds within its body a magnet made of twenty or so opaque, roughly cubic particles, measuring about 500 angstroms on a side (an angstrom is one ten-millionth of a millimeter). These particles are made primarily of the magnetic material Fe_3O_4, called magnetite, or lodestone. Frankel then calculated the total magnetic moment per bacterium and found that each contained enough magnetite to orient itself in the earth's magnetic field against the disturbing influence of Brownian motion. (Particles small enough to be unaffected by the gravitational fields that stabilize us or by the surface forces that affect objects of intermediate size are buffeted in a random manner by thermal energy of the medium in which they lie suspended. The 'play' of dust particles in sunlight provides a standard illustration of Brownian motion.)

The magnetotactic bacteria have built a remarkable machine, using virtually the only configuration that could work as a compass within their tiny bodies. Frankel explains why the magnetite must be arranged as particles and why the particles must be about 500 angstroms on a side. To work as an efficient compass, magnetite must be present as so-called single domain particles, that is, as bits with a single magnetic moment, containing opposite north- and south-seeking ends. The bacteria contain a chain of such particles, oriented with their magnetic moments north pole to the next south pole along the row – 'like the elephants head to tail in a circus finale,' as Frankel states. In this way, the entire chain of particles operates as a single magnetic dipole with north- and south-seeking ends.

If the particles were a bit smaller (less than 400 angstroms on a side), they would be 'superparamagnetic' – a big word indicating that thermal energy at room temperature would cause internal reorientation of the particle's magnetic moment. On the other hand, if particles were greater than 1,000 angstroms on a side, separate magnetic domains pointing in different directions would form *within* the particle. This 'competition' would reduce or cancel the particle's overall magnetic moment. Thus, Frankel concludes, 'the bacteria have solved an interesting problem in physics by producing particles of magnetite of just the right size for a compass, of dimension 500 angstroms.'

But evolutionary biology is predominantly the science of 'why,' and we must ask what such a small creature could possibly do with a magnet. Since a bacterium's cruising range is probably a few inches for the few minutes of its existence, I find it hard to believe that oriented motion in a north or south direction can play any role in its repertoire of adaptive traits. But what preferred direction of motion might make a difference? Frankel suggests, quite plausibly in my view, that an ability to move *down* might be crucial for such a bacterium – for down is the direction of sediments in aquatic environments, and down might lead to a region of preferred oxygen pressure. In this instance, 'them of low degree' might wish to debase themselves even further.

But how does a bacterium know which way is down? With the smug prejudices of our enormous selves, we might think the question inane for its obvious answer: all they have to do is stop whatever they are doing and fall. Not at all. We fall because gravity affects us. Gravity – the standard example of a 'weak force' in physics – influences us only because we are large. We live in a world of competing forces, and the relative strength of these forces depends primarily upon the size of objects affected by them. For familiar creatures of macroscopic dimensions, the ratio of surface area to volume is crucial. This ratio decreases continually as an organism grows, since areas increase as length squared and volumes as length cubed. Small creatures, insects, for example, live in a world dominated by forces acting on their surfaces. Some can walk on water or hang upside down from a ceiling because surface tension is so strong and the gravitational force that might pull them down so weak. Gravitation works on volumes (or, to be more precise, upon masses that are proportional to volumes in a constant gravitational field). Gravitation rules us with our low ratio of surface to volume. But it troubles an insect very little – and a bacterium not at all.

The world of a bacterium is so unlike our own that we must abandon all our certainties about the way things are and start from scratch. Next time you see *Fantastic Voyage* on the tube, take your eyes off Raquel Welch and the predaceous white blood corpuscle long enough to ponder how the miniaturized adventurers would really fare as microscopic objects within a human body (they behave just like regular folks in the film). They would, first of all, be subject to shocks of the Brownian motion, thus making the film something of a random blur. Also, as Isaac Asimov pointed out to me, their ship could not run on its propeller, since blood is too viscous at such a scale. It should have, he said, a flagellum – like a bacterium . . .

In asking why bacteria might build magnets within their bodies, Frankel speculated cogently that swimming north could make little difference to such a tiny creature, but that swimming *down* (another consequence of life around a compass at mid to high latitudes in the northern hemisphere) could be very important indeed. This led me to predict that if Frankel's explanation be valid, magnetic bacteria in the southern hemisphere should swim *south* in order to swim down – that is, their polarity should be reversed relative to northern hemisphere relatives.

In March 1980, Frankel sent me a preprint of a paper with colleagues R. P. Blakemore and A. J. Kalmijn. They traveled to New Zealand and Tasmania in order to test the magnetic polarity of southern hemisphere magnetic bacteria. Indeed, they all swam south and down – an impressive confirmation of Frankel's hypothesis and the basis of my essay.

Source: Stephen Jay Gould, 'Natural Attraction: Bacteria, the Birds, and the Bees', in *The Panda's Thumb* (New York: W. W. Norton, 1980), pp. 306–14.

EXPLODING ANTS

Edward O. Wilson may have abandoned his teenage ambition of classifying the ants of Alabama (see p. 174), but myrmecology – the study of ants – nevertheless went on to occupy much of his professional life. Wilson (b. 1929), a prolific and widely admired science writer, set off a storm of controversy with his 1975 book *Sociobiology: The New Synthesis*, which offered an evolutionary explanation for all kinds of social behaviour, including altruism and aggression in humans as well as in animals. The following account of ants, from a later book of essays entitled *In Search of Nature* (1996), again touches on the subject of animal altruism, and culminates in an astonishing description of the self-detonating worker ants that patrol the Malaysian rainforest:

The question I'm asked most often about ants is 'What do I do about the ones in my kitchen?' And my answer is always the same: 'Watch where you step.' Be careful of little lives. Feed them crumbs of coffeecake. They also like bits of tuna and whipped cream. Get a magnifying glass. Watch them closely. And you will be as close as any person may ever come to seeing social life as it might evolve on another planet. The evolutionary line that gave rise ultimately to ants and other social insects separated more than 600 million years ago from the line that gave rise to human beings. Insect social systems are completely independent of our own and differ from it in many profound ways. They're another grand experiment in evolution for our delectation. The study of their unique traits has already proved very rewarding in several fields of biology.

At present there are about 9,500 described species of ants; this is the number so far given a scientific name. I'd venture a guess that there are in actual existence two or three times that many, and there is immense diversity within this group of hymenopterous insects. A colony of the world's smallest ant could dwell comfortably inside the braincase of the world's largest ant. One genus of ants that I've been studying, *Pheidole*, contains 285 named species from the New World alone. In the collection at Harvard's Museum of Comparative Zoology I have about 600 species; in other words, some 315 are new to science. More pour in from collectors every few months.

Ants are the dominant little-sized organisms of the planet — that is, intermediate in size between bacteria and elephants. My rough estimate is that at any given moment there are about 10^{15}, or a million billion, ants in the world. In terms of overall biomass, measured as dry weight, they are truly formidable. For example, in forests near Manaus, in the central Brazilian Amazon, ants and termites together make up more than one-quarter of the biomass — which includes everything from very small worms and other invertebrates to the largest mammals. Ants alone weigh four times as much as the birds, amphibians, reptiles, and mammals combined. This proportion of ants is approached or exceeded in most other major types of land habitat around the world. When we consider insect biomass alone, we find that the ants and termites, the most highly social of all organisms, plus the social wasps and social bees, which rival them in colonial organization, make up about 80 percent of the biomass. These insects dominate the insect world from the Arctic Circle to Tierra del Fuego and Tasmania. In fact, ants are the principal predator of small animals roughly their own size. They are the 'cemetery squad,' scavenging and removing the corpses of more than 90 percent of the small animals. They are movers and enrichers of the earth, more so than the earth-worms. Indeed, although the social insects as a group make up only 2 percent of all of the known described species of insects in the world, they probably make up most of the biomass . . .

Ants and other social insects are dominant because their social organization gives them competitive superiority over solitary insects. Wherever you go in the world, from rain forest to desert, social insects occupy the center — the stable, resource-rich parts of the environment. Solitary insects, although they also exist in great abundance, are specialists of the fringes — the ephemeral part of the habitat. They are concentrated on outer foliage, deep in wood, in tiny crevices of the soil, and in other sites not preempted by the social insects. A colony of ants can be regarded as a kind of superorganism — a gigantic, amoebalike entity that blankets the foraging field, collecting food and launching forays to engage enemies before they can approach the nest. At the same time they care for the queen and the immature ants — ranging from eggs through larvae to pupae — that are sequestered with her in the nest. They accomplish all these things with high efficiency by a division of labor. Most important, they do them simultaneously. No task is left undone for more than a very short time. No enemy is left unchallenged; no hapless caterpillar fallen from a tree is left uncollected. Also, individuals are able to risk or even to sacrifice their lives in suicidal ventures on behalf of the colony without greatly reducing its productivity. Through close identity with their common mother, the queen, they are able to take far greater risks, in a Darwinian sense, than are solitary insects — and they often do so through mass defense and recruitment to the battlefield, using tactics whose sophistication is worthy of a von Clausewitz. Ant societies are the most warlike of all known animal groups, solitary or social. Most species of ants engage in frequent territorial battles, during which kamikazelike assaults by sterile workers turn the tide. In the deserts of the Southwest, for example, scouts of the genus *Dorymyrmex*, upon discovering a nest of their rivals in the genus *Myrmecocystus*, recruit fellow colony members, who then surround the next entrance, take bits of gravel to the edge of the nest, and drop them in. Any *Myrmecocystus* that continue to resist eventually get buried under the rubble, which at least temporarily closes off their exit to the outside. And in the Malaysian rain forest, worker ants of certain species of *Camponotus* have grotesquely hypertrophied paired glands that open out at the base of the mandibles and fill a large part of the body. These receptacles are loaded with a sticky toxic chemical. When confronted by enemies, and under extreme duress, the ants are able to contract their abdominal muscles and explode in the face of the enemy, rather like walking grenades. One of these ants can trade its life for those of several enemies. In Darwinian terms this is an excellent tactic.

Source: Edward O. Wilson, *In Search of Nature* (Washington, D.C.: Island Press, 1996), pp. 47–51.

THE HUMAN RESERVE

'Primates stand at the hinge of evolution', as evolutionary biologist Alison Jolly (b. 1937) writes in the preface to the second edition of her best-known book *The Evolution of Primate Behavior*, from which this thoughtful account of the future of biodiversity is taken. Writing partly in response to the erosion and desertification that she witnessed during her years of fieldwork in Madagascar and elsewhere – most of which was caused by forest clearance – Jolly considers the long-term impact of habitat loss on primate survival, while rightly resisting the temptation to condemn the local peoples responsible; after all, as she points out, a Malagasy peasant farmer who sets off with his axe to clear an acre of forest is doing so only to feed his family, 'as mankind has done everywhere since the Neolithic Revolution'.

Jolly is also an astute observer of the field habits of her fellow primatologists, and she ends this extract with a salutary vignette of a typical working day spent watching a troop of howler monkeys sitting doing nothing up a tree:

Man is a primate. We know this in every detail of our physical form, from our flat fingernails to our blunt big toes, and in the colors we see out of our eyes. Linnaeus (1758) classed man with monkey, lemur, and bat. When T. H. Huxley (1863) drew the anatomy of man and ape, he challenged his hearers to deny their kinship with the chimpanzee.

Does man behave as a primate? No. Chimpanzees, like people, probe twigs into holes; we also analyze moon rock. Chimpanzees, like people, have destroyed their own kind, but one of our species wrote *Medea*. However, the creationist who looks for absolute differences between ourselves and the ape is likely to be disappointed. Our love and hate and fear and curiosity, our birth and death, all our emotions, and even some of our logic stem from the primate past.

On present reckoning we have been placental mammals for more than 100 million years and have been monkeylike, large-brained, possibly group-living creatures for 40 million years. Sometime in the past 5 or 10 million years we became terrestrial hunter-gatherers. Only for the final millennia have some tribes tilled fields, built villages, constructed human life as we now conceive it.

All written history is the chronicle of a *nouveau riche*, abruptly risen to dictatorship of his planet – not what we are, but what we have made ourselves. Our earlier history, the part without writing or artifacts, can only be deduced from what we do and what our primate relatives do – the hieroglyphics of the chimpanzee's whimper and the baby's smile.

The more urgent question in primate evolution is not what it reveals about the past. The real question is what happens in the future. How many kinds of plants and animals will continue to share our planet? As an order of magnitude, perhaps one wild species now becomes extinct every day. By the end of this century, it may be one every hour, and half the world's present wild species will already be dead.

How many species will continue in something close to their original habitats? Climax habitats, where nature has approached an equilibrium, may take centuries to reestablish once they have been destroyed. Sometimes they never return to their former state, like cleared areas of Amazon jungle that become hard-baked grassland.

We are approaching a point where the planet will be divided into nature reserves and the human reserve. Only a few wild mammals can live in close proximity to man: rats and mice in houses, bush-babies at the bottom of an African garden, baboons about a garbage dump. Most large mammals cannot. We may be horrified at the slaughter and mutilation of mountain gorillas and quick to condemn the Rwandan poachers who kill them. Few of us, however, want gorillas in our own backyard. We would want them less if rich pasture were forbidden to our own cattle, or if the gorillas demolished our own banana trees, calmly stripping off the leaves and chewing the last shred of pith. Even a tolerant Hindu starts throwing rocks, or seeks out an off-caste monkey trapper, when the sacred langurs invade his field and he faces a choice between the monkeys' survival or his children's.

The most marginal lands and the most crucial watersheds are threatened by the needs of the poor and the greed of the rich. Imagine a forest canopy 80 to 120 feet above ground, with emergent trees towering to 150 feet. The canopy is so dense that little light reaches the forest floor, and there is almost no undergrowth, except where a tree has fallen to make a 'clearing' – a solidly braided tangle of vegetation, plants locked together in the light, racing each other upward before the canopy roof shall close again. In the tallest trees, often invisible to a ground-based observer, sit mantled howler monkeys. They sit. And sit. An individual howler rests for 80 per cent of his day. They are black lumps on the branch, without expression, individuals distinguished mainly by the goitrous swellings made by botfly larvae on their throats. They may reward, and threaten, the observer with dramatic howls, or they may simply pick themselves up, drifting from tree to tree by several paths, so it is even difficult to count the animals while the observer stumbles down and up the ravine-slashed forest floor in their wake.

This is not a picture of the selfless heroism of the primatologist. He (or, as likely, she) is usually enjoying himself. It is not even a picture of how he wastes his time, for he ticks 'resting' on his checksheet at two-minute intervals, and builds up a quantitative account of the howlers' lack of activity. He will calculate their energy budget, their choosiness over food, and the food available in their environment, so he may contrast results with the many other howler studies from other sites. He even looks forward to a stand-up argument about howlers at the next conference, for such is the peculiar social life of primatologists.

This is, instead, a picture of the primatologist's luck, that he may escape to the wild. He chooses his study site, if he can, where he will not hear the distant grinding of bulldozers or the nearby toc, toc, toc of the peasant's axe. But primates compete for living space and for human food. They carry human diseases. It is clear that in the future most wild primates will live in natural reserves, islanded in the rising sea of the human reserve. It is even clearer that only protected reserves might preserve the richness of natural communities, the remaining fragments of the ecosystems where our own ancestors evolved.

Source: Alison Jolly, *The Evolution of Primate Behavior*, 2nd edn (New York: Macmillan, 1985), pp. 3–5.

THE EVOLUTION OF THE EYE

Although evolutionary theory is the best tool we currently have for understanding the complexity of life on earth, it remains opposed by a vocal minority of sceptics, most of whom are creationists of one religious persuasion or another. Forever on the lookout for the evidence they hope will undermine the Darwinian edifice, creationists often point to the mammalian eye as an example of an organ that could only have appeared fully formed, and not in a sequence of intermediate stages, since a half-adapted eye is of no use to anyone. Yet, as Richard Dawkins (b. 1941) points out, in the following extract from *Climbing Mount Improbable* (1996), his metaphor-laden survey of the evolution of complex mechanisms, even an organ as 'perfected' as the eye is part of a continuous evolutionary gradient. This gradient is so shallow that there are no steep 'jumps' from one model of the eye to another – from animals with a single light-detecting photocell to animals with highly discriminatory vision – so that Mount Improbable (as Dawkins has named this adaptive landscape) can always be climbed over time.

Charles Darwin was well aware that the evolution of the eye was a challenge to explain, and in *The Origin of Species* he famously admitted that its gradual adaptation 'seems, I freely confess, absurd in the highest possible degree'. This line is often quoted by anti-evolutionists as 'proof' that the eye undermines Darwinian theory, but as Dawkins points out at the end of this extract, they rarely quote the bit that came next:

Darwin famously used the eye to introduce his discussion on 'Organs of extreme perfection and complication':

> To suppose that the eye, with all its inimitable contrivances for adjusting the focus to different distances, for admitting different amounts of light, and for the correction of spherical and chromatic aberration, could have been formed by natural selection, seems, I freely confess, absurd in the highest possible degree.

It is possible that Darwin was influenced by his wife Emma's difficulty with this very point. Fifteen years before *The Origin of Species* he had written a long essay outlining his theory of evolution by natural selection. He wanted Emma to publish it in the event of his death and he let her read it. Her marginalia survive and it is particularly interesting that she picked out his suggestion that the human eye 'may <u>possibly</u> have been acquired by gradual selection of slight but in each case useful deviations'. Emma's note here reads, 'A great assumption / E.D.' Long after *The Origin of Species* was published Darwin confessed, in a letter to an American colleague: 'The eye, to this day, gives me a cold shudder, but when I think of the fine known gradations, my reason tells me I ought to conquer the cold shudder.' . . . Darwin, however, saw his doubts as a challenge to go on thinking, not a welcome excuse to give up.

When we speak of 'the' eye, by the way, we are not doing justice to the problem. It has been authoritatively estimated that eyes have evolved no fewer than forty times, and probably more than sixty times, independently in various parts of the animal kingdom. In some cases these eyes use radically different principles . . . Frogs and squids, for instance, both have good camera-style eyes, but these eyes develop in such different ways in the two embryos that we can be sure they evolved independently. This does not mean that the common ancestor of frogs and squids totally lacked eyes of any kind. I wouldn't be surprised if the common ancestor of all surviving animals, who lived perhaps a billion years ago, possessed eyes. Perhaps it had some sort of rudimentary patch of light-sensitive pigment and could just tell the difference between night and day . . .

As I have already remarked, no animal ever made a living by being an intermediary stage on some evolutionary pathway. What we may think of as a way station up the slope towards a more advanced eye may be, for the animal itself, its most vital organ and very probably the ideal eye for its own particular way of life. High-resolution image-forming eyes, for instance, are not suitable for very small animals. High-quality eyes have to exceed a certain size – absolute size not size relative to the animal's body – and the larger the better in absolute terms. For a very small animal an absolutely large eye would probably be too costly to make and too heavy and bulky to carry around. A snail would look pretty silly if its eyes had the seeing power of human eyes. Snails that grew eyes even slightly larger than the present average might see better than their rivals. But they'd pay the penalty of having to carry a larger burden around, and therefore wouldn't survive so well. The largest eye ever recorded, by the way, is a colossal 37 cm in diameter. The leviathan that could afford to carry such eyes around is a giant squid with 10-metre tentacles.

Accepting the limitations of the metaphor of Mount Improbable, let's go right down to the bottom of the vision slopes. Here we find eyes so simple that they scarcely deserve to be recognized as eyes at all. It is better to say that the general body surface is slightly sensitive to light. This is true of some single-celled organisms, some jellyfish, starfish, leeches and various other kinds of worms. Such animals are incapable of forming an image, or even of telling the direction from which light comes. All that they can sense (dimly) is the presence of (bright) light, somewhere in the vicinity. Weirdly, there is good evidence of cells that respond to light in the genitals of both male and female butterflies. These are not image-forming eyes but they can tell the difference between light and dark and they may represent the kind of starting point that we are talking about when we speak of the remote evolutionary origins of eyes . . .

It may be that living cells are more or less bound to be somewhat affected by light – a possibility that makes the butterfly's light-sensitive genitals seem less strange. A light ray consists of a straight stream of photons. When a photon hits a molecule of some coloured substance it may be stopped in its tracks and the molecule changed into a different form of the same molecule. When this happens some energy is released. In green plants and green bacteria, this energy is used to build food molecules, in the set of processes called photosynthesis. In animals the energy may trigger a reaction in a nerve, and this constitutes the first step in the process called seeing, even in animals lacking eyes that we would recognize as eyes. Any of a wide variety of coloured pigments will do, in a rudimentary way. Such pigments abound, for all sorts of purposes other than trapping light. The first faltering steps up the slopes of Mount Improbable would have consisted in the gradual improvement of pigment molecules. There is a shallow, continuous ramp of improvement – easy to climb in small steps.

Dawkins goes on to narrate some of the many stages by which a simple photocell might evolve, from a cell that captures photons with a single layer of pigment that translates light signals into nerve impulses, to curved, multi-layered pigment cells that can detect the direction of light and shadow, all the way up to the kind of compound eye that many insects have. At the end of this long and impressive account, he returns to the Darwin quote with which the previous extract began:

Nothing is as difficult to evolve as we humans imagine it to be. Darwin gave too much when he bent over backwards to concede the difficulty of evolving an eye. And his wife took too much when she underlined her scepticism in the margin. Darwin knew what he was doing. Creationists love the quotation that I gave at the beginning of this chapter, but they never complete it. After making his rhetorical concession, Darwin went on:

> When it was first said that the sun stood still and the world turned round, the common sense of mankind declared the doctrine false; but the old saying of *Vox populi, vox Dei*, as every philosopher knows, cannot be trusted in science. Reason tells me, that if numerous gradations from an imperfect and simple eye to one perfect and complex, each grade being useful to its possessor, can be shown to exist, as is certainly the case; if further, the eye ever slightly varies, and the variations be inherited, as is likewise certainly the case; and if such variations should ever be useful to any animal under changing conditions of life, then the difficulty of believing that a perfect and complex eye could be formed by natural selection, though insuperable by our imagination, cannot be considered real.

Lucretius

An interesting precursor to the Darwinian rejection of the argument from design appeared in Lucretius's great poem *De rerum natura* (*On the Nature of Things*), in which the poet warns the reader to avoid the fallacy of believing that organs such as eyes were purpose-built from scratch:

> Condemne we here their false opinion
> And evermore their erring doctrines shun
> Who say the splendid eies were made for sight,
> The feete and leggs created that we might
> March on those props, the strong arms hung
> On able shoulders, unto which belong
> The labouring hands, all in their frames decreed
> For those employments which supply the need
> Of human life, so all the rest were made

For severall uses: all perversly sayd.
Nature did not the limbs for use compose,
But th' uses out of their creations rose.
For sight had not a being, before the eies;
Nor speech before the tongue, but contrary,
The tongue long time preceeded words, the eare
Was before sounds, and all the members were
Before their severall uses long producd,
And not created that they might be used.

Finally, I couldn't resist ending this section with the courtroom outburst that appears in chapter 34 of Dickens's *The Pickwick Papers* (1836–37). Mr Pickwick's valet, Sam Weller, is being cross-examined as a witness: 'Have you a pair of eyes, Mr. Weller?' 'Yes, I have a pair of eyes', replied Sam, 'and that's just it. If they was a pair o' patent double million magnifyin' gas microscopes of hextra power, p'raps I might be able to see through a flight o' stairs and a deal door; but bein' only eyes, you see, my vision's limited.'

Sources: Richard Dawkins, *Climbing Mount Improbable* (London: Viking, 1996), pp. 126–79; Lucretius, *De rerum natura*, trans. Lucy Hutchinson (British Library Add. MS 19333, ff. 93^v–94^r [Bk. IV, 871–89]).

LICHENOMETRICS

Like Abraham Trembley's polyps (see p. 121), lichens confound the animal/vegetable hierarchy, being the symbiotic composite of fungi and blue-green algae. They are extraordinary creatures, tough and resilient – 'the toughest organisms on earth', as Richard Fortey (b. 1946) of London's Natural History Museum describes them, and what's more, they cover much of the surface of our planet with an easily readable layer of atmospheric evidence. There are nearly three thousand species of lichen in Britain alone, and undisturbed churchyards are the best places to see them, especially since (as Fortey points out), a lichen-covered gravestone also contains the single most important piece of information that a naturalist can have: a date:

Lichens are particularly important as indicators of pollution because they readily absorb heavy metals into their tissues. They simply mop up elements like lead and cadmium. Lead was formerly present in appreciable quantities in petroleum spirit. In Britain it was practical to assay the damage done to the environment by mapping lichen species; most are unable to tolerate lead pollution for long, but those that can proliferate at the expense of others. The species plot out the state of the environment. It is extraordinary to see how precisely the lead pollution traced the course of trunk roads, forming a series of 'corridors' with low lichen diversity criss-crossing the landscape. However, at least the lead content falls away laterally from the roads themselves. It is a wise precaution not to eat mushrooms picked by main roads, as high lead content has been associated with loss of brain function. The introduction of lead-free fuel received a boost from such findings. Far wider regional pollution across industrial areas is also faithfully recorded by these humble living patches. Inner cities can be very low in species as a result of the pollution from the 'dark, satanic mills' of the last 150 years. As you move westwards across the British Isles the number of lichens increases. The dominant weather systems move in from the clean Atlantic Ocean, flushing most pollutants eastwards, and the lichens revel in the moister atmosphere in the west. In the old oak forests in North Wales every boulder is dappled with lichens, while the twisted oak branches are heavily draped with leafy and feathery forms. Some of these lichens yield vegetable dyes with lovely natural colours, yellows especially, so in addition to dressing the trees, lichens might well help to dress us as well. Around the old mines on the hillsides near by other lichens take up noxious elements or pollutants, for which they seem to have a particular affinity. These can now be accurately assessed using modern mass spectrometry. They are useful things, these unspectacular lichens, for diagnosing the health of the planet. As we have seem so often before, an apparently arcane expertise in an organism that would be passed by unnoticed by many relates to important contemporary issues.

Many lichens also grow very slowly, some just a few millimetres a year, and some even less than a millimetre. It is an interesting problem how to determine this rate of growth. One way to estimate the average growth rates is to use gravestones. These *memento mori* are a favourite habitat for lichens — indeed, there are some species that are now hardly found outside old churchyards. This is partly because many of these sacred acres have escaped chemical spraying or artificial fertilizer for decades or longer. As we have seen, lichens are unusually sensitive to all pollutants, even those of alleged benefit to farmers. Some churchyards are little patches of medieval habitat. Lie back beneath an old yew tree, and imagine priests and squires and villeins going about their Sunday business. A gravestone includes that very useful piece of information — a date. When erected they are pristine, but soon time and lichens make their mark. Lichens on flat gravestones tend to grow outwards in a regular circle, so the diameter of the circle is proportionate to its age. The largest circle found on a gravestone of a given date will give an approximation to the maximum rate of growth. There will be a certain range of variation as a consequence of local conditions, and variation in the time of first colonization. Furthermore, as time passes, new species of lichen will join the gravestone habitat — and younger rings will 'cut' through older ones as they grow, so revealing the order of succession of colonization. A good gravestone will accordingly yield a complex narrative, and many gravestones will provide usable statistics. Rates can then be applied to other sites. The use of lichens in dating is known as lichenometrics. Although not without its critics, lichenometrics has revealed some interesting figures. It seems that lichen growth rate has speeded up in high latitudes since the industrial revolution, and that this may be connected to global warming. Because they can be found nearly everywhere and grow so slowly lichens are potentially a ubiquitous biological 'diary' that records changes to the environment and atmosphere on the century scale . . .

Molecular evidence shows that fungi have become 'lichenized' on many occasions: to put it another way, they have repeatedly made a contract with algae to enter into their special partnership. The great majority of such fungi are 'waxy cup' ascomycetes; this is easy to see on those lichen species that bear little red or yellow cups on their surfaces, because they are not very different in appearance from the spore-bearing fruit bodies of their relatives, which can be found on forest paths or compost heaps. Next time you are in a country grave-yard peer closely at a headstone and you will be almost sure to see these little coloured cups or plates on an encrusting lichen. A few gilled, mushroom-like fungi have also taken this curious evolu-tionary path. On the other hand, the 'algal' partners of the lichen symbiosis are identical to common species living free in nature. It seems that it must have been the fungal partner that hijacked the algae into the collaboration to become lichen. The fungi obtain their carbohydrates from the algal, or blue-green bacterial, partner. The latter employ photosynthesis to manufacture carbohydrates in their cells, at the same time 'fixing' atmospheric carbon dioxide. For their part the fungi access mineral salts like phosphates that would not otherwise be available to the algae. They live on rainwater and dust. Working together, these two organisms produce a collaboration that is remarkably tough. No doubt lichens will outlast us all.

Source: Richard Fortey, Dry Store Room No. 1: The Secret Life of the Natural History Museum (London: HarperCollins, 2008), pp. 163–6.

THE TURTLE'S SHELL

The mechanism of evolution works through the uptake of successful variations, among the most unusual of which is the turtle's lack of ribs. The only four-limbed animal on earth to forgo the protection of a rib cage, the turtle instead grows a bony box into which it can withdraw its limbs and head when under threat. Although the shell seems to have developed initially as a means of bodily protection, the anomaly affected other aspects of turtle evolution, leading to the creature's transformation into one of the oddities of the natural world. But as Mark S. Blumberg notes in the introduction to *Freaks of Nature* (2009), the book from which the following extract comes, our ideas about what constitutes 'normal' and 'natural' tend to overlook the many routes down which evolution has travelled over the millennia. 'Left to its own devices', he writes, 'nature always takes exception to the rule':

As weird as turtles are, they have become so familiar to us that we may forget why they are weird. Of course, the shell has a lot to do with it. So here we focus on the shell, its unique place in evolutionary history, and its curious connection to limb buds.

About 200 million years ago, turtles suddenly (by evolutionary standards) emerged, sporting a unique four-legged body boxed within a bony shell – or more precisely, *shells* comprising the *carapace* above and the *plastron* beneath the turtle's torso. Such rapid innovations are a puzzle to biologists and laypeople alike and have traditionally inspired searches for 'missing links.' Such searches have since lost much of their allure: Biologists may still seek intermediate forms among ancestors, but today we understand that innovative anatomies – such as that of the turtle – can and often do arise suddenly.

Other animals, such as armadillos, use tough, scaly skin as external protection. But the turtle's protective shell stands alone in its shape and its bony composition. Why do turtles, among all four-legged animals, need this distinctive form of protection? Look closely at the turtle skeleton below and you see why: *There is not a rib in sight*.

If we wished to weave an evolutionary story about turtles, we might conjure a tale about ancestors losing the rib cage and adding the shell as an alternative means of protection. But no one has found a turtle ancestor that lacks both ribs and shell – and no one will, because the shell represents a novel way of using ribs, not a novel response to their disappearance. Thus, the carapace embodies both cause and cure for the vulnerability of the turtle's torso.

The truth about turtles can be found in the embryo. The evolutionary innovation that is the shell begins as little more than a swelling on the embryo's flank, between the forelimb and hindlimb buds. This swelling is called the *carapacial ridge*. Similar to the *apical ectodermal ridge*, the AER, that occupies the outer edge of the limb bud, the carapacial ridge also encloses mesenchymal cells that, through inductive interactions, contribute to the construction of a complex appendage.

When you and I were embryos, our ribs grew outward from the spinal column, curving downward to form the cage. Unlike our limbs, which lie *outside* our ribs, turtles' limbs lie *inside* their ribs. For this form to develop, the ribs must grow outward from the vertebral column, but never downward. In the process, they provide supportive struts for the canopy-like shell, sheltering the limbs like an open umbrella resting on a shoulder. Once again, the key to this

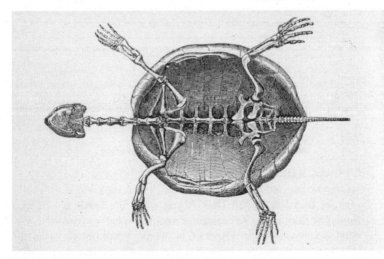

22. As can be seen in this engraving of a turtle skeleton from Sarah Cooper's *Animal Life in the Sea and On Land* (1887), turtles do not have ribs!

unique relationship between ribs and limbs is the carapacial ridge. Specifically, it is the ability of signalling molecules within the carapacial ridge to divert, attract, and ensnare the ribs that sends the turtle down its distinctive developmental path. Disrupt the carapacial ridge and a nonturtle-like rib cage – one more like yours and mine – will emerge.

Beginning as only a small bud, the carapacial ridge expands outward, away from the body, and lengthwise, toward each of the other limbs. Rib cartilage inhabits the carapacial ridge and, as in other limb buds, the cartilage hardens into bone and the ridge continues to expand, ultimately forming the outer edge of the carapace. In this way, we can view the shell as a novel limb form, but one that is produced using ancient mechanisms and found parts.

Source: Mark S. Blumberg, *Freaks of Nature: And What They Tell Us about Development and Evolution* (Oxford: Oxford University Press, 2009), pp. 172–4.

HEBRIDEAN SEAL SURVIVAL

Another evolutionary oddity, this time caused by the actions of Mesolithic man, whose arrival in coastal Scotland some ten thousand years ago spelled catastrophe for the Atlantic grey seal. As a valuable source of meat, skins and oil, as well as bones and sinews for making thread, the grey seal was hunted with such enthusiasm that it soon faced elimination from Scottish waters. As the environmentalist Sir John Lister-Kaye (b. 1946) explains in this extract from his book *Nature's Child* (2004), what saved the grey seal in Scotland was an unusual evolutionary strategy known as delayed implantation, an adaptation that allows healthy embryos to cease development for a while, free-floating in the uterus until a more advantageous moment for implantation occurs. In this case, the seals appear to have discovered, over many years of trial and error, that delaying pupping from May until September meant that sea-going hunters, however rapacious, could no longer reach the newborn pups with ease:

It is a plausible hypothesis that the grey seal assisted early man in his settlement of Highland Scotland, principally throughout the islands. Seals could have been just that extra advantage nature sometimes provides to clinch the successful colonisation of a niche – good for Stone Age man, less so for the seals. They faced a problem. Their biology required them to come ashore to mate and to give birth to helpless pups on the beach. For a month every year the pups were immobile, defenceless and unable to escape to sea – so much easy meat. To make matters worse they were programmed to rut in the autumn and give birth in the spring – in good weather, long daylight and calm seas – all facilitating man's predatory access. To these early men the seal harvest was, quite simply, a bonanza.

But grey seals, *Halichoerus grypus* (the Roman-nosed sea-pig), are intelligent mammals. After several thousand years of Arcadian isolation in these islands the seals now found themselves severely disadvantaged by this unexpected turn of events; they had no alternative but to abandon their favourite beaches and haul-outs for remoter, safer islands, of which there are many off the west coast of the Highlands. Predictably, man followed. It can be argued that by this process and aided by the continuous supply of seal products,

Mesolithic man dispersed himself throughout the Small Isles and the Hebrides, perhaps Orkney and Shetland too. As the seals moved further and further out, so did the hunters. With their vessels they could continue to exploit the skins, meat and bones, and look forward to a seasonal harvest at pupping time. And, as is man's interminable wont, our Stone Age predecessors overdid it. Had marine biologists and conservationists been around at the time, they would probably have forecast the ultimate extermination of the grey seal in Scotland.

But the seal's innate intelligence is not merely perceptual; it is also biological. Deep within its genes lay a trump card known as 'delayed implantation' (actually not uncommon in mammal biology). This is a physiological process by which the brain and the body come together to say, 'We're in deep trouble here and if we are to survive, we must do something about it.' After mating the normal process is for the embryo pup to implant in the uterus wall and steadily develop throughout the nine months of grey-seal gestation until birth. For most of the grey seals in the world that is what happened then, and still does today – but not for the Scottish and Hebridean populations. They have done something else.

Newborn grey-seal pups can't survive for long in water. There is nothing the adults can do about that in the short term, although it is interesting to note that the common seal, *Phoca vitulina*, has evolved to give birth in the sea, in sheltered water, the pup never needing to come ashore. This is a significant evolutionary step forward that will have taken thousands of generations to perfect. Oppressed by Stone Age man all those years ago, the grey seals chose a different and rather more drastic adaptation that may have provided a faster evolutionary response to their crisis. What they could do within a few generations – what they *did* achieve in the Scottish population – was to delay the implantation of the embryo so that the pup was *not* born in the late-May/early-June Highland spring, vulnerable and accessible, but at precisely that moment in the year when man was least likely to be able to get to the outlying islands. Simultaneously, they sensibly migrated their breeding grounds to the most far-flung, storm-battered, uninhabited islands they could find, where many of them have remained to this day.

The real success of this device seems to have been the timing of the pupping. The autumnal equinox falls on 23 September. At this moment in the year there is a reliable pattern of winds and spring tides emerging from the currents and prevailing winds in lower lati-tudes, which build up and lash their way round the northern hemisphere. In the Hebrides these sudden and often fierce squalls make seafaring particularly hazardous – in fact, the very weather in which you *don't* want to be putting to sea in your coracle or dugout canoe. The synchronism of the equinoctial gales and the new season of grey-seal pupping, late September to early October, was arrived at by the continuous process of trial and heart-rending failure – by natural selection, the only way that nature knows how to work. The hunting pressure must have been so great and so constant that the wretched seals were forced to try every option available to them. Only those that succeeded in delaying implantation would survive to breed; all the rest perished. Those that failed to move to pup on out-lying islands were either killed or were perpetually prevented from breeding.

Just how long it took for the selected solution to establish we cannot know. What we do know is that Hebridean grey seals still pup in the autumn – a delay of five months – and then, immediately after giving birth, the cows go straight into rutting harems to mate. They have opted to pit their survival chances against the angry seas, rather than face total extermination at the relentless hand of man, while birth for other populations (in Pembrokeshire, for instance, where fertile soils were far more important to early man than seals), remained in the spring, unchanged, with a separate haul-out to mate in the autumn. To this day, long after the hunting pressure has disappeared, you still have to run the gauntlet of Hebridean storms and high seas to see grey-seal pups in their far-flung colonies.

Source: John Lister-Kaye, *Nature's Child: Encounters with Wonders of the Natural World* (London: Little, Brown, 2004), pp. 67–70.

THE MATHEMATICS OF THE ISBN

Sarah Flannery (b. 1982) shot to fame in her native Ireland when, in 1999, at the age of sixteen, she was named EU Young Scientist of the Year for her work on the development of public-key cryptography (a branch of applied mathematics that deals with data-encryption technology). Her subsequent book, *In Code* (2000), co-written with her father, David Flannery, told the story of her childhood passion for mathematics, while also shedding light on some of the complex numerical codings that underpin the everyday technologies that surround us; in this case, the secrets of the International Standard Book Number – the ISBN – that appears on the back of every published book, including this one:

Every modern book carries an identifying number on its back cover, normally near the bottom right-hand corner. That is true of the hardback edition of this book, *The Art of Science*, which displays the thirteen-digit number 978-0-330-49075-7.* This number is the book's ISBN (International Standard Book Number). Its first twelve digits are 'information digits' because they code information about the book; in this case, the first three digits, 978, denote a code for the book publishing industry and the following digit, 0, identifies the book as being published in an English-speaking country; the next eight digits identify the publishing house as well as the book itself. The thirteenth and final digit is known as a *check digit* and though, in general, it is redundant in the information sense, it is very useful in checking whether or not a purported ISBN is valid.

Apparently the two types of error most commonly committed by people when transcribing long strings of digits are a single transcription error in any position, and an error due to the transposition of any two digits (i.e. '34' instead of '43').

---

* The original published version of this essay analysed the ten-digit ISBN for Sarah Flannery's own book, *In Code* (1-86197-222-9). Since January 2007, however, ISBNs have been made up of thirteen digits, with a different algorithm used to calculate their checksums. In the light of this upgrade, Sarah kindly agreed to rewrite her original essay, which is why it is about the ISBN of the hardback edition of the book you are holding in your hands right now.— R.H.

If an error of either of these types occurs during the transcription, then the thirteenth digit calculated by the recipient of this supposed ISBN will not match the thirteenth digit of the submitted number. Thus a bookseller, for example, need not waste time searching for a book using an invalid ISBN. Let me describe to you how this check digit is calculated, without explaining why computing it in this way provides a 'flag' for the presence of one of the errors just mentioned.

Briefly the first 12 digits in order are alternately multiplied by 1 or 3. The results of all these multiplications are added up to get what is known as *the weighted check sum*, S, say. That sum, S, is then taken modulo 10 and the result subtracted from 10. A zero replaces a ten so that in all cases a single check digit results. (Before the numbering system changed in 2007, an X replaced a ten, so that some ISBNs ended with an isolated X.)

To calculate the check digit in the ISBN for this book, the first twelve information digits of which are 978033049075, compute:

$S = (9\text{x}1) + (7\text{x}3) + (8\text{x}1) + (0\text{x}3) + (3\text{x}1) + (3\text{x}3) + (0\text{x}1) + (4\text{x}3) + (9\text{x}1) + (0\text{x}3) + (7\text{x}1) + (5\text{x}3)$
$= 9 + 21 + 8 + 0 + 3 + 9 + 0 + 12 + 9 + 0 + 7 + 15$
$S = 93$

Then 10 – 93 (modulo 10) = 7. Appending the check digit 7 to the twelve information digits gives the ISBN of *The Art of Science* as: 978-0-330-49075-7.

Interestingly, if the difference between two adjacent digits is 5, the check digit will not catch their transposition!

BRAIN POWER

A large, complex network of rapidly firing neurons, the brain remains the least understood organ in the human body, although recent advances in brain physiology have revealed much about its chemical nature. In this extract from her 'guided tour' of the human brain, the physiologist and former Director of the Royal Institution, Susan Greenfield (b. 1950), describes some of the imaging techniques that allow us to see the various constellations of the brain at work, from the prefrontal cortex, that helps determine thought and personality, to Broca's area (named after Pierre Paul Broca, 1824–80), a region thought to be associated with the development of speech and language. Of course, knowing about the chemical topography of the brain is not the same as knowing about the mysteries of mind and consciousness – areas of research in which some of the most exciting future discoveries are likely to be made:

Of all the organs of the body, the brain is the greediest in its fuel consumption. It burns oxygen and glucose at ten times the rate of all other body tissues at rest. In fact, the brain uses up so much energy that it dies if deprived of oxygen for only a few minutes. Even though the brain is less than 2.5 per cent of our total body weight, it is responsible for 20 per cent of energy consumption at rest. But what happens to this energy? It enables the brain to 'work.'

When a brain region is working, it uses up much more fuel. The fuel for the brain is the carbohydrates in the food that you eat and the oxygen in the air that you breathe: when carbohydrates react with oxygen, they generate carbon dioxide, water, and most important of all, heat. In the body, all the energy from food is not released immediately in a simple combustion, because it would not be very helpful if there was no energy left for any of the functions of the brain and body. Thus, although some heat is needed to keep us warm, there is a chemical in the body that prevents the immediate release of all the energy from food we have eaten. Through its formation we are able to store up this energy for the mechanical, electrical, and chemical work that the body and the brain has to do. The energy-storing chemical adenosine triphosphate (ATP) is produced from the food we eat for as long as we are alive. ATP stores energy and has the potential to liberate it like a compressed spring upon release.

If brain regions are active during a particular task, they are working hard and using more energy; they are making great demand on ATP stores, and hence more carbohydrates, the simplest form of which is glucose, as well as oxygen are required. It follows that if we could trace the increased demand for oxygen or for glucose by certain parts of the brain, we would be able to say what brain areas were most active or working hardest during any particular task. This is the principle of the two particular techniques used to visualize the brain actually at work.

One technique is known as positron emission tomography (PET). The basic requirement in PET is for either oxygen or glucose to be labelled so that it can be easily tracked. The 'label' in this case is a radioactive atom, in the sense that it contains an unstable nucleus that ejects positrons at very high speed. Positrons are fundamental particles similar to electrons except that they have a positive charge. Radioactive oxygen atoms incorporated into either glucose or water molecules are injected intravenously. The radioactive label is then carried by the blood into the brain. The emitted positrons collide with electrons in other molecules within the brain and mutually annihilate each other. The burst of energy that results forms a gamma ray that is of sufficiently high energy to penetrate through the skull and be detected outside of the head.

Because these high-energy gamma rays can travel a long way, they pass right out of the head and strike sensors, the signal from which is then used to build up an image of the brain at work. The glucose or oxygen accumulates in the brain regions that need it most, namely those working the hardest. With PET, it is possible to show different active areas according to tasks as subtly different as saying words compared with reading words.

A second imaging technique, functional magnetic resonance imaging (MRI), is like PET in that it relies on the differential expenditure of energy by whatever brain regions are working hardest; however, this time no injections are involved. Because there is no problem with ascertaining exactly when the injected label reaches the brain, imaging with MRI has the potential to give an even more faithful reflection of what is going on at a given moment. MRI, like PET, also measures changes in blood oxygen concentration serving brain areas that are more active; however, the method of detection is different. Oxygen is carried by the protein haemoglobin. MRI exploits the fact that the actual amount of oxygen present affects the magnetic properties of haemoglobin: these properties can be monitored in the presence of a magnetic field, where the nuclei of the atoms line up as though they were themselves miniature magnets. When bombarded and pushed out of alignment by radio waves, these atoms emit radio signals as they spin back into line. The radio signal is unique to the amount of oxygen carried by haemoglobin in the sample and therefore gives a very sensitive measure of the activity of different regions of the brain. This technique can pinpoint an area as small as 1 to 2 millimetres and measure events taking place over seconds.

With the use of these techniques it is becoming increasingly apparent that during a specific task several different brain regions are working simultaneously. There is not just one brain area for one function but rather several brain areas appear to contribute to a particular function. Moreover, if some aspect of the task changes slightly, such as hearing words rather than speaking words, then a different constellation of brain regions appears.

Brain events are monitored over a time scale exceeding several seconds and averaged over, at best, a cubic millimetre of tissue. Another method, magnetoencephalography (MEG), measures the magnetic field generated by differential electrical activity of the brain and has a superior time resolution, but it is only currently accurate for the outer regions of the brain. Although their true potential lies in the future, when the space and time resolutions are more commensurate with the scale of real brain cells, techniques such as PET, MRI, and MEG are already offering windows to the brain at work. Perhaps the most obvious lesson they have taught us so far is that it is misleading to think of one brain region as having one specific, autonomous function, as in the phrenologists' scenario. Instead, different brain regions combine in some way to work in parallel for different functions.

Source: Susan Greenfield, *The Human Brain: A Guided Tour* (London: Weidenfeld & Nicolson, 1997), pp. 34–9.

THE BENDS

Decompression sickness, known to divers as 'the bends', occurs when dissolved nitrogen turns into bubbles inside the body due to a sudden decrease in ambient pressure – typically during a rapid ascent from an underwater dive. The condition can usually be prevented by incorporating a series of 'decompression stops' into a diver's ascent, during which he or she must remain at a fixed depth for several minutes in order to allow absorbed gases to be safely eliminated from the body. The following account of the phenomenon is taken from one of my favourite books, *Reefscape: Reflections on the Great Barrier Reef* (2001) by Rosaleen Love (b. 1940), an Australian science (and science-fiction) writer who took up scuba-diving at the age of fifty-eight, in emulation of one of her heroes, the nature writer Jean Devanny (1894–1962), whose underwater exploits were begun at a similar age: hence the title of the book's first chapter, 'Diving for Oldies':

Diving is a highly unnatural activity, and it can cause death or illness and disability. We have to take our air with us, if we are to stay underwater for more than a few minutes, and we also have difficulty in dealing with the considerable changes in pressure that occur. As we descend, the volume of air in the lungs decreases and nitrogen, a normal component of air, starts acting differently in the body under pressure. While the body makes use of the oxygen in air, so that we exhale less oxygen than we inhale, nitrogen is an inert gas. This means that at sea level, the amount of nitrogen inhaled and exhaled is the same. As the diver descends, both the pressure, and the amount of oxygen and nitrogen inhaled, are greater than on the surface. The body uses most of the extra oxygen inhaled, but the extra nitrogen, under pressure, enters the blood and the tissues of the body, where it dissolves under the increased pressure. The problems known as 'the bends' and 'decompression sickness' arise from this excess nitrogen, which will dissolve out as harmful bubbles in the blood and tissues if the diver ascends too rapidly. The problem known as 'nitrogen narcosis', or the 'raptures of the deep' arises, also from excess nitrogen, if the diver keeps descending, but it is a different thing.

If nitrogen bubbles form in the blood and the tissues from a too-rapid descent, they cause symptoms ranging from severe pain in the joints – 'the bends' – to dizziness, headache, paralysis, and vomiting.

The bends was the major cause of death in the pearl-shelling industry at the beginning of the twentieth century. Now it can be treated by recompression in a hyperbaric chamber, where the nitrogen bubbles are reabsorbed by the body under pressure, and subsequent decompression can take place very slowly. Recreational divers are taught to time their dives so they can 'stage' their ascent, with rests at the various levels, such as at five metres. With a slow, staged ascent, the excess nitrogen may be disposed of more gently by the body, without forming the dangerous bubbles.

Nitrogen narcosis is dangerous for different reasons. The diver who descends below the recreational diver's limit of eighteen metres may find that at about 35 metres a state of euphoria kicks in. Nitrogen crosses the blood-brain barrier and penetrates the lipids or fats of the brain cells. It induces what is called the 'martini effect', as it is like feeling suddenly drunk. Time ceases to matter. The normal survival mechanisms of worry about loss of control no longer work. The literature of diving is full of stories of divers at great depths who are observed to go deeper, and deeper, out of control but not knowing it. The 'raptures of the deep' have seized people who, in their rational moments on land, know about the dangers. Underwater, they may immerse themselves in the flow of the activity, too much, too deeply, and forever. If the diver realises what is happening, and pulls out of it, the 'martini effect' ceases dramatically at around the 35-metre level, unlike a real hangover from too much alcohol.

Canadian diving doctor David Sawatsky has researched the issues of diving and aging, and says bluntly that older divers have an increased risk of dying while diving. There is an increased risk of developing decompression sickness with age, as well as a bone condition known as dysbaric osteonecrosis. The older person is more likely to have increased body fat, impaired circulation, damaged blood vessels and damaged lung function, as well as joint degeneration, and these factors increase the risk of getting 'bent'. Older people are also in greater danger from both hypo- and hyperthermia (becoming too cold or too hot). Sawatsky also says the older diver should have the fitness level required to run 1.6 kilometres in eight minutes. Uh-oh to that one; I'm glad I read the article after the course was over.

Source: Rosaleen Love, *Reefscape: Reflections on the Great Barrier Reef* (Washington, D.C.: Joseph Henry Press, 2001), pp. 35–6.

ROUNDING ERRORS AND THE
BUTTERFLY EFFECT

This illuminating essay is from George G. Szpiro's mathematical miscellany, *The Secret Life of Numbers* (2006), one of the very few maths books that I have read from cover to cover. In it Szpiro, a Swiss mathematician turned journalist (he is currently the Middle East correspondent for the *Neue Zürcher Zeitung*, for which he also writes a monthly mathematics column), points to the serious consequences of apparently trivial errors, amplified, in this case, by the rounding function built in to computer software. As Szpiro notes, minute variations in initial values can lead to wildly anomalous results, with unforeseen impacts in a variety of situations, from missile strikes to election results. One such episode, in the early 1960s, led to the discovery of the so-called butterfly effect, when the American mathematician Edward Lorenz (1917–2008) found that small changes in complex, dynamic systems could have unexpectedly large effects elsewhere in the models. Lorenz's subsequent work, most notably his 1963 paper 'Deterministic Nonperiodic Flow', helped found a branch of applied mathematics that would be known as chaos theory:

Electronic calculators are precise and never, ever, make mistakes. This at least is what we would like to think. But, in fact, such errors occur all the time. It is just that we hardly ever notice. Take a pocket calculator, for example, which has buttons for 'square' and 'square root,' and follow this procedure: Press the number 10, then the square root button, and then the square button. As expected, the number 10 appears on the display screen, since the square number of the square root of 10 is, of course, 10. So far so good. Now try this: Press the number 10, then press the square root button 25 times, and follow up by pressing the square button 25 times. The result, one would expect, should again be 10, but the display shows something like the number 9.9923974. Ordinarily not much thought is given to this rather minor divergence of 0.07 percent. It is an error one can usually live with. But now repeat the experiment by pressing the square root button and the square button 33 times. The resulting number, 5.5732436 or something similar, no longer bears any resemblance to the real answer, which, of course, is 10.

The reason for this phenomenon, which occurs without fail in one way or another with each and every digital calculator, is the fact that a number can have an infinite number of decimals. An example is the fraction $1/3$. Expressing it in decimals results in an infinite number of threes after the decimal point. But – and it is a very big 'but' – calculators can only store a finite amount of numbers. As a general rule, numerical values are truncated after 15 digits by computers. Thus very small errors exist between the true numbers and the stored or displayed values.

In general, we just put up with these inaccuracies since it is not difficult to manage daily life with only two or three digits after the decimal point. There are times, though, when rounding errors can lead to catastrophes. On February 25, 1991, during the Persian Gulf War, an American Patriot missile battery in Dharan, Saudi Arabia, failed to intercept an incoming Iraqi Scud missile. The Scud struck an American army barracks and killed 28 soldiers. The cause for this tragic mishap was an inaccurate conversion of time, measured in tenths of seconds, to the binary values as they are stored in the computer. Specifically, elapsed time was measured by the system's internal clock in tenths of a second and stored in binary numbers. Then the result needed to be multiplied by 10 to produce the time in seconds. This calculation was performed using 24 bits. Hence the value 1/10, which has a nonterminating binary expansion, was truncated after 24 bits, resulting in a minute error.[*] This truncation error, when multiplied by the large number giving the time in tenths of a second, led to what was to be a fatal mistake.

On the evening of the election day of April 5, 1992, the Green Party in Germany's state of Schleswig-Holstein was elated. By a hair's breadth, the party had mastered the 5 percent threshold required for entry into the state parliament. The rude awakening came shortly after midnight. The true election results were published, and the Greens discovered to their dismay that they had actually received only 4.97 percent of the vote. The program that calculated the election results throughout the day had only listed one place after the decimal, and the count had been rounded to 5.0 percent. This particular piece of software had been used for years, but nobody had thought of turning off the rounding feature – if not to

---

[*] Bit is short for Binary digit, the latter being either 0 or 1.

say bug – at the crucial moment. The long and the short of it was that the Greens were unable to occupy any seat in parliament.

On June 4, 1996, an unmanned Ariane 5 rocket was launched off the island of Courou in French New Guinea, but it exploded just 40 seconds after liftoff. The rocket had veered off its flight path and had to be destroyed by ground control. Due to a software error, the guidance system had misinterpreted a rounded figure.

In 1982 the stock market in Vancouver introduced a new index and set the initial value at 1,000 points. After less than two years the index was down by nearly half, even though the average value of the stocks had increased by some 10 percent. The discrepancy was, again, due to rounding errors. While calculating the index, the weighted averages of the stock prices were truncated after too few decimal places.

In one particular instance, however, rounding errors led to a significant discovery. One day in the 1960s, Edward Lorenz, a meteorologist at the Massachusetts Institute of Technology, was busy observing weather simulations on his computer. After a while he felt that he needed a break. Lorenz stopped running the program and jotted down the intermediate results. After finishing his cup of coffee, Lorenz returned to his desk, fed the intermediate results back into the computer, and let the simulation run its course. To his surprise, the weather on his computer took a completely different turn to what he expected based on the previous simulations.

After brooding for a while over this puzzle, Lorenz realized what had happened. Before leaving for the coffee shop, he had copied down the numbers he saw on his computer screen. These numbers were displayed to three places behind the decimal point. But inside the computer, numbers were stored to eight decimal places. Lorenz realized that his computer program had been working with values that had been rounded. Since the weather simulation involved several nonlinear operations, it was not surprising that divergences cropped up rather quickly. Nonlinear expressions – that is, expressions like squaring or taking the square root – have the annoying characteristic of amplifying even minute mistakes very quickly.

Edward Lorenz's discovery set the foundation for so-called chaos theory, which today is a well-known concept. One of the consequences of this theory is the notorious butterfly effect. Basically it says that the movement of a butterfly's wings may unleash a hurricane at the other end of the world. Tiny vortices in the air, caused by a butterfly's flapping wings, may represent no more than a change in the 30th digit behind a decimal point. However, non-linearities in the weather could augment the tiny air movements a billionfold and thus escalate into a hurricane.

But there is another, less sinister, way of looking at things. By flapping its dainty wings, a butterfly could, by the same token, prevent a hurricane from arising. Mathematical models that make use of the reverse butterfly effect have found applications in, for example, cardiology. Minute electrical shocks, released at precisely the right moment, may correct a chaotic heartbeat and prevent a heart attack.

Source: George G. Szpiro, *The Secret Life of Numbers: 50 Easy Pieces on How Mathematicians Work and Think* (Washington, D.C.: Joseph Henry Press, 2006), pp. 102–5.

PULSARS

Pulsars – rapidly rotating neutron stars left over from long-ago supernovae – were discovered in 1967 by a team from Cambridge University that included a twenty-four-year-old PhD student named S. Jocelyn Bell (b. 1943), who was the first to notice the 'scruff' signals appearing on the chart recordings. The announcement of the discovery (with its suggestive talk of 'signals from another civilization') attracted a good deal of press attention, although the treatment of Bell, as described below, added a layer of sexist absurdity to the story. Worse was to come, however, when in 1974, Bell's PhD supervisor, Antony Hewish, shared, with his colleague Martin Ryle, the Nobel Prize in Physics for the discovery of pulsars; the Nobel rules allow for up to three individuals to share each prize, but for unexplained reasons Bell was not included, a scandal that outraged astronomers at the time, including Sir Fred Hoyle, who publicly condemned the conduct of the Royal Swedish Academy of Sciences. Bell, however, has never made an issue of it, and in this extract from an after-dinner speech given in New York in 1977, Jocelyn Bell Burnell (as she became known after her marriage) describes the build-up and aftermath of the momentous discovery, and ends with some thoughtful reflections on the ensuing Nobel controversy:

The story began in the mid-1960s, when the technique of interplanetary scintillation (IPS) was discovered. IPS is the apparent fluctuation in intensity of the radio emission from a compact radio source. It is due to diffraction of the radio waves as they pass through the turbulent solar wind in interplanetary space. Compact radio sources, e.g. quasars, scintillate more than extended radio sources. Professor Tony Hewish realized this technique would be a useful way of picking out quasars, and designed a large radio telescope to do this. I joined him as a Ph.D. student when construction of this telescope was about to start.

The telescope covered an area of $4\frac{1}{2}$ acres — an area that would accommodate 57 tennis courts. In this area we put up over a thousand posts, and strung more than 2,000 dipoles between them. The whole was connected up by 120 miles of wire and cable. We did the work ourselves — about five of us — with the help of several very keen vacation students who cheerfully sledge-hammered all one

summer. It took two years to build and cost about £15,000, which
was cheap even then. We started operating it in July 1967, although
it was several months more before the construction was completely
finished.

I had sole responsibility for operating the telescope and analyzing
the data, with supervision from Tony Hewish. We operated it with
four beams simultaneously, and scanned all the sky between declina-
tions +50° and –10° once every four days. The output appeared on
four 3-track pen recorders, and between them they produced 96 feet
of chart paper every day. The charts were analyzed by hand, by me.
We decided initially not to computerize the output because until we
were familiar with the behavior of our telescope and receivers we
thought it better to inspect the data visually, and because a human
can recognize signals of different character whereas it is difficult to
program a computer to do so.

After the first few hundred feet of chart analysis I could recognize
the scintillating sources, and I could recognize interference. (Radio
telescopes are very sensitive instruments, and it takes little radio
interference from nearby on earth to swamp the cosmic signals;
unfortunately, this is a feature of all radio astronomy.) Six or eight
weeks after starting the survey I became aware that on occasions
there was a bit of 'scruff' on the records, which did not look exactly
like a scintillating source, and yet did not look exactly like man-made
interference either. Furthermore I realized that this scruff had been
seen before on the same part of the records — from the same patch
of sky (right ascension 1919).

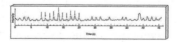

'Scruff'

The source was transiting during the night — a time when interplanetary scintillation should be at a minimum, and one idea we had
was that it was a point source. Whatever it was we decided that it
deserved closer inspection, and that this would involve making faster
chart recordings as it transited. Towards the end of October when we
had finished doing some special test on 3C273, and when we had at

last our full complement of receivers and recorders, I started going out to the observatory each day to make the fast recordings. They were useless. For weeks I recorded nothing but receiver noise. The 'source' had apparently gone. Then one day I skipped the observations to go to a lecture, and next day on my normal recording I saw the scruff had been there. A few days after that, at the end of November '67 I got it on the fast recording. As the chart flowed under the pen I could see that the signal was a series of pulses, and my suspicion that they were equally spaced was confirmed as soon as I got the chart off the recorder. They were 1¹/₃ seconds apart. I contacted Tony Hewish who was teaching in an undergraduate laboratory in Cambridge, and his first reaction was that they must be man-made. This was a very sensible response in the circumstances, but due to a truly remarkable depth of ignorance I did not see why they could not be from a star. However he was interested enough to come out to the observatory at transit-time the next day and fortunately (because pulsars rarely perform to order) the pulses appeared again.

This is where our problems really started. Tony checked back through the recordings and established that this thing, whatever it was, kept accurately to sidereal time. But pulses 1¹/₃ seconds apart seemed suspiciously man-made. Besides, 1¹/₃ seconds was far too fast a pulsation rate for anything as large as a star. It could not be anything earth-bound because it kept sidereal time (unless it was other astronomers). We considered and eliminated radar reflected off the moon into our telescope, satellites in peculiar orbits, and anomalous effects caused by a large, corrugated metal building just to the south of the 4¹/₂ acre telescope.

Then Scott and Collins observed the pulsations with another telescope with its own receivers, which eliminated instrumental effects. John Pilkington measured the dispersion of the signal, which established that the source was well outside the solar system but inside the galaxy. So were these pulsations man-made, but made by man from another civilization? If this were the case then the pulses should show Doppler shifts as the little green men on their planet orbited their sun. Tony Hewish started accurate measurements of the pulse period to investigate this; all they showed was that the earth was in orbital motion about the sun.

Meanwhile I was continuing with routine chart analysis, which was falling even further behind because of all the special pulsar observations. Just before Christmas I went to see Tony Hewish about something and walked into a high-level conference about how to present these results. We did not really believe that we had picked up signals from another civilization, but obviously the idea had crossed our minds and we had no proof that it was an entirely natural radio emission. It is an interesting problem — if one thinks one may have detected life elsewhere in the universe how does one announce the results responsibly? Who does one tell first? We did not solve the problem that afternoon, and I went home that evening very cross — here was I trying to get a Ph.D. out of a new technique, and some silly lot of little green men had to choose my aerial and my frequency to communicate with us. However, fortified by some supper I returned to the lab that evening to do some more chart analysis. Shortly before the lab closed for the night I was analyzing a recording of a completely different part of the sky, and in amongst a strong, heavily modulated signal from Cassiopeia A at lower culmination (at 1133) I thought I saw some scruff. I rapidly checked through previous recordings of that part of the sky, and on occasions there was scruff there. I had to get out of the lab before it locked for the night, knowing that the scruff would transit in the early hours of the morning.

So a few hours later I went out to the observatory. It was very cold, and something in our telescope-receiver system suffered drastic loss of gain in cold weather. Of course this was how it was! But by flicking switches, swearing at it, breathing on it I got it to work properly for 5 minutes — the right 5 minutes on the right beam setting. This scruff too then showed itself to be a series of pulses, this time 1.2 seconds apart. I left the recording on Tony's desk and went off, much happier, for Christmas. It was very unlikely that two lots of little green men would both choose the same, improbable frequency, and the same time, to try signalling to the same planet Earth.

Over Christmas Tony Hewish kindly kept the survey running for me, put fresh paper in the chart recorders, ink in the ink wells, and piled the charts, unanalyzed, on my desk. When I returned after the holiday I could not immediately find him, so settled down to do some chart analysis. Soon, on the one piece of chart, an hour or so apart in right ascension I saw two more lots of scruff, 0834 and 0950. It was another fortnight or so before 1133 was confirmed, and soon after that the third and fourth, 0834 and 0950 were also. Meanwhile

I had checked back through all my previous records (amounting to several miles) to see if there were any other bits of scruff that I had missed. This turned up a number of faintly possible candidates, but nothing as definite as the first four.

At the end of January the paper announcing the first pulsar was submitted to *Nature*. This was based on a total of only 3 hours' observation of the source, which was little enough. I feel that comments that we kept the discovery secret too long are wide of the mark. At about the same time I stopped making observations and handed over to the next generation of research students, so that I could concentrate on chart analysis, studying the scintillations and writing up my thesis.

A few days before the paper was published Tony Hewish gave a seminar in Cambridge to announce the results. Every astronomer in Cambridge, so it seemed, came to that seminar, and their interest and excitement gave me a first appreciation of the revolution we had started. Professor Hoyle was there and I remember his comments at the end. He started by saying that this was the first he had heard of these stars, and therefore he had not thought about it a lot, but that he thought these must be supernova remnants rather than white dwarfs. Considering the hydrodynamics and neutrino opacity calculations he must have done in his head, that is a remarkable observation!

In the paper to *Nature* we mentioned that at one stage we had thought the signals might be from another civilization. When the paper was published the press descended, and when they discovered a woman was involved they descended even faster. I had my photograph taken standing on a bank, sitting on a bank, standing on a bank examining bogus records, sitting on a bank examining bogus records: one of them even had me running down the bank waving my arms in the air. Look happy dear, you've just made a Discovery! (Archimedes doesn't know what he missed!) Meanwhile the journalists were asking relevant questions like was I taller than or not quite as tall as Princess Margaret (we have quaint units of measurement in Britain) and how many boyfriends did I have at a time?

That was how my part in the proceedings ended. I finally finished the chart analysis, measured the angular diameters of a number of radio sources, and wrote my thesis. (The pulsars went in an appendix.) Then I moved out of the field to another part of the country, to get married. It has been suggested that I should have had a part in the Nobel Prize awarded to Tony Hewish for the discovery of pulsars. There are several comments that I would like to make on this: First, demarcation disputes between supervisor and student are always difficult, probably impossible to resolve. Secondly, it is the supervisor who has the final responsibility for the success or failure of the project. We hear of cases where a supervisor blames his student for a failure, but we know that it is largely the fault of the supervisor. It seems only fair to me that he should benefit from the successes, too. Thirdly, I believe it would demean Nobel Prizes if they were awarded to research students, except in very exceptional cases, and I do not believe this is one of them. Finally, I am not myself upset about it — after all, I am in good company, am I not!

Source: S. Jocelyn Bell Burnell, 'Petit Four', *Annals of the New York Academy of Sciences* 302 (1977), pp. 685–9.

THE FIRST THREE MINUTES

This Olympian overview of the birth of our universe by Nobel Prize-winning physicist Steven Weinberg (b. 1933) describes the first appearance of the basic building blocks of matter, the 'zoo' of subatomic particles whose inter-actions with the four fundamental forces of nature – gravity, the strong and weak nuclear forces, and electromagnetism – have occupied the attentions of theoretical physicists since the 1930s.

By the 1970s an entire new picture of these fundamental interactions had emerged; known as the Standard Model, it seeks to explain the history, behaviour and interaction of most forms of matter and energy, while organizing them into a mathematically based – and hauntingly worded – classification (for more on this see p. 459). As one of the architects of the Standard Model, Weinberg is well placed to guide us through its cosmolog-ical implications, which he does with a rare poetic skill that is exemplified by the last line of the book from which this extract has been taken: 'The effort to understand the universe is one of the very few things that lifts human life a little above the level of farce, and gives it some of the grace of tragedy':

In the beginning, there was an explosion. Not an explosion like those familiar on earth, starting from a definite center and spreading out to engulf more and more of the circumambient air, but an explosion which occurred simultaneously everywhere, filling all space from the beginning, with every particle of matter rushing apart from every other particle. 'All space' in this context may mean either all of an infinite universe, or all of a finite universe which curves back on itself like the surface of a sphere. Neither possibility is easy to compre-hend, but this will not get in our way; it matters hardly at all in the early universe whether space is finite or infinite.

At about one-hundredth of a second, the earliest time about which we can speak with any confidence, the temperature of the universe was about a hundred thousand million (10^{11}) degrees Centigrade. This is much hotter than in the center of even the hottest star, so hot, in fact, that none of the components of ordinary matter, molecules, or atoms, or even the nuclei of atoms, could have held together. Instead, the matter rushing apart in this explosion consisted of various types of the so-called elementary particles, which are the subject of modern high-energy nuclear physics . . .

One type of particle that was present in large numbers is the electron, the negatively charged particle that flows through wires in electric currents and makes up the outer parts of all atoms and molecules in the present universe. Another type of particle that was abundant at early times is the positron, a positively charged particle with precisely the same mass as the electron. In the present universe positrons are found only in high-energy laboratories, in some kinds of radioactivity, and in violent astronomical phenomena like cosmic rays and supernovas, but in the early universe the number of positrons was almost exactly equal to the number of electrons. In addition to electrons and positrons, there were roughly similar numbers of various kinds of neutrinos, ghostly particles with no mass or electric charge whatever. Finally, the universe was filled with light. This does not have to be treated separately from the particles – the quantum theory tells us that light consists of particles of zero mass and zero electrical charge known as photons. (Each time an atom in the filament of a light bulb changes from a state of higher energy to one of lower energy, one photon is emitted. There are so many photons coming out of a light bulb that they seem to blend together in a continuous stream of light, but a photoelectric cell can count individual photons, one by one.) Every photon carries a definite amount of energy and momentum depending on the wavelength of the light. To describe the light that filled the early universe, we can say that the number and the average energy of the photons was about the same as for electrons or positrons or neutrinos.

These particles – electrons, positrons, neutrinos, photons – were continually being created out of pure energy, and then after short lives being annihilated again. Their number therefore was not preordained, but fixed instead by a balance between processes of creation and annihilation. From this balance we can infer that the density of this cosmic soup at a temperature of a hundred thousand million degrees was about four thousand million (4×10^9) times that of water. There was also a small contamination of heavier particles, protons and neutrons, which in the present world form the constituents of atomic nuclei. (Protons are positively charged; neutrons are slightly heavier and electrically neutral.) The proportions were roughly one proton and one neutron for every thousand million electrons or positrons or neutrinos or photons. This number – a thousand million photons per nuclear particle – is the crucial quantity that had to be taken from observation in order to work out the standard model of the universe . . .

As the explosion continued the temperature dropped, reaching thirty thousand million (3×10^{10}) degrees Centigrade after about one-tenth of a second; ten thousand million degrees after about one second; and three thousand million degrees after about fourteen seconds. This was cool enough so that the electrons and positrons began to annihilate faster than they could be recreated out of the photons and neutrinos. The energy released in this annihilation of matter temporarily slowed the rate at which the universe cooled, but the temperature continued to drop, finally reaching one thousand million at the end of the first three minutes. It was then cool enough for the protons and neutrons to begin to form into complex nuclei, starting with the nucleus of heavy hydrogen (or deuterium), which consists of one proton and one neutron. The density was still high enough (a little less than that of water) so that these light nuclei were able rapidly to assemble themselves into the most stable light nucleus, that of helium, consisting of two protons and two neutrons.

At the end of the first three minutes the contents of the universe were mostly in the form of light, neutrinos, and antineutrinos. There was still a small amount of nuclear material, now consisting of about 73 percent hydrogen and 27 percent helium, and an equally small number of electrons left over from the era of electron-positron annihilation. This matter continued to rush apart, becoming steadily cooler and less dense. Much later, after a few hundred thousand years, it would become cool enough for electrons to join with nuclei to form atoms of hydrogen and helium. The resulting gas would begin under the influence of gravitation to form clumps, which would ultimately condense to form the galaxies and stars of the present universe. However, the ingredients with which the stars would begin their life would be just those prepared in the first three minutes.

Source: Steven Weinberg, *The First Three Minutes: A Modern View of the Origin of the Universe*, 2nd edn (New York: Basic Books, 1993), pp. 5–8.

BLACK HOLES

The term 'black hole' was first used in print by the science journalist Anne Ewing (1921–2010), who overheard it during a meeting of the American Association for the Advancement of Science in 1964. It describes a region of distorted space-time so dense that nothing, not even light, can escape its gravitational pull. Although the existence of black holes (or 'dark stars') was hypothesized more than two hundred years ago by the brilliant John Michell, who we met earlier in connection with seismic waves (p. 157), it was not until the early 1970s that the first positive identification of a black hole was made. The object in question, Cygnus X-1, famously cost Stephen Hawking (b. 1942) the price of a year's subscription to *Penthouse* magazine (he lost a bet with a colleague at Caltech that the object would turn out *not* to be a black hole). As is apparent from the following account of black hole formation, taken from a scientific biography of Hawking, there is a strangely alluring vocabulary associated with the science of black holes: 'singularity'; 'event horizon'; 'the no-hair theorem'; and, of course, 'spaghettification', the vertical stretching and compression of objects in inescapably strong gravitational fields (though sadly the word does not appear here):

By the 1930s, physicists knew that the nucleus of an atom is made of closely packed particles called protons and neutrons. The protons each carry one unit of positive charge; the neutrons, as their name suggests, are electrically neutral, but each has about the same mass as a proton. In everyday atoms, like the ones this book is made of, each nucleus is surrounded by a cloud of electrons. Each electron carries one unit of negative charge, and there is the same number of electrons as protons, so the atom as a whole is electrically neutral.

But an atom is largely empty space. The nucleus is tiny but very dense, and the cloud of electrons is (by comparison) huge and insubstantial. In proportion to the size of a whole atom, the nucleus is like a grain of sand in the middle of a concert hall. In white dwarf stars, some of the electrons are knocked off their atoms by the high prevailing pressure, and the nuclei are embedded in a sea of electrons that belong to the whole star, not to any particular nucleus. But there is still a lot of space between the nuclei, even though that space contains electrons. Each nucleus has positive charge, and like charges repel, so the nuclei keep their distance from each other.

But quantum theory said that there is a way to make a star denser than a white dwarf. If the star were squeezed even more by gravity, the electrons could be forced to combine with protons to make more neutrons. The result would be a star made entirely of neutrons, and these could be packed together as closely as the protons and neutrons in an atomic nucleus. This would be a neutron star.

Calculations suggested that this ought to happen for any dead star with a mass more than 20 percent larger than that of our Sun (that is, more than 1.2 solar masses). A neutron star would have that much mass packed within a radius of about 10 kilometers, no bigger than many mountains on Earth. The density of the matter in a neutron star, in grams per cubic centimeter, would be 10^{14} – that is, 1 followed by 14 zeros, or one hundred thousand billion. Even an object this dense would not be a black hole, though, for light could still escape from its surface into the Universe at large.

Making a black hole from a dead star would require, as the theorists of the early 1960s were well aware, crushing even neutrons out of existence. The quantum equations said, in fact, that there was no way that even neutrons could hold up the weight of a dead star of 3 solar masses or more and that, if any such object were left over from the explosive death throes of a massive star, it would collapse inward completely, shrinking to a mathematical point called a singularity. Long before the collapsing star could reach this state of zero volume and infinite density, it would have wrapped space-time around itself, cutting off the collapsar from the outside Universe.

Indeed, the equations said that if you squeezed any collection of matter hard enough it would collapse in this way. The special feature of objects more than 3 solar masses is that they will collapse anyway, under their own weight. But if it were possible to squeeze our own Sun down into a sphere with a radius of about 3 kilometers, it would become a black hole. So would the Earth, if it were squeezed down to about a centimeter. In each case, once the object had been squeezed down to the critical size, gravity would take over, closing space-time around the object while it continued to shrink away into the infinite density singularity inside the black hole. But notice that it is much easier to make a black hole if you have a lot of mass. The critical size is not simply proportional to the amount of mass you have; the density at which a black hole forms is larger if you have less mass to squeeze.

For any mass there is a critical radius, called the Schwarzchild radius, at which this will occur. As these examples indicate, the Schwarzchild radius is smaller for less massive objects – you have to squeeze the Earth harder than the Sun, and the Sun harder than a more massive star, in order to make a black hole. Once it had formed, there would be a surface around the hole (a bit like the surface of the sea) marking the boundary between the Universe at large and the region of highly distorted space-time from which nothing could escape. It would be a one-way horizon (unlike the surface of the sea!) across which both radiation and material particles could happily travel inward, tugged by gravity to join the accumulating mass of the singularity, but across which nothing at all, not even light, could travel.

Source: Michael White and John Gribbin, *Stephen Hawking: A Life in Science*, new edn (Washington, D.C.: Joseph Henry Press, 2002), pp. 76–8.

CLIMATE CHANGE

Over the past twenty years, global climate change has emerged as the over-arching narrative of our age, uniting a series of concerns about humanity's place in the natural world, the relative responsibilities of first-world nations, and the obligations owed by present populations to the generations to come. But if the climate-change story started out as a data-driven scientific concept – an hypothesis to be tested like any other – it was soon transformed into a kind of secular prophecy, a grand narrative freighted with powerful, even transcendent, languages and values that influence the terms of the current debate just as much as any scientific content. The following pair of extracts offers a historical snapshot of how the climate-change story developed from its nineteenth-century origins, and features the moment when international attention was first attracted to the subject.

The Greenhouse Effect

In December 1895, the Swedish chemist Svante Arrhenius (1859–1927) read a paper at the Royal Swedish Academy of Sciences, entitled 'On the Influence of Carbonic Acid in the Air upon the Temperature of the Ground'. This now famous paper outlined the likely impact that increased concentrations of atmospheric carbon dioxide would have on the surface temperature of the earth ('carbonic acid' in this context refers to a solution of carbon dioxide in water). Arrhenius calculated that the hypothetical removal of all atmospheric CO_2 would cause the earth's temperature to drop by at least 20–30 °C, and, conversely, that the doubling of atmospheric CO_2 (from its 1890s concentrations of c. 300 parts per million) would cause average global temperatures to rise by around 5 °C, with the greatest increase being seen at the poles.

A few years later Arrhenius expanded on these ideas in his *Worlds in the Making* (1906), a widely translated work of popular science in which he speculated about the likely benefits of an anthropogenically warming world:

The actual percentage of carbonic acid in the air is so insignificant that the annual combustion of coal, which has now (1904) risen to about 900 million tons and is rapidly increasing, carries about one-seven-hundredth part of its percentage of carbon dioxide to the atmosphere. Although the sea, by absorbing carbonic acid, acts as a regulator of huge capacity, which takes up about five-sixths of the produced carbonic acid, we yet recognize that the slight percentage of carbonic acid in the atmosphere may by the advances of industry be changed to a noticeable degree in the course of a few centuries. That would imply that there is no real stability in the percentage of carbon dioxide in the air, which is probably subject to considerable fluctuations in the course of time.

Volcanism is the natural process by which the greatest amount of carbonic acid is supplied to the air. Large quantities of gases originating in the interior of the earth are ejected through the craters of the volcanoes. These gases consist mostly of steam and of carbon dioxide, which have been liberated during the slow cooling of the silicates in the interior of the earth. The volcanic phenomena have been of very unequal intensity in the different phases of the history of the earth, and we have reason to surmise that the percentage of carbon dioxide in the air was considerably greater during periods of strong volcanic activity than it is now, and smaller in quieter periods. Professor Frech, of Breslau, has attempted to demonstrate that this would be in accordance with geological experience, because strongly volcanic periods are distinguished by warm climates, and periods of feeble volcanic intensity by cold climates. The ice age in particular was characterized by a nearly complete cessation of volcanism, and the two periods at the commencement and in the middle of the Tertiary age (Eocene and Miocene) which showed high temperatures were also marked by an extraordinarily developed volcanic activity. This parallelism can be traced back into more remote epochs . . .

Since, now, warm ages have alternated with glacial periods, even after man appeared on the earth, we have to ask ourselves: Is it probable that we shall in the coming geological ages be visited by a new ice period that will drive us from our temperate countries into the hotter climates of Africa? There does not appear to be much ground for such an apprehension. The enormous combustion of coal by our industrial establishments suffices to increase the percentage of carbon dioxide in the air to a perceptible degree. Volcanism, whose devastations – on Krakatoa (1883) and Martinique (1902) – have been terrible in late years, appears to be growing more intense. It is probable, therefore, that the percentage of carbonic acid increases at a rapid rate. Another circumstance points in the same direction; that is, that the sea seems to withdraw carbonic acid from the air. For the carbonic acid percentage above the sea and on islands is on an average 10 per cent less than above the continents.

If the carbonic acid percentage of the air had kept constant for ages, the percentage of the water would have found time to get into equilibrium with it; but the sea actually absorbs carbonic acid from the air. Thus the sea-water must have been in equilibrium with an atmosphere which contained less carbonic acid than the present atmosphere. Hence the carbonic acid percentage has been increasing of late.

We often hear lamentations that the coal stored up in the earth is wasted by the present generation without any thought for the future, and we are terrified by the awful destruction of life and property which has followed the volcanic eruptions of our days. We may find a kind of consolation in the consideration that here, as in every other case, there is good mixed with the evil. By the influence of the increasing percentage of carbonic acid in the atmosphere, we may hope to enjoy ages with more equable and better climates, especially as regards the colder regions of the earth, ages when the earth will bring forth much more abundant crops than at present, for the benefit of rapidly propagating mankind.

'It is changing our climate now'

Complacency about the effects of increased atmospheric CO_2 dwindled over the course of the twentieth century, to be replaced by a slowly growing unease. In 1975 the oceanographer Wallace S. Broecker (b. 1931) published a paper in the journal *Science*, in which he pointed out that, although the northern hemisphere was then experiencing a cooling phase, it was already apparent that the trend was about to go into rapid reverse, and that, as he put it, 'we may be in for a climatic surprise'. Broecker's paper, entitled 'Climatic Change: Are We on the Brink of a Pronounced Global Warming?', is notable for having popularized the term 'global warming', as well as heralding a now-familiar note of climatic alarm.

A decade later, scientists were beginning to worry about the climate in earnest, and James E. Hansen's deposition to the United States Congress on 23 June 1988 proved to be a watershed in science communication, marking the moment when anxiety over climate change entered the public realm: 'Global Warming Has Begun' was the next morning's headline in the *New York Times*, and Hansen (b. 1941), Director of NASA's Goddard Institute for Space Studies, became the first in a long line of climate-change celebrities. Media interest was helped by the fact that the United States was then experiencing a severe heat wave: it was 98 °F (36.7 °C) on the day of Hansen's appearance, and even the delegates in the air-conditioned conference room were finding it hard to keep cool. Note, by the way, the unavailability of PowerPoint at the time – Hansen has to ask for each of his slides to be put up by a technician. Apart from that minor historical intrusion, everything else that Hansen said remains a familiar fixture of the story we live with today:

Mr. Chairman and committee members, thank you for the opportunity to present the results of my research on the greenhouse effect which has been carried out with my colleagues at the NASA Goddard Institute for Space Studies.

I would like to draw three main conclusions. Number one, the earth is warmer in 1988 than at any time in the history of instrumental measurements. Number two, the global warming is now large enough that we can ascribe with a high degree of confidence a cause and effect relationship with the greenhouse effect. And number three, our computer climate simulations indicate that the greenhouse effect is already large enough to begin to effect the probability of extreme events such as summer heat waves.

My first viewgraph, which I would like to ask Suki to put up if he would, shows the global temperature over the period of instrumental records which is about 100 years. The present temperature is the highest in the period of record. The rate of warming in the past 25 years, as you can see on the right, is the highest on record. The four warmest years, as the Senator mentioned, have all been in the 1980s. And 1988 so far is so much warmer than 1987, that barring a remarkable and improbable cooling, 1988 will be the warmest year on the record.

Now let me turn to my second point which is causal association of the greenhouse effect and the global warming. Causal association requires first that the warming be larger than natural climate variability and, second, that the magnitude and nature of the warming be consistent with the greenhouse mechanism. These points are both addressed on my second viewgraph. The observed warming during

the past 30 years, which is the period when we have accurate measurements of atmospheric composition, is shown by the heavy black
line in this graph. The warming is almost 0.4 degrees Centigrade by
1987 relative to climatology, which is defined as the 30 year mean,
1950 to 1980 and, in fact, the warming is more than 0.4 degrees
Centigrade in 1988. The probability of a chance warming of that
magnitude is about 1 percent. So, with 99 percent confidence we can
state that the warming during this time period is a real warming
trend.

The other curves in this figure are the results of global climate
model calculations for three scenarios of atmospheric trace gas
growth. We have considered several scenarios because there are
uncertainties in the exact trace gas growth in the past and especially
in the future. We have considered cases ranging from business as
usual, which is scenario A, to draconian emission cuts, scenario C,
which would totally eliminate net trace gas growth by year 2000.

The main point to be made here is that the expected global warming is of the same magnitude as the observed warming. Since there is
only a 1 percent chance of an accidental warming of this magnitude,
the agreement with the expected greenhouse effect is of considerable
significance. Moreover, if you look at the next level of detail in the
global temperature change, there are clear signs of the greenhouse
effect. Observational data suggests a cooling in the stratosphere
while the ground is warming. The data suggest somewhat more
warming over land and sea ice regions than over open ocean, more
warming at high latitudes than at low latitudes, and more warming
in the winter than in the summer. In all of these cases, the signal is at
best just beginning to emerge, and we need more data. Some of these
details, such as the northern hemisphere high latitude temperature
trends, do not look exactly like the greenhouse effect, but that is
expected. There are certainly other climate change factors involved
in addition to the greenhouse effect.

Altogether the evidence is that the earth is warming by an
amount which is too large to be a chance fluctuation and the similarity of the warming to that expected from the greenhouse effect
represents a very strong case. In my opinion, the greenhouse effect
has been detected, and it is changing our climate now.

Sources: Svante Arrhenius, *Worlds in the Making: The Evolution of the Universe*, trans. H. Borns (London and New York: Harpers, 1908), pp. 54–63; James E. Hansen, 'Statement of Dr. James Hansen, Director, NASA Goddard Institute for Space Studies', *The Hearing of the Committee on Energy and Natural Resources, U.S. Senate, 100th Congress, June 23, 1988* (Washington, D.C.: U.S. Government Printing Office, 1988), II, pp. 39–41.

FUZZY SETS AND BEAUTY MAPS

Fuzzy set theory was introduced in 1965 by the Azerbaijan-born mathematician Lotfi Zadeh (b. 1921) as a means of incorporating the imprecision of human thought processes in a systematic algorithmic framework. Fuzzy-ism rests on the notion that 'the key elements in human thinking and human decision-making are based not on numbers but on fuzzy sets – classes of objects in which transition from membership to nonmembership is gradual rather than abrupt', as Zadeh explained. He later claimed that the more we learn about human cognition, the more we come to appreciate that our ability to manipulate fuzzy concepts is an asset rather than a liability, and it is this ability, above all, that constitutes a key to understanding the nature of human intelligence.

A particularly good example of a fuzzy set is a subjective concept such as love or beauty, for which there are no units or numerical scales (though has Zadeh not heard of the 'millihelen', the quantum of beauty required to launch one ship?). The following extract is from a paper published in 1976, by which time Zadeh's original concept had evolved and migrated beyond set theory to embrace fuzzy logic, fuzzy systems and fuzzy linguistics:

The high standards of precision which prevail in mathematics, physics, chemistry, engineering and other 'hard' sciences stand in sharp contrast to the imprecision which pervades much of sociology, psychology, political science, history, philosophy, linguistics, anthropology, literature, art and related fields. This marked difference in the standards of precision is due, of course, to the fact that the 'hard' sciences are concerned in the main with the relatively simple mechanistic systems whose behavior can be described in quantitative terms, whereas the 'soft' sciences deal primarily with the much more complex non-mechanistic systems in which human judgement, perception and emotions play the dominant role.

Although the conventional mathematical techniques have been and will continue to be applied to the analysis of humanistic* systems, it

---

* By a humanistic system we mean a non-mechanistic system in which human behavior plays a major role. Examples of humanistic systems are political systems, economic systems, social systems, religious systems, etc. A single individual and his thought processes may also be viewed as a humanistic system.

is clear that the great complexity of such systems calls for approaches that are significantly different in spirit as well as in substance from the traditional methods – methods which are highly effective when applied to mechanistic systems, but are far too precise in relation to systems in which human behavior plays an important role.

In the *linguistic approach* which represents one such departure from conventional methods – words or sentences are used in place of numbers to describe phenomena which are too complex or too ill-defined to be susceptible of characterization in quantitative terms. For example, if the probability of an event is not known with precision, then it may be characterized linguistically as, say, *quite likely, not very unlikely, highly unlikely*, etc., where *quite likely, not very unlikely* and *highly unlikely* are interpreted as labels of fuzzy subsets of the unit interval. Such subsets may be likened to ball-parks without sharply defined intervals which serve to provide an approximate rather than exact characterization of the value of a variable.

The use of the linguistic approach in the case of humanistic systems is dictated by the fact that as the complexity of a system increases, our ability to make precise and yet significant statements about its behavior diminishes until a threshold is reached beyond which complexity, precision and significance can no longer coexist. The essence of the linguistic approach, then, is that it sacrifices precision to gain significance, thereby making it possible to analyze in an approximate manner those humanistic as well as mechanistic systems which are too complex for the application of classical techniques.

A key feature of the linguistic approach has to do with its use of the notion of a *primary fuzzy set* as a substitute for the basic notion of a unit of measurement. More specifically, much of the power of mathematical techniques for dealing with mechanistic systems derives from the existence of a set of units for such basic parameters as length, area, weight, force, current, heat, etc. In general, such units do not exist in the case of humanistic systems, and it is this fact that contributes significantly to the difficulty of analyzing humanistic systems through the use of techniques which depend so essentially on the existence of units of measurement.

In the linguistic approach, a role comparable to that of a unit of measurement is played by one or more primary fuzzy sets from which other sets can be generated through the use of linguistic modifiers such as *very, quite, more or less, extremely, essentially, completely*, etc. To illustrate, consider a property, say *beautiful*, for which we have neither a unit nor a numerical scale. The meaning of this property may be defined via *exemplification* by associating with each member, *u*, of a subset of objects in a given universe of discourse, U, the grade of membership of *u* in the fuzzy subset labeled *beautiful*. For example, the grade of membership of Fay in the class of beautiful women might be 0·9, that of Gillian 0·85, of Helen 0·8, etc. This set of women, then, would constitute a primary fuzzy set which serves as a reference for defining the meaning of *very beautiful, quite beautiful, more or less beautiful, extremely beautiful*, etc. as fuzzy subsets of U. Thus, in terms of these subsets, an assertion of the form '*Nora is very beautiful*', may be interpreted as the assignment of a linguistic rather than a numerical value to the beauty of Nora. In this way, the linguistic values *beautiful, very beautiful, quite beautiful*, etc. which are generated from the primary fuzzy set *beautiful*, play a role which is roughly similar to that of the multiples of a unit of measurement, when such a unit exists.

Our main purpose in the present paper is to apply the linguistic approach to the definition of concepts which are too complex or too imprecise to be susceptible of exact definition. In general, such concepts are fuzzy in the sense that they correspond to classes of objects or constructs which do not have sharply defined boundaries. For example, the concepts of *oval, in love, young* and *masculine* are fuzzy whereas those of *straight line, married, brother* and *male* are not. Note that *oval* is a more complex concept than *straight line, in love* is more complex than *married, friend* is more complex than *brother*, and *masculine* is more complex than *male*. Indeed, most complex concepts tend to be fuzzy, and it is in this sense that fuzziness may be regarded as a concomitant of complexity.

Emerging at around the same time as fuzziness, fractal geometry was also concerned with the mathematical representation of intuitive forms of complexity. The field is most closely associated with the French mathematician Benoît Mandelbrot (1924–2010), who coined the term 'fractal' in 1975 in order to categorize amorphous or irregular geometric shapes such as clouds, rivers, coastlines, or trees, which when divided tend to form a reduced-size copy of the former whole. One of the most celebrated examples of such

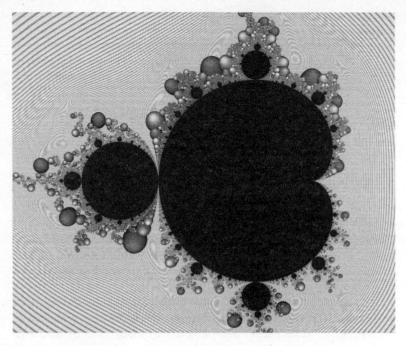

23. This computer-generated fractal, derived from a Mandelbrot set, displays the repeating self-similarity associated with complex geometric forms in nature, such as coastlines, rivers and trees.

fractal self-similarity reveals itself whenever an attempt is made to measure the coastline of an island such as Britain or Ireland. Cartographers have long been aware that the shape of bays and inlets remains similarly jagged no matter how close one's view-point zooms in: so at what point on the scale does one decide to stop measuring? The microscopic? Coastlines, it transpires, are potentially infinite in length, and they approximate more closely to a fuzzy set than to a precise or well-defined measurement.

Mandelbrot's term fractal, which derives from the Latin for 'break' or 'fracture', has also proved linguistically useful to mathematicians, particularly within chaos theory, for as Mandelbrot pointed out in the new word's defence, 'scientists will (I am sure) be surprised and delighted to find that not a few shapes they had to call grainy, hydralike, in between, pimply, pocky, ramified, seaweedy, strange, tangled, tortuous, wiggly, wispy, wrinkled, and the like, can henceforth be approached in rigorous and vigorous quantitative fashion'.

The Beauty Map

The quantification of beauty might seem to be the ultimate collision between the mathematical sciences and the world of subjective personal judgement; perhaps this is why scientists have returned to it so often. The Victorian polymath Francis Galton (1822–1911), who we met earlier in connection with the measurement of boredom (see p. 290), published his memoirs in 1908, and in them he described his voyeuristic attempts to quantify the attractiveness of women he passed in the street:

I may here speak of some attempts by myself, made hitherto in too desultory a way, to obtain materials for a 'Beauty-Map' of the British Isles. Whenever I have occasion to classify the persons I meet into three classes, 'good, medium, bad,' I use a needle mounted as a pricker, wherewith to prick holes, unseen, in a piece of paper, torn rudely into a cross with a long leg. I use its upper end for 'good,' the cross-arm for 'medium,' the lower end for 'bad.' The prick-holes keep distinct, and are easily read off at leisure. The object, place, and date are written on the paper. I used this plan for my beauty data, classifying the girls I passed in streets or elsewhere as attractive, indifferent, or repellent. Of course this was a purely individual estimate, but it was consistent, judging from the conformity of different attempts in the same population. I found London to rank highest for beauty; Aberdeen lowest.

Sources: Lotfi A. Zadeh, 'A Fuzzy-Algorithmic Approach to the Definition of Complex or Imprecise Concepts', *International Journal of Man-Machine Studies* 8 (1976), pp. 249–91; Francis Galton, *Memories of My Life* (London: Methuen, 1908), pp. 315–16.

THE HUMAN GENOME

The Human Genome Project (HGP) was launched in 1990 with the aim of identifying every gene stored in human chromosomes, as well as sequencing the billions of chemical base pairs that make up our DNA. The project involved several hundred scientists and volunteers across three continents at a cost of nearly $3 billion. In June 2000, a 'working draft' of the sequence was published, and three years later, during the fiftieth anniversary of the discovery of the structure of DNA, the complete first draft of the human genome was finally published. It was, in the words of the Nobel laureate Sir John Sulston, 'an extraordinary enterprise, one of the notable achievements of late twentieth-century science'. Sulston (b. 1942) was Director of the Sanger Institute in Cambridge, one of the research laboratories most closely involved in the Genome Project, and has also been one of the most vocal defenders of the consortium's principle that their collective results – 'the heritage of humanity' – should be made freely available to the world. (In 1998 a rival organization, Celera Genomics, headed up by the biologist and entrepreneur J. Craig Venter (b. 1946), had entered the race to sequence the human genome, much to the disquiet of the HGP consortium, who feared that this would lead to commercial exploitation of human genetic data.)

In this extract from *The Common Thread* (2002), a first-hand account of the genome project that Sulston co-wrote with the award-winning science writer Georgina Ferry (b. 1955), the team has just made the surprising discovery that there are only between 20,000 and 25,000 genes in the human array, far fewer than the 100,000-plus genes that most biologists had been expecting:

At first sight, the conclusion that it takes only twice as many genes to make a human as it does to make a tiny worm or a fly seems to lead very naturally to the conclusion that the genes are fairly boring and don't really have a lot to do with the essence of being a human. After all, runs the argument, a human is obviously so much more complex than a fly that twice as many genes just won't be enough. When I hear this argument I tend to hear also a subtext to the effect that humans, and the speaker in particular, are so much more important than flies that twice as many genes won't do.

What tends to be forgotten is management. Many of the extra

genes that are added in going from worm or fly to human appear to be control genes, and they come in hierarchies which are only just beginning to be worked out. So in principle, by elaborating the control mechanisms, a huge range of tissue types can be specified and a very complex structure can be built up. It's a bit like the expansion of an organization: although some of us wish it wasn't so, an essential part of building up a large organization is the introduction of more complex management structures and the employment of more executives. The control genes are the executives of biological development, and they allow complex and diverse structures to be built from units that are fundamentally quite similar. Many control genes operate by switching groups of other genes on or off; in addition, a single gene typically gives rise to a variety of products. One gene can have two or more differently spliced RNA transcripts, acting as templates for different proteins, and enzymes can further modify the protein after it has been synthesized; all of these processes offer further opportunities for control.

But the most remarkable thing is the power of genes in combination. Consider a gene that can exist in two variants, or alleles: A and B. This single gene will then allow us to specify two cell types, A and B. Now add a second gene, that can exist as C or D. Together the two genes will allow us to specify four cell types: AC, AD, BC and BD. Three genes can specify eight types, four, sixteen types, five, thirty-two types; ten genes, over 1,000 types, twenty genes, over a million. The conclusion is that just a few dozen genes, if applied in a hierarchical executive fashion can provide an immense amount of additional complexity. So the addition of an extra 15,000 over the worm's allowance allows plenty of room for manoeuvre in the construction of a human being. In real life things will not be as neat as this simple discussion suggests, but by thinking in this way we can avoid being trapped by falsely limiting assumptions.

By the same line of argument, we can tackle the concern that an extra 15,000 genes are too few to explain the range of human inheritance. Indeed they would be too few if each gene were solely responsible for the specification of one recognizable human characteristic. But we have long known that this is the exception rather than the rule. In particular, many of the subtle human attributes about which we care most – intelligence, athleticism, beauty, wisdom, musicality and so on – are clearly not heritable in the same way as hair or eye colour, for example, leading some to the conclusion that they are not heritable at all.

But think again about the power of different alleles combining in different ways. Extending our range to thirty-three genes allows us to generate over 8 billion different types, enough to give every living person a unique label. Three hundred genes will provide as many different types as there are particles in the universe or seconds since time began – vastly more than will ever be needed uniquely to identify every person that has ever lived or will live. And this is on the parsimonious assumption that each gene comes in only two forms, whereas in fact there are numerous alleles of every gene. No wonder identical twins are so special: we shall never see two identical human genomes by chance, but only by the splitting of the fertilized ovum.

Taken together, combinations and hierarchical control allow us to see in principle how both the complexity and the diversity of humans can be specified by a relatively small number of genes. All this should give us optimism in moving forward to find out exactly how it all works; but it should also caution us against jumping to quick conclusions. The complexity of control, overlaid by the unique experience of each individual, means that we must continue to treat every human as unique and special, and not imagine that we can predict the course of a human life other than in broad statistical terms.

The genes are the starting point for a human being, and we should think of them as offering potentials rather than exercising constraints. Many fear that genetic information about individuals will be used to discriminate against them, and this is a concern that has to be taken seriously. Insurers are pressing to be allowed to use the results of genetic tests taken by their clients in deciding whether or not to issue policies; in future both insurers and employers, if the law permitted, might make genetic testing a condition of issuing a policy or offering a job. It is immensely important that we do not make presumptions about a person's health or ability on the basis of their genotype, but rather look to see what they can actually achieve. It is a matter of fundamental human rights: rights that are broadly accepted, at least in principle, in Western society as far as sex discrimination and race discrimination are concerned. The same rights must now encompass all forms of genetic discrimination, because we are acquiring the ability to measure a vastly greater range of characteristics than before. Although the correlation of genetic characteristics with physical and mental outcomes will in most cases be purely statistical, there will be a temptation – there already is a temptation – for those to be used in actuarial prediction, very possibly to the detriment of some individuals' opportunities. This we must oppose.

Source: John Sulston and Georgina Ferry, *The Common Thread: A Story of Science, Politics, Ethics and the Human Genome* (Washington, D.C.: Joseph Henry Press, 2002), pp. 248–51.

SWIMMING WITH *E. COLI*

Life from the point of view of an *E. coli* bacterium sounds enviably warm and comfortable, swimming through the nutrient-rich regions of a mammal's lower intestine, although, as the American neurobiologist Debra Niehoff points out, there are plenty of hazards to watch out for on the journey. Bacteria such as *E. coli* have evolved impressive life skills with which to negotiate their chosen terrain, including the ability to navigate at will, as well as to communicate through a complicated chemical 'language'. In this extract from Niehoff's second book, *The Language of Life* (2005), she takes a bacterium's-eye view of the pleasures and challenges of life in a mammalian gut:

In broad daylight, you are floating in darkness. You swim in silence because you cannot hear. On your right, a cloud of sugar molecules drifts just a few centimetres away – dinner, but how will you find it? On your left, a noxious chemical seeps towards you – but unless you know that you're in danger, how can you escape? If you are the bacterium *Escherichia coli*, sharing space in the human gut with other indigenous flora and fauna, this is your world: unpredictable, at times even inhospitable. But don't dwell on your limitations. Your kind came to life when the earth was still hot and sulfurous and has used the intervening eons to craft all the tools and techniques you need to survive in a capricious environment.

You have many talents. For one thing, you can move. Let the bottom dwellers and the stone huggers ache to be noticed by some passing current; evolution taught you to waltz – chasse, pause, spin, glide, ONE-two-three, ONE-two-three left, right, zigzag. Your dancing shoes are a skein of protein filaments, or flagella, powered not by muscles and tendons but by a gearbox of proteins that operate as a rotor, twirling the flagella at more than 100 revolutions per second. When the rotor turns counterclockwise, the flagella spiral into a single tail and you glide in a smooth, straight line. When it spins clockwise, the flagella unfurl and stroke, each to its own beat, and you spin and tumble in place. Reverse again and you resume swimming in a new direction. The rotor switches back and forth as regularly as a metronome, spinning one way for a few seconds, then the other. Spinning and swimming, you meander along to its rhythm, improvising a daydream of a dance that textbooks refer to as the 'random walk'. . .

By repeatedly adjusting the proportion of clockwise-to-counter-clockwise rotation, you can forego your wandering ways in favor of a one-step-sideways shuffle – not exactly what a more advanced creature might think of as purposeful movement but a 'biased random walk' that hitches determinedly, if somewhat erratically, toward satisfaction or away from catastrophe. Microbiologist Ann Stock explains: 'If the cell finds itself moving in the proper direction, it suppresses tumbling and moves further in that direction. Then it randomly reorients and heads off in a new direction, one that might still be good or may now be bad. If it's going in the wrong direction, it goes back to tumbling. It seems tortuous, but the same patterns are used by grazing animals to find better pastures. You do a little sampling in all directions and keep going when things are good and reorient when they're not.' But how did you determine which direction was the 'right direction'? And how did you use that information to change the rotation of your flagella?

I look and listen and touch; if I could swim with you, I would know the human gut as wet and warm, a muffled rush of murky water. But there are other ways to experience life. Your world is vivid with chemicals as well as colors, molecules as well as sounds. They could be your connection to the outside world, the cues that could chart your path. Even this solution presents difficulties, however. Like other cells, you are surrounded by a protective membrane of lipids and proteins isolating your fastidiously composed internal fluids from the unpredictable excesses of your surroundings. Nothing that might upset the delicate balance critical to life can penetrate this barrier – but neither can the chemical signals bearing news about current events. Inside your hermetically sealed bubble, the rotor proteins controlling your flagella are waiting for direction; outside, messengers with the information needed by the rotor proteins mill restlessly in front of closed doors.

Troublesome organism, isn't there any end to your problems? Evolution should have just given up on you, allowed you to starve in silence. Instead it has crafted words to describe your world and rules for combining them, the building blocks of a language that tells you how to move in step with the world around you and gives your

crooked walk a glamorous name: 'chemotaxis,' movement directed
by chemicals.

As Niehoff goes on to explain, cells communicate through a set of complex
chemical interchanges, with instructions and questions being despatched to
receptors through the chemical chatter of enzymes and proteins. This chemical language allows cells to talk to one another, to cooperate as a group yet
to behave as single organisms – it is, she says, 'one of the most extraordinary
achievements in the history of life'. Yet it is a relatively recent discovery,
because bacteria have tended to be studied in laboratory cultures, far from
the stresses of their natural environments (such as inside our guts). In the
well-fed comfort of the Petri dish, cells quickly begin to forget their survival
skills, and thereby lose some of their most interesting habits – including the
ability to talk:

In the laboratory incubator it is never winter – but it's not home
either. 'We've tended to use bacteria as model systems in the laboratory,' says Ann Stock. 'But we realize now that the conditions that
you keep them under in a laboratory don't even come close to
scratching the surface of the kinds of environments they find themselves in when they're living in the real world.' For some species, like
marine microbes, the difference between their native environment
and the lab environment is more than their fragile physiology can
handle. Others are more resilient; a few, like E. coli, adapt so well,
Stock notes, that it's easy to forget they were once wild creatures.
Unless culture conditions demand some effort, these domesticated
bacteria find it easy to forget survival skills that were essential in the
wild; after all, why waste effort on rotors and receptors when safety
is a given and food is delivered regularly to your doorstep? If you
don't use it, it's best to lose it. 'There's a tremendous selection for
efficiency,' Stock observes. 'We see all the time in the laboratory that
cells will lose genes or the expression of genes and behaviors that
you'd see in the wild-type environment because you don't maintain
selection for them.'

Shielded from the demands of the real world, many domestic
bacteria have not only lost their ambition but also mothballed a
talent that until recently no one knew they even had – the ability to
talk. Speechless in agar, bacteria at liberty in soil and pond water,
milk and meat, the open sea, or the human body are not mute, but

chatter relentlessly to each other, in chemical dialects that follow the same organizational principles as chemotaxis. They trade gossip. They sound alarms and plan invasions. And, most surprising of all, they form alliances, primitive communities that allow them to enjoy benefits – more efficient use of resources, access to environmental niches unsuited to single cells, safety from predators – once thought to be the sole province of 'traditional' multicellular organisms like plants and animals.

'Certainly bacteria are the best-studied organisms,' says microbiologist Bonnie Bassler. 'But they have always been considered to live these sorts of individualistic, asocial lives. And it's just not true.' In their native environment, bacteria are social creatures, eager to collaborate and able to do so because, like more complex organisms, they have discovered that the key to civilization is communication.

Source: Debra Niehoff, *The Language of Life: How Cells Communicate in Health and Disease* (Washington, D.C.: Joseph Henry Press, 2005), pp. 9–12; 20–21.

COFFEE STAINS

Jay Ingram (b. 1945) is a Canadian broadcaster and journalist, known for his enthusiastic popularizations of the science of everyday life. This piece, from his pleasingly titled book of essays, *The Velocity of Honey* (2003), describes the complicated patterns that gather on the surface of a steaming cup of coffee, as well as on the table after you've spilled it:

If you can position a cup of hot, black coffee so that light strikes it at an angle, you should see a whitish sheen on the surface. (This works even better with a cup of clear tea.) There's something more to this sheen than first meets the eye. It makes a flagstone pattern on the surface of the coffee, with patches of this lighter colour separated from other patches by dark lines. The patches are usually a centimetre or so across.

These patches are what scientists call convection 'cells,' areas where warm fluid is rising and cold is sinking. Convection is what the weather is all about, not to mention ocean currents, and the same thing in miniature happens in your coffee. As the surface layer cools from contact with the air above it, it becomes denser and sinks, forcing warmer, less dense coffee up to the surface. But this doesn't happen in a haphazard or confusing way. Rather, the areas of up-flow and downflow organize themselves into roughly similar sized columns, one beside the other. In the coffee cup the areas with the whitish sheen are rising columns of hot coffee, and it's the heat of that coffee that creates the sheen, although saying it in that straight-forward way misses the point: a drama is being played out at the surface of your coffee.

The sheen is actually a thin layer of tiny water droplets, droplets that have condensed just above the surface of the coffee and are hovering there, less than a millimetre above the surface. It's whitish because so much light reflects from the surfaces of the droplets. The droplets form because as the water evaporates from the hot surface of the liquid, it cools suddenly, condenses and coalesces. The drops that form do not fall back onto the surface of the coffee because they are buoyed up by the trillions of water molecules still rising up underneath them. Held there, suspended above the surface, they are clouds on a scale so minute that only careful lighting reveals them. It would be an incredible experience to be there in the tiny space under the droplets but above the liquid coffee. It would be hellish hot for one thing, but you'd also be buffeted by stuff evaporating from the surface, and concerned all the while about slipping into the downstream convection (the black lines separating the clouds) and vanishing into the blackness of the coffee below. Even from our mundane perspective (simply looking down on the cup) it should have been apparent from the start that the drops were hovering — you would have noticed that a breath scatters them instantly, like clouds before the wind, but they form again just as quickly.

The only place where you can see right down to the coffee surface is along the black lines, as if you are seeing the surface of Venus through a sudden break in its impenetrable clouds. The cool coffee sinks in those black lines, completing the convection cell . . .

Less beauteous than evanescent clouds or churning convection cells, but certainly more common, is the dark ring left behind when coffee spills. Even the ring presented a puzzle for physicists to solve. Funnily enough, when the puzzle was solved, the processes involved turned out to be the same as seen in coffee when it was still in the cup: the flagstone pattern and the clouds, the movement of fluid from one place to another, and evaporation.

The puzzle is this: why, when a drop, or half a cup, of coffee spills and then dries, does it form a ring, with almost all of the dark coffee stuff in the ring and the centre almost empty? Why shouldn't it dry and leave a uniform beige stain on the counter?

Here are some clues: you can show that it doesn't have anything to do with gravity by throwing your cup of coffee onto the ceiling, then watching as it dries. Each individual drop will still form a dark ring at its perimeter. On the other hand, it must have something to do with evaporation, the process by which the water molecules move into the air, leaving the solids behind. A couple of early experiments by Sid Nagel and his colleagues at the James Franck Institute at the University of Chicago tested this by interfering with the normal evaporative process. In one, a drop was placed under a tiny glass lid that had only a minute hole over the very centre of the drop. You would expect under these conditions that the only evaporation possible would be from the centre of the drop, not from the edge. In this special circumstance drops did *not* leave a ring behind. So evaporation from the edge of the spill must have something to do with the formation of the ring.

In a second experiment the scientists placed drops on Teflon, to which, as you know, nothing sticks. Drops left on Teflon didn't leave a ring either. In this case you'd have to suppose that the smoothness of Teflon would be the key, suggesting that a second factor in ring formation is the surface on which the drop is sitting. Add to these the fact that if you use a microscope to watch the behaviour of tiny particles in the drop as it is drying, you'll see that the particles are streaming headlong out to the edge of the drop. Sid Nagel described it as being like watching rush hour in New York. Evaporation, the surface, the streaming — those are the things you need to know to be able to account for the ring . . .

Of course, it's not just the water that vacates the centre of the drop for the edges. With it goes all the dissolved and particulate matter that exists in a cup of coffee. It is carried along, then finally dumped at the edge of the drop when all the water has evaporated.

The Teflon experiment worked because the surface is virtually free of irregularities, so the drop can contract as it evaporates, maintaining its preferred shape to the bitter end. The lid experiment worked because the water could not evaporate from the edges of the drop, only from the centre, so there was no need for the liquid to migrate out to the edges, no transport of particulate matter from the centre and therefore no ring. In that case the particulate matter was simply left where it was, forming a smudge.

Sip your coffee, gulp it, even spill it, but above all, take a second or two to check it out. After all, a glance at an apple stimulated great thoughts in Isaac Newton's head. It's true there aren't very many Newtons, but a few moments at the breakfast table can serve as a reminder that yes, our lives are under the influence of forces beyond our control: forces like surface tension, viscosity, evaporation and gravity.

Source: Jay Ingram, *The Velocity of Honey; and More Science of Everyday Life* (London: Aurum Press, 2004), pp. 18–23.

THE GOD PARTICLE

The Standard Model of particle physics is a classification with a difference, in that one or two of its major constituents may turn out not to exist.* The Higgs boson is probably the most famous of these fugitives, a hypothetical elementary particle that was first proposed in 1964 by the British physicist Peter Higgs (b. 1929), and that has so far failed to turn up. But as science writer Anil Ananthaswamy describes in the following extract from his cosmological travelogue, *The Edge of Physics* (2010), the hunt for the Higgs is on at Geneva's Large Hadron Collider, a 27-kilometre-long tunnel down which beams of protons are accelerated towards each other at close to the speed of light. When the particles smash together, all their energy is concentrated

* In the early 1930s the only known subatomic particles were protons, neutrons, electrons and photons, but since then many others, such as quarks and bosons, have been detected. By the mid-1960s physicists realized that a new model of the subatomic world was needed to explain the interactions of this flotilla of new particles with the four fundamental forces of nature: gravity, the strong and weak nuclear force, and electromagnetism. Thus arose the so-called Standard Model, an elegant classification of elementary and composite particles, whose mysterious names have drifted into the wider culture as if from some impenetrable modernist poem:

ELEMENTARY PARTICLES

a) **Fermions:** the building blocks of matter:

   Quarks (of which there are three groups or 'flavours': up & down; charm & strange; top & bottom)

   Leptons (of which there are three groups: electron & electron neutrino; muon & muon neutrino; tau particle & tau neutrino)

b) **Bosons:** the carriers of forces:

   Gauge Bosons (carriers of force: these include photons, gluons, and W & Z gauge bosons)

COMPOSITE PARTICLES (HADRONS)

a) **Fermions:** building blocks of matter:

   Baryons (made of three quarks, bound by gluons: these include protons & neutrons)

b) **Bosons:** carriers of forces:

   Mesons (quark–antiquark pairs: these include pions and kaons)

The model is not complete, however, and there is plenty of room for new particles to join, such as the projected Higgs boson, the discovery of which would answer one of the most pressing questions in science: how do the elementary particles acquire their mass? — R.H.

into a space thousands of times smaller than the width of a human hair, creating the energy densities necessary for spawning new particles that, if spotted and identified, could fill some of the gaps in the Standard Model. The Higgs boson, the most significant of these hypotheticals, is thought to be responsible for giving all the other particles their mass; no wonder it has been nicknamed 'the God particle':

I was on my way to see the 7,000-ton ATLAS, a particle detector being built for the Large Hadron Collider (LHC), the world's largest particle accelerator. The LHC smashes protons together at energies never before achieved on Earth, creating particles that could have existed only in the infant universe, when the cosmos was hot enough for them to pop in and out of existence. And then detectors like ATLAS sift through the debris of particles, hoping to find answers to a multitude of questions. How do elementary particles get their mass? What is dark matter? Are there extra dimensions? Is there a whole new set of particles besides the ones we already know about? Will these particles lead us toward a theory of quantum gravity? . . .

On the day of my visit, the place was humming with the efforts of physicists and engineers striving to put the finishing touches on the LHC and its various detectors. The LHC is a once-in-a-generation machine. But before it could be built, an older accelerator, the Large Electron-Positron (LEP) collider running in the same tunnel, had to be torn down. (Such machines are built underground not for safety reasons but because it is cheaper to dig gigantic tunnels than to pay for the real estate on the surface – especially in prime locations like Geneva.) By 1999, the LEP was nearing the end of its life, but it refused to die a quiet death. The machine had come tantalizingly close to finding an elusive particle called the Higgs boson, the one undetected member of the so-called zoo of particles described by the standard model of particle physics. The Higgs is thought to give elementary particles their mass. Finding it is essential for a complete understanding of the material world. So, teased by the promise of the LEP, CERN (the European Organization for Nuclear Research) officials gave it more time to find the Higgs while the construction of the LHC proceeded apace . . .

To understand just how important the LHC is to particle physi-cists and cosmologists, one needs to understand the problems with the standard model of particle physics. The model, which divides the material world into fermions (the building blocks of matter) and bosons (the carriers of fundamental forces), has been amazingly successful at explaining subatomic particles and their interactions. But over the past few decades, there have been strong hints that it is incomplete.

One glaring omission in the standard model is that it does not incorporate gravity and hence does not predict a particle for this fourth fundamental force. But that's just one of its problems. Take the Higgs boson, a particle required by the model. This particle is crucial to explaining the origin of mass – not the universe's mass but the mass of elementary particles, such as electrons and quarks. Most of the mass of atoms in the universe consists of protons and neutrons (electrons contribute less than one part in a thousand to the mass of atoms). Protons and neutrons, in turn, get their mass from their constituents, quarks. The quarks are held together by particles called gluons, the carriers of the strong nuclear force. It is the 'binding energy' of this force that gives protons and neutrons their mass, and it accounts for 90 percent of the mass of normal matter. Except for one big problem: The standard model forbids quarks and gluons (and, indeed, all other elementary particles, including the electron) from having any mass if it weren't for the Higgs boson.

The explanation goes something like this: The Higgs boson is associated with a Higgs field, just as the electron is associated with an electromagnetic field, but with one major difference. An electromagnetic field is a vector field, which means that at every point in space it has an associated direction. The Higgs field, however, is scalar – it has a value but no direction, not unlike temperature in a room. Think of the Higgs field as a thick, gooey background against which all other particles have to move. Every other fundamental particle interacts with the Higgs field in its own specific manner, and the strength of this interaction gives each particle its mass. By all accounts, the Higgs should have been found by now. That hasn't happened.

Then there is something called the hierarchy problem. The Higgs boson gets some of its mass from its interactions with every other particle. Add up these contributions, and the Higgs mass turns out to be nearly 10^{17} greater than what experiments say it should be. To rein in the mass of the Higgs, the various contributions – some positive, some negative – have to cancel out exactly, and for that to happen some parameters of the standard model have to be fine-tuned to an astounding precision of one part in 10^{34}. 'We believe that fine-tuning is an undesirable property,' said Dan Tovey of the University of Sheffield, UK, a member of the ATLAS team. 'Obviously we don't want the mass of particles to be what we observe just because of some remarkable coincidence.'

Among the more notable visitors who came to view the collider's tunnels in the months before they were closed to the public was the man whose name has become so closely associated with the search for the missing boson:

Of all the people who were waiting for the LHC to start up, one man could rightfully lay claim to being the most eager: Peter Higgs. In April 2008, Higgs, a seventy-eight-year-old British physicist, came to CERN for the first time in decades, just before the caverns and tunnels were closed to the public, to see for himself the machine that could find his eponymous particle. It was in the 1960s that Higgs had invented the Higgs mechanism, which endows elementary particles with mass, and after decades of hype and hoopla the particle is now close to being found. There's little doubt in his mind that the Higgs particle (or particles) will be discovered soon.

The day after he toured the LHC, the usually reclusive Higgs faced a gaggle of reporters in Geneva. He is famously modest, refusing to refer to the Higgs particle by name, preferring to call it the 'boson named after me,' and remains careful to point out that he wasn't the only physicist to come up with the idea. Two physicists in Belgium, Robert Brout and François Englert, discovered it almost simultaneously; it is now officially called the Brout-Englert-Higgs mechanism. But Higgs did make one unique contribution: he predicted a particle that could be found by experiments. 'The only reason I suggest that maybe I deserve to have the particle named after me is because I drew attention to it,' said Higgs.

The Higgs boson gained wider notoriety when Nobel laureate Leon Lederman wrote a 1993 book about it called *The God Particle*. Lederman had wanted to call it *The Goddam Particle*, but his publisher's commercial instincts won out and the book ended up with the more grandiose title. An atheist, Higgs finds it extremely embarrassing that the boson has such a nickname. Still, what if the LHC finds it? 'I shall open a bottle of something,' said Higgs. 'Whisky or champagne?' quipped a media wag. 'Champagne,' said Higgs. 'Drinking a bottle of whisky takes a little more time.' More poignantly, Higgs was asked if he thought the LHC would find the boson before he turned eighty. 'I hope so,' he said. '[Or] I'll just have to ask my GP to keep me alive a bit longer.'

The Large Hadron Collider was fired up on 10 September 2008, and nearly a billion people around the world tuned in to bear witness to the first high-speed journey of protons around the tunnel. Two days later a power transformer failed and the collider was shut down for a week; as soon as it reopened an electrical fault blew a hole in the plumbing, leading to a build-up of pressure that damaged some of the equipment. The collider was shut down again for repairs and did not reopen until November 2009. The following year saw rumours of the first major breakthrough in the search for the Higgs boson, when scientists at the Tevatron particle accelerator at Fermilab in the United States – the main American rival to CERN's Large Hadron Collider – were reported by an Italian physics blogger to have detected 'a light Higgs boson signal'. These rumours were quickly denied by Fermilab, whose Twitter feed on 13 July 2010 read: 'Let's settle this: the rumours spread by one fame-seeking blogger are just rumours. That's it.'

In January 2011 it was announced that the Large Hadron Collider will run continuously until the end of 2012, rather than shutting down for a year at the end of 2011, as previously planned. The search for the elusive boson continues.

Source: Anil Ananthaswamy, The Edge of Physics: Dispatches from the Frontiers of Cosmology (London: Duckworth, 2010), pp. 222–0, 245–6.

AFTERWORD

When I first told my friends and colleagues that I was working on a compilation of readable science writing, one of them looked at me in surprise and said, 'Doesn't "readable" just mean that you leave all the science out?' 'Very funny,' I replied, but the question turned out to be more pertinent than I'd realized, for, as I mentioned in the introduction, I wasn't about to fill nearly five hundred pages with stuff that hardly anyone could understand; but at the same time I was reluctant to restrict myself to populist accounts of Great Scientific Breakthroughs, especially since such pieces tend to be fairly similar in tone ('little did Doctor Brilliant suspect, as he strode into his laboratory that fateful spring morning, that he was about to make a world-changing discovery . . .'). I knew from the outset that the collection needed to be diverse in character as well as in content, that it was important to reflect the wide range of times and places in which the stories of science have been told. Thus, from the long-vanished cities of ancient Mesopotamia to the newly dug site of the Large Hadron Collider – the world's biggest and costliest scientific instrument – this anthology has taken us on a virtual grand tour of the places where science has been made.

But in spite of the historical and geographical scope, a number of persistent themes and preoccupations seem to have recurred in many places over time, and it seemed to me, as I came towards the end of the selection process, that they might have something interesting to say about our ongoing efforts to understand and interpret the world. Take, for example, the habit of self-examination, both literal (in the case of medical scientists such as Avicenna, or John Dalton, whose self-diagnosis of his own colour-blindness is one of the most remarkable episodes in the book), and figurative, in the form of a series of related anxieties about what one is doing when one is engaged in science, what it really means to make new discoveries, who owns the knowledge that is thereby created, and to what ends it might reasonably be put. Right from the start, from the Presocratics onwards, scientists (or rather, natural philosophers) sought to put themselves in the picture, to understand what it was they were doing even as they sought to understand the workings of the physical world; and, looking back over my choices for this anthology, I am struck by how thoughtful and reflective so many of them are. This may, of course, be down to editorial bias, informed by my own fondness for authorial introspection, but it seems to me that scientists throughout history have rarely imagined that what they are doing is philosophically unproblematic. How on earth could they, given that they are routinely faced with some of the biggest questions there are: What is life? How did it arise? Why is there something rather than nothing? And though scientists – along with everyone else – have not yet found the answers to these questions, and perhaps they never will, I find it reassuring to know that they are working on them as I write.

This habit of self-scrutiny is also apparent in the close attention paid to scientific language, and once again, as I came to the end of the editing process I was struck by how many of my chosen extracts were concerned, directly or indirectly, with the creation of new kinds of narrative. The wish to circumscribe an exclusively 'scientific' language goes back a long way, Thomas Sprat's conception of 'a close, naked, natural way of speaking . . . bringing all things as near the Mathematical plainness as they can', being only one of many concerted attempts to cordon off the discourse of science from the surrounding intellectual culture. It took several centuries, but, unfortunately, it worked in the end, and some of my favourite pieces are those that explore the linguistic consequences of this cultural separation, from William Whewell's campaign on behalf of the new word 'scientist' (p. 216) to Lotfi Zadeh's didactic yet curiously winning attempt to quantify the idea of beauty (p. 443). The same is also true of some of the pieces that I ended up leaving out of my final selection: among those that I most regretted losing was John McPhee's awed encounter with mineralogical nomenclature from his 1981 book *Basin and Range*, a taxonomic *wunderkammer* that opens with a delirious stream of hardcore geology-speak, tinged with the poetry of unknowability: 'Metakirchheimerite, phlogopite, katzenbuckelite, mboziite, noselite, neighborite, samsonite, pigeonite, muskoxite, pabstite, aenigmatite, Joesmithite.' A dazzling array, although McPhee's enchantment was not with the language of science so much as the language of specialization, in the form of an esoteric jargon used by geological initiates as a means of avoiding ambiguity. Yet this kind of language is forever on the move, and as McPhee pointed out, 'what had previously been described as the granite of the world turned out to be a large family of rock that included granodiorite, monzonite, syenite, adamellite, trondhjemite, alaskite, and a modest amount of true granite . . . the enthusiasm geologists show for adding new words to their conversation is, if anything, exceeded by their affection for the old. They are not about to drop granite. They say granodiorite when they are in church and granite the rest of the week', an observation that goes some way towards countering the allure of the 'two cultures' pessimism that was advanced by C. P. Snow.*

But what has really surprised me, looking back over this selection, is the emphasis on what might be called the science of everyday life, from wet towels to coffee stains, via rusty nails, housework, boredom, and the barcode on the back of this book. I hadn't originally planned on making a feature of this kind of material, but I'm glad that I did, for it has served to counterbalance some of the more abstract entries about things that none of us will ever actually see (black holes, the God particle, the fourth dimension ...). For as I discovered in the course of making my selections, I have lost all interest in the 'gee-whizz-just-imagine-that' variety of popular science writing that so beguiled me when I was younger, and find that I much prefer the more down-to-earth kind in which human stories weave in and out of the wider scientific narrative. Some of these stories really do read like novels: Robert Wood's account of the notorious 'N-rays' delusion, for example, resembles a hard-boiled detective yarn, complete with a skulking accomplice who glares from the laboratory shadows as our hero uncovers the deception through a series of cunning ruses: 'The assistant commenced to turn the wheel, and suddenly said hurriedly to Blondlot in French, "I see nothing; there is no spectrum. I think the American has made some *dérangement*." Whereupon he immediately turned up the gas and went over and examined the prism carefully. He glared at me, but I gave no indication of my reactions. This ended the séance, and I caught the night train for Paris.' This is science by way of Dashiell Hammett. And what about Alfred Russel Wallace's description of his antics in pursuit of rare Malaysian butterflies? The image of him lunching in a jungle

* The many other 'lost' pieces that I wish I could have kept could easily fill another volume. They include Karl von Frisch's classic account of bee communication through dance; Dr John Harlow's eye-watering case history of Phineas Gage, a nineteenth-century railway worker who suffered severe frontal brain injury when an iron tamping rod was blasted through his skull; Dorothy Crawford's bravura description of the microbe responsible for seventeenth-century 'Tulipomania', from her book *The Invisible Enemy* (2000); Peter and Jean Medawar's essay on the biology of water from their unique collaboration, *Aristotle to Zoos: A Philosophical Dictionary of Biology* (1984); a scene from Michael Frayn's 'atomic' play *Copenhagen* (1998) in which Niels Bohr and Werner Heisenberg debate the properties of uranium-235; or Frank Close's dazzling account of the Alice-in-Wonderland world of antimatter. For me, perhaps the saddest omission of all was Roy Herbert's joyous essay, 'On First Encountering Sodium Benzoate', published in the *New Scientist* in 1978: this mini-masterpiece of chemical prose is not just one of my favourite pieces of science writing, but one of my favourite pieces of writing, full stop. It was left out solely because we couldn't afford the permission fee – that unpredictable stumbling-block over which every anthologist must trip.

clearing as his hired man clambers among the rocks with a specimen net is like something out of Saki or Somerset Maugham, with Wallace cast as a perspiring British nabob in the grip of a private mania.

'Oh, what a fool I've been to neglect science all these years,' as the French novelist Érik Orsenna recently remarked, 'for natural history is the mother of every form of history, every sort of story, the novel of all novels.' And as this anthology has set out to show, such stories of science really are among the greatest stories ever told: tales of wonder in which 'our imagination is stretched to the utmost, not, as in fiction, to imagine things which are not really there, but just to comprehend those things which are there', as the physicist Richard Feynman characterized the nature of scientific testimony. It was Feynman who told the well-known story of an artist friend of his who accused him of undermining the beauty of a flower, claiming that he, as an artist, could see and appreciate its beauty, while Feynman, as a rational scientist, saw only one more object to dissect. Feynman's response – that scientific knowledge of a flower only adds to its excitement and mystery and awe, 'it only adds; I don't understand how it subtracts' – remains, in my view, one of the finest defences of science ever mounted; in fact, it ought to be printed on a T-shirt. Maybe it already is. At any rate, it would be an improvement on the one that I recalled at the beginning of this book, the one that claimed that there are only 10 (that is to say, two) types of people in the world: maths geeks and everybody else. What I hope is that this collection has shown such notions, however wittily applied, to be false. I hope, too, that it has served to illuminate some of the richness of long-vanished times and minds, and that it has shown science to be a truly collaborative cultural endeavour. I hope, above all, that it has given a fair hearing to the myriad people and places that have contributed over the centuries to what I hold to be the greatest invention of the human imagination: the art of scientific thinking.

FINIS

Further Reading

I have read and consulted many books in the course of compiling *The Art of Science* and, apart from the ones which went on to supply its raw materials (the details of these can be found at the end of each section), I found the following works of reference and reflection particularly useful when it came to making those difficult decisions about what and what not to include:

Adler, Robert, *Science Firsts: From the Creation of Science to the Science of Creation* (Hoboken, N J: John Wiley, 2002)

Al-Khalili, Jim, *Pathfinders: The Golden Age of Arabic Science* (London: Allen Lane, 2010)

Atkins, Peter, *Galileo's Finger: The Ten Great Ideas of Science*, new edn (Oxford: Oxford University Press, 2004)

Bolles, Edmund Blair (ed.), *Galileo's Commandment: An Anthology of Great Science Writing* (New York: W. H. Freeman, 1997)

Bonta, Marcia Myers (ed.), *American Women Afield: Writings by Pioneering Women Naturalists* (College Station: Texas A&M University Press, 1995)

Bragg, Melvyn, *On Giants' Shoulders: Great Scientists and Their Discoveries from Archimedes to DNA* (London: Hodder & Stoughton, 1998)

Bronowski, Jacob, *The Ascent of Man* (London: British Broadcasting Corporation, 1973)

Byers, Nina, and Gary Williams (eds), *Out of the Shadows: Contributions of Twentieth-Century Women to Physics* (Cambridge: Cambridge University Press, 2006)

Bynum, W. F. and Roy Porter, *The Oxford Dictionary of Scientific Quotations*, new edn (Oxford: Oxford University Press, 2006)

Calder, Nigel, *Magic Universe: A Grand Tour of Modern Science* (Oxford: Oxford University Press, 2005)

Carey, John (ed.), *The Faber Book of Science* (London: Faber, 1995)

Coley, Noel G., and Vance M. D. Hall (eds), *Darwin to Einstein: Primary Sources on Science and Belief* (Harlow: Longman/Open University, 1980)

Dawkins, Richard (ed.), *The Oxford Book of Modern Science Writing* (Oxford: Oxford University Press, 2008)

Dixon, Bernard (ed.), *From Creation to Chaos: Classic Writings in Science* (Oxford: Blackwell, 1989)

Eastwood, W. (ed.), *Science and Literature: The Literary Relations of Science and Technology; An Anthology* (London: Macmillan, 1957)

Fara, Patricia, *Scientists Anonymous: Great Stories of Women in Science* (Cambridge: Icon Books, 2005)

—, *Science: A Four Thousand Year History* (Oxford: Oxford University Press, 2009)

Gribbin, John, *Science: A History 1543–2001* (London: Penguin, 2003)

Gullberg, Jan, *Mathematics: From the Birth of Numbers* (New York: W. W. Norton, 1997)

Hart-Davis, Adam *et al.* (eds), *Science: The Definitive Visual Guide* (London: Dorling Kindersley, 2009)

Hawley, Judith *et al.* (eds), *Literature and Science, 1660–1832*, 8 vols (London: Pickering & Chatto, 2003–4)

Huff, Toby E., *The Rise of Modern Science: Islam, China, and the West* (Cambridge: Cambridge University Press, 1993)

Jardine, Lisa, *Ingenious Pursuits: Building the Scientific Revolution* (London: Little, Brown, 1999)

Jennings, Humphrey (ed.), *Pandaemonium, 1660–1886: The coming of the Machine as Seen by Contemporary Observers* (London: André Deutsch, 1985)

Kirk, G. S., J. E. Raven and M. Schofield, *The Presocratic Philosophers: A Critical History with a Selection of Texts*, 2nd edn (Cambridge: Cambridge University Press, 1983)

Mackay, Alan L. (ed.), *The Harvest of a Quiet Eye: A Selection of Scientific Quotations* (Bristol and London: Institute of Physics, 1977)

Nasr, Seyyed Hossein, *Science and Civilization in Islam*, 2nd edn (Cambridge: The Islamic Texts Society, 1987)

Otis, Laura (ed.), *Literature and Science in the Nineteenth Century: An Anthology* (Oxford: Oxford University Press, 2002)

Phillips, Patricia, *The Scientific Lady: A Social History of Women's Scientific Interests 1520–1918* (London: Weidenfeld and Nicolson, 1990)

Pickover, Clifford A., *From Archimedes to Hawking: Laws of Science and the Great Minds Behind Them* (New York: Oxford University Press, 2008)

Porter, Roy (ed.), *The Faber Book of Madness* (London: Faber, 1991)

Riordan, Maurice, and Jon Turney (eds), *A Quark for Mister Mark: 101 Poems About Science* (London: Faber, 2000)

Sagan, Carl, *Cosmos* (London: Random House, 1980)

Waller, John, *Fabulous Science: Fact and Fiction in the History of Scientific Discovery* (Oxford: Oxford University Press, 2002)

Weber, Alan S. (ed.), *Nineteenth-Century Science: An Anthology* (Peterborough, ON: Broadview Press, 2000)

Weber, Robert L. (ed.), *A Random Walk in Science* (Bristol and London: The Institute of Physics, 1973)

—, *More Random Walks in Science* (Bristol and London: The Institute of Physics, 1982)

Whitfield, Peter, *Landmarks in Western Science: From Prehistory to the Atomic Age* (London: British Library, 1999)

Acknowledgements

It is a pleasure to thank the friends and colleagues who have helped in the preparation of this volume, whether by suggesting pieces for inclusion, offering advice about the overall structure, or just listening while I blathered on about the amazing things I'd been reading. Thanks, then, to Markman Ellis, Michael Newton, Gregory Dart, and David Hamblyn; also to Jon Adams, Nicholas Alfrey, Giles Bergel, Nicholas Blake, Emma Bravo, Martin John Callanan, Tom Dunkley-Jones, William Fiennes, Angela Foster, Michael Griffiths, Dan Grimley, Paul Hardaker, Judith Hawley, Megan Hiatt, Steven Hiatt, Claudia Jessop, Gavin Jones, Sarah Lancaster, Joshua Li, Anthony and Paula Lynch, Rob McSweeney, Maria Majsa, Mark Maslin, Adrian Simpson, and Jem Southam.

The staff of the British Library; the Senate House Library, University of London; the Wellcome Collection Library; the Royal Society Library and Archives; and the Science Museum Library have been helpful and efficient as always. I would particularly like to thank Barbara Wolff and the staff of the Albert Einstein Archives, Hebrew University of Jerusalem, for their kindness and generosity. Special thanks are also due to Sarah Flannery for agreeing to rewrite her piece on the ISBN (see p. 414); and to Mahmud Mirza for his help in translating the Arabic of Ibn al-Haytham. Thanks also to Peter Straus for commissioning the book, and to Paul Baggaley for taking it on so willingly; while Sam Humphreys was, as always, an insightful and supportive editor. The biggest thanks of all, though, are to my family, Jo, Ben and Jessie Hamblyn, for making life, the universe and everything so infinitely worthwhile.

For permission to reprint copyright material the editor and publishers gratefully acknowledge the following:

ANIL ANANTHASWAMY: from *The Edge of Physics*, by Anil Ananthaswamy. By permission of Gerald Duckworth & Co. Ltd; AVICENNA (IBN SINA): from *The Canon of Medicine*, adapted by Laleh Bakhtiar (Great Books of the Islamic World, 1999). Reprinted by permission of Kazi Publications, Inc.; JOHN R. BAKER: from 'English Style in Scientific Papers'. Reprinted by permission of Macmillan Publishers Ltd: *Nature* vol. 176, copyright © 1955; Mark S. Blumberg: from *Freaks of Nature* by Mark S. Blumberg, 2009. By permission of Oxford University Press; George Bradley: 'About Planck Time', from *Terms to be Met* (1986). Copyright © 1986 by George Bradley. By permission of Yale University Press; EDWARD C. BULLARD: from *Proceedings of the Royal*

Society of London, Series A, 342 (1975). By permission of HighWire Press; JOCELYN
BELL BURNELL: from The Annals of the New York Academy of Sciences. Copyright ©
1977, John Wiley and Sons; RACHEL CARSON: from The Sea Around Us, by Rachel
Carson. Copyright © 1950 by Rachel L. Carson. Used by permission of Frances
Collin, Trustee. All copying, including electronic, or redistribution of the this text,
is expressly forbidden; UGO CERLETTI: from The Great Physiodynamic Therapies in
Psychiatry: An Historical Reappraisal (1956). Reproduced with permission of Lippincott
Williams & Wilkins/Wolters Kluwer; ERWIN CHARGAFF: reproduced from Heraclitean
Fire: Sketches from a Life Before Nature. Copyright © 1978 The Rockefeller University
Press; MARCUS CHOWN: from The Quantum Zoo: A Tourist's Guide to the Neverending
Universe by Marcus Chown. Reproduced with permission of Joseph Henry Press via
Copyright Clearance Center; RICHARD DAWKINS: from Climbing Mount Improbable
(Viking, 1996). Copyright © Richard Dawkins, 1996. Reproduced by permission of
Penguin Books Ltd; ALBERT EINSTEIN AND LEOPOLD INFELD: from The Evolution of
Physics (1938). Reproduced by permission of the Albert Einstein Archives, Hebrew
University of Jerusalem, and Prof. Eryk Infeld; SARAH FLANNERY: Copyright © 2011
by Sarah Flannery. Used by kind permission of Sarah Flannery; RICHARD FORTEY:
from Dry Store Room No. 1: The Secret Life of the Natural History Museum (2008).
Reprinted by permission of HarperCollins Publishers Ltd. © 2008 Richard Fortey
Dry Store Room No. I; OTTO FRISCH: from What little I remember (1979). Copyright
© Cambridge University Press 1979, reproduced with permission of the family of
Otto Robert Frisch and the publisher; STEPHEN JAY GOULD: from The Panda's Thumb:
More Reflections in Natural History by Stephen Jay Gould. Copyright © 1980 by
Stephen Jay Gould. Used by permission of W. W. Norton & Company, Inc.; SUSAN
GREENFIELD: from The Human Brain: A Guided Tour (1997) by Susan Greenfield.
Reprinted by permission of Weidenfeld & Nicolson, a division of The Orion Publish-
ing Group, London; J. B. S. HALDANE: from Possible Worlds and Other Essays (Chatto &
Windus, 1927). By permission of the Random House Group Ltd; JAY INGRAM: from
The Velocity of Honey by Jay Ingram, reprinted by permission of Aurum Press; ALISON
JOLLY: from Jolly, Alison, The Evolution of Primate Behavior, 2nd Edition, © 1985, pp. 3–5.
Reprinted by permission of Pearson Education, Inc., Upper Saddle River, NJ; ARTHUR
KOESTLER: from The Sleepwalkers by Arthur Koestler (© Arthur Koestler, 1958),
reproduced by permission of PFD (www.pfd.co.uk) on behalf of The Estate of Arthur
Koestler; JOHN LISTER-KAYE: from Nature's Child: Encounters with Wonders of the
Natural World (Little, Brown, 2004). By permission of Little, Brown Book Group;
ROSALEEN LOVE: from Reefscape: Reflections on the Great Barrier Reef (2001), by
Rosaleen Love. Reprinted with permission from the National Academies Press,
Copyright © 2001, National Academy of Sciences; TODD MCEWEN: from Arithmetic
by Todd McEwen, published by Jonathan Cape. Reprinted by permission of the

Random House Group Ltd; DEBRA NIEHOFF: from *The Language of Life: How Cells Communicate in Health and Disease*, by Debra Niehoff. Reprinted with permission from the National Academies Press, Copyright © 2005, National Academy of Sciences; C. P. SNOW: from *The Two Cultures and a Second Look: An Expanded Version of the Two Cultures and the Scientific Revolution*, 1964, © Cambridge University Press, reproduced by permission of the Hon. Philip Snow and the publisher; GEORGE G. SZPIRO: from *The Secret Life of Numbers* (2006) by George G. Szpiro. Reprinted with permission from the National Academies Press, Copyright © 2006, National Academy of Sciences; JOHN SULSTON AND GEORGINA FERRY: from *The Common Thread* (2002), by John Sulston and Georgina Ferry. Reprinted with permission from the National Academies Press, Copyright © 2002, National Academy of Sciences; JAMES. D. WATSON: from *The Double Helix: A Personal Account of the Discovery of the Structure of DNA* (1968), by James D. Watson. Reprinted by permission of International Creative Management, Inc. Copyright © 1968 by James D. Watson; ALFRED WEGENER: from *The Origin of Continents and Oceans*, 4th edn (Dover Publications, 1966). By permission of Dover Publications Inc.; STEVEN WEINBERG: from *The First Three Minutes*, Copyright © 1993 Steven Weinberg. Reprinted by permission of Basic Books, a member of the Perseus Books Group; MICHAEL WHITE AND JOHN GRIBBIN: from *Stephen Hawking: A Life in Science*, new edn, 2002, by Michael White and John Gribbin. Reprinted with permission from the National Academies Press, Copyright © 2002, National Academy of Sciences; EDWARD O. WILSON: from *In Search of Nature* by Edward O. Wilson. Copyright © 1996 Edward O. Wilson. Reproduced by permission of Island Press, Washington, DC; ELLIS L. YOCHELSON: *Proceedings of the Biological Society of Washington*, Copyright © 1969 by Biological Society of Washington. Reproduced with permission of Biological Society of Washington; LOTFI A. ZADEH: 'A Fuzzy-Algorithmic Approach to the Definition of Complex or Imprecise Concepts', *International Journal of Man-Machine Studies* 8 (1976), pp. 249–91; Copyright © 1976 by Academic Press Inc. (London) Ltd.

Index

Addison, Joseph 133
Adrian, Edgar 375
Aesop 15
Agassiz, Louis 138, 141
Agrippa, Heinrich Cornelius 367–8
air pump 84–7
Alexander, William 32–3
Algarotti, Francesco 98, 110–12
Alhazen (Ibn al-Haytham) 38–40, 56, 80
Alphonso X of Castile ('the Wise') 34
Alzheimer, Alois 322
American Association for the Advancement
    of Science 434
Amontons, Guillaume 101
Ananthaswamy, Anil 459–63
Anaxagoras of Clazomenae 12
Anaximander of Miletus 8, 9–10
ants xvii, 174, 348, 394–6
Arabic science 7, 34–7, 38–40, 41–2, 48–51,
    56, 115–16, 118
Arago, François 260–3
Archimedes 429
Aristotle xx, 8, 11, 12, 13, 15–16, 22, 34, 56, 57,
    62, 63, 64 5, 67, 304
Arrhenius, Svante 437–9
Asimov, Isaac 393
astronomy xix, 1–3, 9, 34–7, 52, 61, 63–6,
    67–70, 71–5, 98–100, 111–14, 185–92,
    213–15, 253, 278–81, 330, 349, 369–71,
    425–30, 434–6
atoms xv, 8, 13, 19–23, 90, 105, 210, 212,
    304–8, 337, 350–3, 354–9, 362–4,
    431–3, 434, 459–63
Aubert, Alexander 189–90, 191–2
Avery, Oswald T. 377

Avicenna (Abu Alī al-Husain ibn Adballah
    ibn Sīnā) 48–51, 57, 464
Ayrton, Hertha 136, 325–9

Babbage, Charles 134–5, 219, 222–7
Babylonians 1–3, 4, 5, 8, 34, 101, 335, 464
Bacon, Francis xvii, 38, 56, 80–3, 94, 361
Bacon, Roger 38, 56–8
bacteria 91, 260–3, 377, 391–3, 452–5
    E. coli 452–5
Bailly, Jean-Sylvain 178, 179–81
Bainbridge, Kenneth 360
Baker, John R. xviii, 129–32
Banks, Sir Joseph 156, 189, 190–2
Barrington, Daines 166–8
Bassler, Bonnie 455
Becquerel, Henri 309–10, 315
Becquerel, Jean 315, 316
Behn, Aphra 98–100, 136*n*
Bell, Alexander Graham 287
Bentley, Wilson A. 330–4
binary numbers xiii, 467
biology xvi, 94–7, 163–6, 174, 253–6, 372,
    377–9, 391–3, 411–13, 448–51
black holes 434–6, 466
Blackett, Patrick 352
Blagden, Charles 186
Blake, William xiv
Blakemore, Richard 391–3
Bleuler, Eugen 319
Blondlot, Prosper–René 315–18, 466
blood 76–9, 116–20, 165, 393
Blumberg, Mark S. 408–10
Bohr, Niels 342, 350, 354, 356, 357–8, 466*n*
boredom 290–2, 447, 466

botany 88–91, 121–6, 163–6, 182, 193–8, 210–11, 245, 273–7, 405–7

Boyle, Robert 84–7, 199

Bradley, George 340–1

Bragg, Sir William 362

British Association for the Advancement of Science xviii, 216, 218, 316

Broca, Pierre Paul 416

Broecker, Wallace xviii, 439

Bronowski, Jacob 360

Brout, Robert 462

Brown, Robert 210–12

Brownian motion 20–1, 210–12, 391–3

Buckland, Revd William 228, 230

Bullard, Sir Edward C. 364–6

Burnell, Jocelyn Bell 137, 425–30

butterflies 250–2, 423–4, 466–7

butterfly effect 421

Byron, George Gordon, 6th Baron 219, 241

California Institute of Technology 159, 346, 350, 378, 434

Carey, John xv

Carpenter, James 278–81

Carroll, Lewis (Charles Lutwidge Dodgson) 107–9

Carson, Rachel xx, 16–18

Carter, Elizabeth 110–12, 136n

Catlow, Agnes 253–6

Cavendish, Margaret 91–3

Cavendish Laboratory, University of Cambridge 350, 364–5, 377, 378

cells xviii, 88–91, 377–9, 391–3, 401–3, 416–18, 448–51, 453

Celsius, Anders 102–03

Cerletti, Ugo 322–4

Chaldea 2–3

chaos theory 421, 423–4, 446

Chargaff, Erwin 30, 360–2

charlatans 41–7, 56, 178, 315–18

Charpentier, Augustin 316

Charpentier, Jean de 141

chemistry xvii–xviii, 8, 17–18, 30–3, 199–203, 204, 293, 298–303, 304–8, 309–14, 336, 355, 416–18, 452–5, 466n

Chermock, Ralph 174

Chiao, Raymond 348

China 52–5, 115, 119, 125, 160–2, 163, 330

Chiswell, Sarah 116

Chown, Marcus xvi, 346–8

classification 163–74, 193, 243, 245, 304–8, 319, 397, 447

Clifford, David 166

climate change xviii, 54, 138–43, 232, 406–7, 437–42

Close, Frank 466n

clouds 10, 17, 24–5, 27, 28–9, 152, 169–74, 445

coffee 134, 423, 456–8

Coleridge, Samuel Taylor xvii–xviii, 218

Collinson, Peter 144

colour-blindness 182–4, 464

comets 67, 185–92, 369

computers 219–27, 421–4

contraception 338–9

convection cells 456–8

Cooper, Sarah 409

Copernicus, Nicholas xix, 34, 35, 63–6, 75, 98

cosmology 1–3, 9, 14, 34–7, 63–6, 104–7, 340–1, 425–30, 431–3, 434–6, 459–60

Crawford, Dorothy H. 466n

creationism xx, 236, 248, 335, 400–4

Crick, Francis xvi, 377–9

Crookes, William 43–7

Curie, Marie xvii, xviii, 136, 309–14, 315

Curie, Pierre xvii, 309–14, 315

Dalton, John 182–4, 464

Darwin, Charles 76, 166, 227, 236, 242–9, 250, 273, 276, 400–3

Darwin, Emma 242, 400–1, 403

Darwin, Prof. George 326–8

Davy, Humphry xvii–xviii, 199, 204–9, 234, 362

Dawkins, Richard xx, 400–4

de Beaumont, Élie 230

de Fontenelle, Bernard 98–100, 110

Democritus of Abdera 8, 13, 15, 19, 22, 56

Dennett, Daniel 242

d'Eslon, Charles 178–81

Devanny, Jean 419

Dewar, Sir James 362

diatoms 17, 253–6

Dickens, Charles 375, 404

Digges, Leonard 63

Digges, Thomas xix, 63–6

dinosaurs xviii, 268–72

Diogenes Laertius 14

Diogenes of Apollonia 8, 14

Dirac, Paul xiii–xiv, 375

    Dirac equation xiv

diving 419–20

DNA xvi, 377–9, 448–51

Dobbs, Arthur 273

Döbereiner, Johann 304, 305

Dollond, John 191

earthquakes 157–62

ecology 244–7, 250, 298, 394–6, 397–9, 405–7, 411–13

Eddington, Arthur 105–7, 346, 375

Edison, Thomas 287–9

Egypt 1–2, 4–7, 8, 34, 335

Einstein, Albert xv, 11, 105, 210, 212, 337, 340, 342–5, 346–8, 354, 357

    $E = mc^2$ xiii, xiv n, 354, 357

electricity 23, 32–3, 144–56, 157, 287, 288–9, 322–4, 344, 372

elements xvii, 8, 9, 13, 14, 52, 199–203, 204, 304–8, 309–14, 315–18, 350–3, 354–9, 362–4, 405–6, 431–3

Empedocles of Acragas 8, 12, 15

Englert, François 462

Epicurus 19, 90

Euclid of Alexandria 9

eugenics 290

evolutionary biology 174, 391–3, 394–6, 397–9, 400–4, 407, 408–10, 411–13, 452–5

evolutionary theory xx, 166, 174, 242–9, 250, 400–4

    opposition to xx, 247, 248–9, 400–4

Ewing, Anne 434

Fahrenheit, Daniel 101–3, 138–9

Fantastic Voyage (film) 393

Faraday, Michael 46, 199, 362

Fermi, Enrico 355

Fermilab (National Accelerator Laboratory), Illinois 463

Ferry, Georgina 448–51

Feynman, Richard 467

Fielding, Henry 124–5

Finley, John P 76–9

Finnegans Wake (Joyce) xiv

Flamsteed, John 191

Flannery, David 414

Flannery, Sarah 414–15

Flinders, Matthew 210

Forel, F. A. 328

Fortey, Richard 405–7

fossils 8, 10, 54, 143, 228–31, 237, 253, 268–72, 384–90

fractals 445–6

Frankel, Richard B. 391–3

Franklin, Benjamin 144–7, 152–6, 178, 181, 234

Franklin, Rosalind xvi, 377, 379

Frayn, Michael 466n

French Academy of Sciences 101, 136, 179, 315, 316, 318

Frisch, Karl von 466n

Frisch, Otto 354–9

fuzzy set theory 443–6

Gage, Phineas 466n

Galen of Pergamon 48, 76

Galileo Galilei 71–5, 76, 98, 278

Galton, Francis 290–2, 447

Gamow, George 356

Geikie, Sir Archibald 228–31

genes 182, 248–9, 377–9, 412–13, 448–51

geology 10, 30, 48, 138–43, 157–62, 204, 228–31, 236–41, 268–72, 282, 380–3, 384–90, 465–6

Gilbert, Sir Humphry 233

glaciers 138–43

Glaisher, James 334

Goethe, Johann Wolfgang von 217

Goldacre, Ben 367

Goldsmith, Oliver 124, 125–6

Gombrich, Ernst 38

Gould, Stephen Jay 391–3

gravity xvi, 104–9, 111–12, 213, 215, 340, 342, 346–8, 392, 431–3, 434–6, 459

Gray, Asa 243, 273, 276

Greenfield, Susan 416–18

greenhouse effect 437–42

grey seals 411–13

Gribbin, John 434–6

Gumperson's Law 389

Haeckel, Ernst 298

Hahn, Otto 354, 355–6, 357

Haldane, John Burdon Sanderson ('J. B. S.') 335–8

Hamilton, Emma 110

Hammett, Dashiell 466

Han Ying 330

Hansen, James E. 439–42

Hardy, G. H. 374–5

Hardy, Thomas 113–14

Harlow, Dr. John 466n

Harvey, William 76–9

Hawking, Stephen 434–6

Hawkins, Admiral Sir Richard 175

Heisenberg, Werner 466n

Herbert, Roy 466n

Herschel, Caroline xx, 136, 185–92

Herschel, William 136, 185–92

Hewish, Antony 425–30

Higgs, Peter 459, 462–3

Higgs boson 459–63

Hill, 'Sir' John 133–4

Hippolytus of Rome 10

Hoare, Sarah 193–8

Hobbes, Thomas 217

Hodgkin, Dorothy 137

Hooke, Robert xviii, 84, 88–91

Hooker, Joseph 243, 244, 276

Hopkins, Gerard Manley 285–6

housework 293–7, 298–303, 466

Howard, Luke 169–74

Hoyle, Sir Fred 373, 425, 429

Human Genome Project 448–51

Humboldt, Alexander von 205, 232–5

Humboldt Current 232–5

Hutchinson, Lucy 19–21, 136n

Huxley, Thomas Henry 397

hydra 121–6, 405

Ibn Sīnā (Avicenna) 48–51, 57, 464

Ibn al-Haytham (Alhazen) 38–40, 56, 80

Ibn al-Nafis 76
Infeld, Leopold 342–5
Ingram, Jay 456–8
International Standard Book Number
    (ISBN) 414–15
J. E. Ridgeway (ship) 282–5
Jacquard, Joseph Marie 222, 223–4
Jenner, Edward 119–20
Johnson, Samuel xviii
Jolly, Alison 397–9

Kalmijn, A. J. 393
Kelvin, Lord (William Thomson) 103
Kepler, Johannes 34, 35, 38, 330
Koestler, Arthur 1–3, 34
Kraepelin, Emil 319–22

Langmuir, Irving 293
Laplace, Pierre-Simon 213
Large Hadron Collider 459–63, 464
Lavoisier, Antoine 178, 181
Lawrence, D. H. xiv, 371–2
Leavitt, Henrietta Swan 136
Lederman, Leon 462
Leeuwenhoek, Antoni van 91, 253
Leibniz, Gottfried 244
Leucippus of Elea (or Miletus) 8, 12, 19
lichens 405–7
light xv, 20–1, 38–40, 105, 110–12, 182–4,
    258–9, 315, 340–1, 342–5, 346, 350,
    432–3, 434–6
lightning 8, 10, 12, 152–6
    lightning rods 144–5, 152–6
Lind, James xix, 175–7
Linnaeus, Carl 134, 163–6, 169, 174, 193, 194,
    210, 397
Lippershey, Hans 71
Lister-Kaye, John 411–13
Logue, Lionel 264

Lonsdale, Kathleen 136, 362–4
Lorenz, Edward 421, 423–4
Los Alamos Scientific Laboratory 354,
    358–9, 360
Love, Rosaleen 419–20
Lovelace, Augusta Ada 219, 222–7
Lucretius (Titus Lucretius Carus) xv, xx,
    19–21, 22, 212, 403–4
Lyell, Charles 229, 236–41, 243, 244

McEwen, Todd 384–6
McMillan, Edwin 357
McPhee, John 465–6
magnetism 44, 52, 54, 178–81, 315, 346, 364,
    391–3
Malthus, Thomas 242
Mandelbrot, Benoît 445–6
Manners, Lord John 287
Mantell, Gideon 268–72
Mantell, Mary Ann 268
Marcet, Jane xx, 199–203, 213
Martel, Pierre 138–40
Maskelyne, Nevil 190, 191
Massachusetts Institute of Technology 298,
    374, 391, 423
mathematics xiii–xiv, xix, 1, 4–7, 9, 11, 34–7,
    48, 57, 60–2, 103, 104, 159–60, 219–27,
    375, 330, 336–7, 340, 374, 414–15,
    421–4, 443–7
Maugham, W. Somerset 467
mayflies 94–7
Mayr, Ernst 174
Medawar, Jean 466n
Medawar, Peter 76, 466n
medicine 41–2, 48–51, 61, 76, 115–20, 175–7,
    178–81, 264–7, 319–24, 338–9, 416–18,
    419–20
Meitner, Lise 137, 354–8
Menabrea, Luigi Federico 222–6

Mendeleev, Dmitri 304–08
Mesmer, Franz Anton 178–81
  mesmerism 44, 178–81
meteorites 53, 213–5
meteorology 1, 10, 24–9, 56–9, 103, 138–40,
    152–6, 169–74, 234, 282–6, 290, 330–4,
    406–7, 423–4, 437–42
Metrodorus 10
Meyer, Lothar 305–06
Michael of Ephesus 12
Michell, John 157–9, 434
microbiology 91, 137, 253–6, 260–3, 377–9,
    391–3, 452–5
microscopes xviii, 88–93, 123, 124–6, 130,
    198, 210–11, 253–6, 260–1, 274, 283,
    332, 404, 458
mining 18, 204–09, 369, 439
Mitchell, Maria 369–71
Møller, Christian 357–8
moon, the 3, 9, 36, 65, 71, 99, 112, 215,
    278–81
moons of Jupiter 71–5, 98, 99
Murchison, Sir Roderick Impey 228–31

Nagel, Sidney 457, 458
NASA Goddard Institute for Space Studies
    440
Nasmyth, James 278–81
Nature (journal) 129, 285–6, 290–2, 293–7,
    315, 318, 429
Needham, Joseph 52, 160
neutron stars 425–30
New Scientist (journal) 466n
Newlands, John 304, 305
Newton, Isaac xix, 76, 104–7, 110–12, 144,
    183, 344, 346, 458
Nickson, Edward 152–3
Niehoff, Debra 452–5
Nietzsche, Friedrich 361

Nobel Prizes xvi, 21, 124, 136, 137, 211, 293,
    308, 309, 342, 425, 429–30, 448, 462
N-rays see under X-rays
nuclear physics 309–14, 350–3, 354–9,
    360–6, 431–3, 459–62
nuclear weapons 354, 358–9, 360–6

oceanography xx, 10, 15–18, 232–5, 236, 325,
    364, 380–3, 419–20, 439
Oppenheimer, J. Robert 360
optics 38–40, 91–3, 182–4, 258–9
Orsenna, Érik 467
Owen, Richard xviii, 268, 272

Parmenides of Elea 12
particle physics xv, xviii, 21–3, 210–12,
    309–14, 315–18, 338, 340–1, 342,
    350–3, 354–9, 431–3, 434–6, 459–63
  Standard Model 431–3, 459–62
Pascal, Blaise 219–21
Pasteur, Louis 260–3
Peacock, Thomas Love 368–9
Peet, Thomas Eric 4, 5, 7
Pepys, Samuel 88
Pericles 12
Perrin, Jean 210, 211–12
physics xiv, xv, 13, 21–3, 84–7, 101–3, 210–12,
    257–9, 293–7, 315–18, 325–9, 330, 337,
    340–1, 342–5, 346–8, 350–3, 354–9,
    360–6, 431–3, 434–6, 456–8
physiology 41–2, 48–9, 76–9, 147–51, 165–6,
    178–81, 182–4, 264–7, 268–72, 408–10,
    411–13, 416–18, 419–20
Piazzi, Giuseppe 186
Pilkington, John 427
Planck, Max 340–1, 342, 344
plate tectonics 157, 380–3
Plato 4, 9, 62
Pliny (the Elder) 30–2

Plutarch 12

Pockels, Agnes 293–7, 298

Poett, J. H. Ayres 264–7

presocratic philosophy 8–14, 15–16, 464

Priestley, Joseph 144, 147–51

primatology 247, 397–9

Pringle, Sir John 155–6

Proclus Diadochus 9

psychiatry 319–24

Ptolemy (Claudius Ptolemy) 34–7, 38, 56, 63

pulsars 425–30

quantum mechanics xiv, 340–1, 342–5,
  434–6, 459–63

radioactivity xvii, xviii, 309–14, 315, 354–9

Ramsay, William 304–8

Rayleigh, John W. Strutt, 3rd Baron Rayleigh
  293–7, 304, 306, 308

Recorde, Robert 60–2

religion and science 1–3, 12, 34, 43–7, 62,
  67, 96–7, 214, 242, 244–5, 335–9,
  367–8, 371–2, 400–3, 462

Rhazes (Muhammad ibn Zakariyyā al-Rāzī)
  41–2, 56, 115–16

Rhind, Alexander Henry 4

Richards, Ellen H. 298–303

Richter, Charles 157, 159–60

Rilke, Rainer Maria 375

ripplemark 325–9

Rømer, Ole 101

Romilly, Sir Samuel 217

Röntgen, Wilhelm Conrad 23

Rootsey, Samuel 193, 196n

Rousseau, Jean-Jacques 367

Royal Astronomical Society 136, 185, 213

Royal Geographical Society 290–2

Royal Institution 199, 204, 257, 287, 362,
  416

Royal Irish Academy 136, 213

Royal Society of London 80, 84, 88, 91, 101,
  124, 127–9, 133–5, 136, 137, 144, 152–6,
  166, 179, 185, 186, 189, 213, 228, 325,
  362, 364

  Philosophical Transactions (journal) 101,
    127–9, 166

Royal Swedish Academy of Sciences 136,
  425, 437

rust 30–3, 466

Rutherford, Ernest 298n, 350–3, 354, 364,
  365, 375

Ryle, Martin 425

Sabin, Florence 136

Sabine, Elizabeth 136n, 232–5

Salisbury (ship) xix, 175–7

Sawatsky, David 420

schizophrenia 319–24

Schrödinger, Erwin xiv, 377

Schwarzchild radius 436

Science (journal) 439

scurvy xix–xx, 175–7

Sedgwick, Adam 140–3, 229

seismology 157–62

Seneca (Lucius Annaeus Seneca) 9, 10, 12,
  24–6, 57, 58–9

Servetus, Michael 76

Shakespeare, William 60, 63, 287, 376

Shapin, Steven 292

Shelley, Mary xv, 367

  Frankenstein xv, 363, 367

Shên Kua 52–5, 56

Siegbahn, Manne 355

Simplicius of Cilicia 8, 13

Simpson, Sir James Young 338–9

smallpox 115–20, 338–9

  inoculation 115–20, 338–9

Smith, A. L. 374

Smithsonian Institution 386
Smollett, Tobias 124
Snow, C. P. xiv, 373–6, 466
   'The Two Cultures' xiv, 373–6, 466
snowflakes 330–4
Somerville, Mary xx, 136, 185, 213–15,
   216–18
sound 80–3, 166–8, 264–7, 287–9, 316
Speliotopoulos, Achilles 348
Sprat, Thomas 127–9, 465
SS *Great Britain* (ship) 278
stammering 264–7
Standen, Anthony 372
Steele, Richard 133
Stephenson, Marjory 136, 362
Stock, Ann 453, 454
Stopes, Marie 338–9
Strassmann, Fritz 355–6
Street, Arthur 32–3
Strindberg, August 163
Stukeley, William 104–05
Sulston, John 448–51
sunspots 98–100
supernovae 67–70, 425, 429
surface tension 293–7
Swammerdam, Jan 94–7
Swinburne, Algernon 325
Szpiro, George G. 421–4

telephones 287–9
telescopes 67, 71–5, 100, 185–92, 278–81,
   425–30
Thales of Miletus 8–9, 56
thermometers 27, 101–03, 138–40, 234
Thomson, Sir Joseph John ('J. J.') 21–3, 365
Ticknor, George 226
tornadoes 24–9
Tovey, Dan 462
Travers, Morris 307–8

Treat, Mary 273–7
Trembley, Abraham 121–6, 405
Tsu Kêng-Chih 53
Turing, Alan 225
Turkey 116–19
turtles 8, 408–10
Tycho Brahe 67–70
Tyndall, John 257–9, 287–9, 362

Vallery-Radot, René 260–3
Venter, J. Craig 448
vivisection 84–7, 124, 147–51, 322–4, 372
volcanoes 18, 30, 100, 197, 215, 229, 282–6,
   438–9
Voltaire (François-Marie Arouet) 117–19

Wakefield, Priscilla 193
Wallace, Alfred Russel 243, 250–52, 466–7
war 30, 31–2, 62, 153, 358–9, 360–6, 369, 375,
   380, 422
Watson, James D. xvi, 377–9
Wegener, Alfred 380–3
Weinberg, Steven 431–3
Welch, Raquel 393
Westminster Abbey xiii–xiv
Wharton, Joseph 282–5
Wheeler, John A. 346, 358
Whewell, William 213, 216–18, 465
White, Michael 434–6
Wilkins, Maurice 377–9
Wilson, Benjamin 152–6
Wilson, Edward O. xvii, 174, 394–6
Withering, William 193, 195n, 196n
women in science xvi, xx, 19, 98–100,
   110–14, 116–20, 136–7, 185–92, 193,
   199–203, 213, 216–17, 222, 273–7,
   293–7, 298–303, 309, 325, 338–9,
   354–8, 362–4, 369–71, 399, 425–30
Wood, Robert W. 315–18, 466

Wordsworth, William 140–1
Wortley Montagu, Lady Mary 116–19
Wright, Joseph (of Derby) 84

Xenophanes of Colophon 8, 10, 12
X-rays 23, 309–10, 315, 350, 378, 379
    N-rays 315–18, 466

Yeats, William Butler 367

Yochelson, Ellis L. 386–90

Zadeh, Lotfi A. 443–6, 465
Zeno of Elea 8, 11
Zhang Hêng 160–2
Zim, Herbert S. 384–6
zoology xvii, 12, 94–7, 121–6, 163–8, 174,
    250–2, 273–7, 372, 388, 394–6, 397–9,
    405–7, 408–10, 411–13

picador.com

blog
videos
interviews
extracts